T0327645

# Mobile Radio Network Design in the VHF and UHF Bands

# Mobile Radio Network Design in the VHF and UHF Bands

## A Practical Approach

**Adrian W. Graham, Nicholas C. Kirkman and Peter M. Paul**

*All of Advanced Topographic Development and Images (ATDI) Ltd, UK*

John Wiley & Sons, Ltd

Telephone (+44) 1243 779777

Email (for orders and customer service enquiries): cs-books@wiley.co.uk
Visit our Home Page on www.wiley.com



***Other Wiley Editorial Offices***

John Wiley & Sons Inc., 111 River Street, Hoboken, NJ 07030, USA

Jossey-Bass, 989 Market Street, San Francisco, CA 94103-1741, USA

Wiley-VCH Verlag GmbH, Boschstr. 12, D-69469 Weinheim, Germany

John Wiley & Sons Australia Ltd, 42 McDougall Street, Milton, Queensland 4064, Australia

John Wiley & Sons (Asia) Pte Ltd, 2 Clementi Loop #02-01, Jin Xing Distripark, Singapore 129809

John Wiley & Sons Canada Ltd, 6045 Freemont Blvd, Mississauga, ONT, Canada L5R 4J3

Wiley also publishes its books in a variety of electronic formats. Some content that appears in print may not be available in electronic books.

***British Library Cataloguing in Publication Data***

A catalogue record for this book is available from the British Library

ISBN-13 978-0-470-02980-0 (HB)
ISBN-10 0-470-02980-3 (HB)

Typeset in 10/12 pt Times by Thomson Digital.

# Contents

# Foreword

In 1687, Sir Isaac Newton published his *philosophiae naturalis principia mathematica* or *Mathematical Principles of Natural Philosophy* postulating gravity, the very general law that held the universe together. Throughout the Enlightenment and ever since, engineers and scientists have sought the universal laws with which to explain the natural world in which we live. This desire to explain things spawned the idea that we can build and exercise models of our world that will describe how the systems that comprise it might function. Once proven to be useful in example cases, engineers and scientists can hypothesise, seek evidence and prove them generally.

Progress in modelling the radiocommunications domain effectively started with Kenneth Bullington's landmark paper in 1946 proposing the single knife-edge diffraction model. The body of knowledge has grown steadily since to a point where engineers can explain their world in high fidelity using computer simulation. The accuracy with which this can be done is astounding to the point that no-one should consider radiocommunications network deployment without simulation of their plans against these laws. In our risk-averse times it is the only way to effect risk control without huge expense in equipment, effort and sometimes lives.

That sets the scene for this book. Using 60 years of authoritative knowledge one can model the radiocommunications world. And yet two, maybe three, schools still exist today. Like the positivists of the Enlightenment, there are those who view that everything can be explained and it is just a question of mankind finding and using the necessary models. On the contrary there are those engineers who, like the Enlightenment's interpretivists, consider that radiocommunications is all a 'black art' with too many independent variables to permit anything but a 'feel' to be gained using very broad models. The third group fall into the disbelievers, retorting 'it's only a model' and hence not to be considered useful in any way. Let us be under no illusion. Modelling radiocommunications networks is now a science that is of age and those engineers who lean towards positivism can have their day. The laws exist. The error bands in these laws are known and when technology shifts the goal posts, the techniques exist to re-calibrate the laws. Be sure: every radiocommunications network can be modelled.

Models seek to give engineers the truth. Within this book there are two truths. The first is the positivist truth that the laws exist with known accuracy to afford engineers real, tangible benefit from modelling and from the parallel functions of spectrum regulation, network planning and measurement. The second is that Adrian Graham and his colleagues have put forward the most appropriate, specific and complete models spanning every part of the radiocommunications domain and every part of the radiocommunications project life cycle. In achieving this they have achieved the 'Holy Grail' of authorship: writing something that is actually useful to practicing radiocommunications engineers. By embodying their own experience, they have achieved a post-modern 'Enlightenment'; a lifting of the myths that doomed radio-network planning and other simulation activities to be described as 'black art'.

Now, engineers are the world's worst project managers. Most fall into the Myers-Briggs 'judgemental' personality preference type. The result is a high desire to make early decisions and to foresee the result before the problem has been completely explored though simulation. The opposite would be to develop a process in which the project team can trust to lead them to success; something that in abstract everyone would agree makes more sense. This book identifies the project nature of radiocommunications regulation, modelling, planning and measurement. Several sections are devoted to the elicitation of requirements, discussing how one might describe the desires of radiocommunications users. Adrian, Nick and Peter have identified the criticality of this user requirement in setting the tone for the whole project leading to performance, acceptance and ultimately payment. Get the requirements wrong and the project will fail to satisfy these users; and yet many radiocommunications projects fall short of their final aim suggesting that engineers have much to learn. This book provides a framework for that learning.

Finally, I personally welcome this book. It is a most useful addition to company bookshelves wherever radiocommunications engineers work. Poor communications were cited in the report on the emergency services' response to the bombings in London in 2005. Poor communications contributed hugely to the defeat of the British 1 Airborne Division at the Battle of Arnhem in 1944 and since then poor communications has been cited the world over whenever there is a critical need for human performance. And yet we engineers can model radiocommunications performance. We do need to impress on our management, accountant or economist colleagues that we have this competence and that we can tell them the risks surrounding the various options that they propose. That said, we need to get our story right. I welcome the book wholeheartedly because it defines specification parameters including one word that is the key to every wide area radiocommunications network, that is, coverage. This word of all words is the least well understood by the engineering fraternity. If you need definition or clarification of this and other parameters in the fixed, mobile, broadcast and satellite simulation world, read on.

John Berry BSc DIS DMS MBA CEng FIET
Managing Director
ATDI

# Preface

Mobile radio networks have risen in prominence over the last few years, primarily by the rise in popularity of cellular phones. It is important to recognise however that mobile radio technology fulfils a far wider range of applications that meet the demands of the modern world. These include the networks that allow police and emergency services to serve the public, military networks for operations and humanitarian support, and the mobile technologies that are vital to the safety of aircraft. These and many others beside make up the panoply of the mobile radio network domain. In this book, we look at the design of mobile radio networks in the VHF and UHF band. In each case, the design approach taken must reflect the unique requirements of the specific project, but there are design methods that can be applied to a range of mobile technologies and we have described some of these in this book. Throughout the book, we focus on the radio-related aspects of design; the location and configuration of base stations, tools and techniques for the analysis of network coverage and interference, design for capacity in the mobile to base station link and so on. We have included a chapter on fixed microwave links that are used to route mobile traffic to its destination, but again this focuses on the radio aspects of design rather than anything above the physical layer.

We have not addressed aspects such as traffic switching, access protocols or anything to do with the line aspects or the services carried by the networks. These subjects have received a great deal of attention in a variety of good books on the subject of radio technology. These books provide vital background to the reader in terms of understanding the technology and how it achieves the objectives of being a bearer for the services it is designed to cover. In writing this book, the authors wanted to avoid repeating this work, but instead to provide a practical tool for the radio network design engineer. This means focusing on those aspects that the designer has influence over, and spending less time on those aspects that the designer cannot change. We have tried to include material that radio engineers will need on a day-to-day basis, but not to spend time on derivations of complex mathematical formulae. Such derivations are generally already well covered in books also available and are not generally needed for day-to-day work. We have also looked at radio network projects from a wider perspective, so that the engineer can gain an understanding of other groups of people that are involved in the process. This is not filler material; more projects fail from business reasons than from technical ones, and the designer has a role in supporting the whole project.

We contend that modern radio network design implicitly depends on the use of computer-based radio planning tools; so much of the book is based on their use. We have provided guidance on features we believe any state-of-the-art planning tool should have, and how they are used in the network design process. In addition, we also focus on the data that must be used with planning tools for them to be effective. We have tried not to focus on individual radio technologies as far as possible, but rather to identify design concepts that can be tailored to many different technologies; we feel that learning these techniques is more useful than being able to crib from a prescribed process. We have been forced to omit many detailed design aspects and to provide no more than overviews on others, simply due to the space available in the book and the time available to create it, so we do not intend this book to replace others but rather to be a companion to other data sources; where possible, we have identified suitable sources of data such as ITU recommendations.

Our guiding principle has been to provide the sort of information we would have liked to avail when we first entered the world of radio network design, all in one place. We also intend to provide it to engineers that join our company, and we will expect them to refer to the book as they develop their careers in this domain. We have also provided information for those at a more senior level, such as project managers and others who are involved in radio projects or who need to be able to interpret the results produced by the techniques covered.

In writing this book, we have enjoyed the support of our colleagues and friends. We would like to thank the following people for their contribution and ideas:

John Berry, Cyprien de Cosson, Colin Desmond, Robin Hughes, Peter Johnson, Kevin Morrison, Sarah Roddis and John Talbot (of Emas Consulting).

Our particular thanks go to Ed Douglas of RSI (www.rsi-uk.com) for his invaluable contribution to the survey section, to M. Philippe and David Missud of ATDI SA (www.atdi.com) for their permission to use images of the ICS Telecom radio-planning tool, and to ATDI Ltd (www.atdi.co.uk) for permission to use partial screen shots of IMP Calc and other radio-planning tools and software components.

The authors hope you enjoy the book and find it useful. Readers may like to visit www.atdi.co.uk, where they will be able to find a range of useful radio planning calculators and tools that can be used in conjunction with the methods described in this book to solve many engineering problems.

# Glossary

| | |
|---|---|
| AGA | Air-Ground-Air communications (aeronautical) |
| AM | Amplitude Modulation |
| ASRP | Arc Standard Raster Product – NATO file format for map image data |
| CBS | Common Base Station |
| CDMA | Code Division Multiple Access |
| C/I | Carrier-to-Interference ratio in dB |
| CIS | Command Information System |
| CNR | Combat Net Radio: military mobile radio network system |
| COTS | Commercial-Off-The-Shelf |
| CRP | Compressed Raster Product – geographic data file format |
| DDD | Detailed Design Document |
| DEM | Digital Elevation Model |
| DF | Direction Finding |
| DME | Distance Measuring Equipment (aeronautical) |
| Downlink | Link from base station to mobile |
| DRDF | Digitally Resolved Direction Finding |
| DTED | Digital Terrain Elevation Data – NATO file format for DTM data |
| DTM | Digital Terrain Model |
| DVOR | Doppler VHF Omni-directional Radio range (aeronautical) |
| EA | Electronic warfare Attack – jamming |
| EP | Electronic warfare Protection – protection from jamming, DF etc |
| ES | Electronic warfare Support – detection, direction finding etc |
| FAA | Federal Aviation Authority |
| FDMA | Frequency Division Multiple Access |
| FFZ | First Fresnel Zone |
| FH | Frequency Hopping |
| FM | Frequency Modulation |
| FS | Functional Specification |
| FSL | Free Space Loss |
| GIS | Geographic Information System |
| GPS | Global Positioning System |
| GTD | Geometric Theory of Diffraction propagation modelling method |
| GUI | Graphical User Interface |
| ICAO | International Civil Aviation Organisation |

| | |
|---|---|
| IF | Intermediate Frequency |
| IFF | Identification Friend or Foe |
| IMP | Inter-Modulation Product |
| IRF | Interference Rejection Factor of a radio receiver |
| ITT | Invitation To Tender |
| ITU | International Telecommunications Union |
| MTBF | Mean Time Between Failure |
| MCFA | Most Constrained First Assigned |
| MER | Message Error Rate |
| MGRS | Military Grid Reference System map projection |
| MMOFS | Minimum Median Operating Field Strength (dB$\mu$V/m) |
| MMOL | Minimum Median Operating Level (dBm) |
| MLS | Microwave Landing System (aeronautical) |
| MOS | Mean Opinion Score |
| MOTS | Mostly-Off-The-Shelf |
| MS | Method Statement |
| MTTR | Mean Time To Repair |
| NDB | Non-Directional Beacon (aeronautical) |
| Net | Military shortening of the term 'network' |
| NFD | Net Filter Discrimination |
| NGR | National Grid Reference map projection (e.g. UK NGR) |
| NM | Nautical Miles |
| ODBC | Open DataBase Connectivity standard for database |
| OFDM | Orthogonal Frequency Division Multiplexing |
| PAMR | Public Access Mobile Radio |
| PDH | Plesiosynchronous Digital Hierarchy (backhaul term) |
| PESQ | Perceptual Evaluation of Speech Quality |
| PM | Pulse Modulation |
| PMR | Private Mobile Radio |
| PSD | Power Spectral Density of a transmission |
| PSO | Probability of Successful Operation |
| PSTN | Public Switched Telephone Network |
| RFQ | Request For Quote |
| RSSI | Received Signal Strength Indication |
| Rx | Shorthand for radio receiver |
| SDH | Synchronous Digital Hierarchy (backhaul term) |
| SMM | Simplified Multiplication Method (multiple interferer calculation) |
| SNR | Signal-to-Noise Ratio |
| SSR | Secondary Surveillance Radar (aeronautical) |
| System value | Maximum permissible loss in a system (normally military) |
| TACAN | TACticial Aid to Navigation (aeronautical) |
| TDMA | Time Division Multiple Access |
| TETRA | TERrestrial Trunked Radio Access |
| TPC | Tactical Pilotage Chart |
| TS | Test Specification |
| Tx | Shorthand for radio transmitter |

| | |
|---|---|
| UAV | Unmanned Airborne Vehicle |
| UHF | UHF Band 300–3000 MHz |
| UMTS | Universal Mobile Telecommunications System |
| Uplink | Link from mobile to base station |
| URS | User Requirements Specification |
| UTD | Uniform Theory of Diffraction propagation model |
| UTM | Universal Transverse Mercator map projection |
| VDF | VHF Direction Finding |
| VOR | VHF Omni-directional Radio range (aeronautical) |
| VORTAC | Combined VOR & TACAN site (aeronautical) |
| WGS | World Geodetic System – geographic datum |

# PART ONE

# 1

# Introduction

## 1.1 Mobile Radio Network Design in the Modern World

Over the last few years, mobile radio has emerged from being a niche technology to becoming ubiquitous throughout much of the world. Most people come directly in touch with the technology through mobile phones and the networks that serve them, but this is only one aspect of mobile radio. Mobile radio is, however, used for many other applications in networks that span the air, land and sea, and although not immediately familiar to most people, their use is essential to the standard of life expected in the modern world. It would be difficult for airline passengers to travel to their holiday destination without the dedicated aeronautical networks that aid smooth management of aircraft and their navigation. Police and other emergency services would be unable to serve the public effectively without the ability to communicate rapidly and securely. Table 1.1 shows some users, applications and technologies within the broader mobile radio sphere. The list is by no means exhaustive.

In this book, we will be looking at the design and optimisation of mobile radio networks operating in the VHF and UHF bands, with a brief look at microwave links that are sometimes used for backhaul (the fixed infrastructure used to direct calls to their destination). We have chosen not to include LF, MF and HF; not because they are unimportant, but rather that their design processes are somewhat different from that employed at frequencies above HF. The aim is to provide the mobile radio network designer or engineer with a practical toolkit of knowledge and techniques that can be brought to bear to a wide range of radio network design tasks. We will focus on the aspects of network design that are under the control of the radio network designer, rather than the nuts and bolts of the underlying technology. The reason for this is that while it is essential that engineers understand the basic principles of the technology, aspects such as the interface protocols and frame structure are not under the control of the typical radio network designer. Since there is no control, these aspects represent constraints rather than design factors in most mobile radio network design activities. Additionally, our focus is almost exclusively at the physical layer of the radio network design – in other words – the RF aspects from the output of the transmitter part of one radio to the input of a receiving

Mobile Radio Network Design in the VHF and UHF Bands: A Practical Approach
*Adrian W. Graham, Nicholas C. Kirkman and Peter M. Paul* 

**Table 1.1.** Some typical mobile radio applications.

| Category | Typical users | Type of application | Typical technologies |
|---|---|---|---|
| Public cellular | Public | Mobile phone analog voice | AMPS, NMT-450. NMT-900, ETACS |
| | | digital voice and data | GSM, EDGE, GPRS, UMTS, IS-95, IS-136 |
| PMR | Business Outdoor activities Mountain rescue | Analog voice | PMR 460, MPT 1327 |
| PAMR | Public safety Business | Mobile analog voice and data | APCO 16, MPT 1327 |
| | | Digital voice and data | TETRA, TETRAPOL, APCO 25, APCO 25/34 |
| Aeronautical | Civil aviation | Voice | VHF AGA analog voice, UHF ground movements |
| | Military aviation | Data | VHF Data Link Mode 2, SSR Mode 2 |
| | | Navigation | VOR, DVOR, VDF, DME, TACAN, NDB, GBAS |
| | | Landing systems | ILS, MLS |
| | | Radar systems | Primary RADAR, secondary RADAR |
| Maritime | Coastguard | Analog voice | Marine VHF |
| | Port authorities | Search and rescue | Marine VHF, GMDSS |
| | Vessels | | |
| | Pilots | | |
| | Customs | | |
| Military | Military forces | Combat net radio | Clansman, Bowman, JTRS |
| | Paramilitary | Trunk radio (combat) | Parakeet, RITA, Autoko, Sotrin, Zodiac, Ptarmigan, MSE |
| | | Trunk radio | TETRA, TETRAPOL |
| | | Air-land-sea tactical data link | JTIDS, LINK 11-A, LINK 11-B, LINK 16 (TADIL-J), EPLRS |
| | | Telemetry/command | UAV command links |

radio. This includes the effects of feeders, the antennas used for transmission and reception, and the radio channel environment. We will extend this to include dimensioning for traffic, and we will examine design of backhaul for mobile radio networks. We will not be covering access protocols, switching or other line aspects of the design, as these are adequately covered in a wide variety of other books currently available. This leaves us with the physical positioning and orientation of antennas, selection of antenna types, radio equipment configuration, interference management, frequency assignment and dimensioning for mobile traffic. These aspects of design form a major part of the design of any mobile radio network, and they are usually under the control of the radio network designer, albeit with a wide array of technical and nontechnical constraints.

This book is also intended to be a *practical* guide to mobile radio network design. It is split into two parts. Part One covers some essential background, and Part Two looks at the practical aspects of network design. The emphasis on the practical aspect is intended to convey the fact that understanding a technology, being able to perform coverage simulations, frequency assignment, and so on, is, although vital in itself, not sufficient to bring a network from conception all the way through to implementation, rollout and use. More projects are doomed by poor understanding of customer's requirements, poor project structuring, poor communications and *not listening* than are ever limited by the technology being used – even though it is common to blame the technology as a convenient scapegoat. We, therefore, focus on these additional aspects throughout the book, mostly in Part Two.

A useful method of understanding the competing entities within a radio network design process is to consider the project in terms of 'stakeholders'. A stakeholder is an entity or person that can exert influence over the implementation of the network. The probability of successfully completing a project is maximised if the concerns of as many as possible stakeholders are addressed and minimised for the converse condition. We also focus on the necessity for the design team to continually work to meet the needs of the customer – where 'customer' may be any organisation on whose behalf the network is being implemented. Without that focus, the delivered network design may not meet the original requirements. We make no apologies for including these aspects, which some may not regard as 'pure engineering', but we contend that in the sense that engineering is applying technology to deliver a service to a customer, they are utterly essential. Part Two also examines practical methods of exploiting the radio planning tools and data to achieve specific objectives within the overall plan. Again, the emphasis is to allow the reader to be able to go off and perform these tasks and understand what they are doing. What we will not do is spend a major portion of the book describing specific technologies in great depth. Other books provide this type of information. We will only look at the most important aspects of mobile radio planning and we will try to provide information and skills that can be adapted for as wide a range of mobile technologies as possible. We believe it is more important to get across adaptable methods rather than focus on the planning of a single technology that will quickly lose its currency – although naturally we will include as much technical detail as is necessary and will illustrate the principles by describing how they are applied to specific technologies.

Part One of the book introduces essential background information that radio network planners must be aware of, and some aspects such as propagation models and constructing link budgets that must be understood in as much depth as possible. Much of this can be found in other publications, but we intend to present it in a way that has proved very successful during many of the training courses we have run on the subject. The second part moves on to the practical aspects of mobile radio network design. Before going on to do this, however, we will spend the rest of this chapter looking in outline at the 'business' environment in which the activity of radio mobile network design resides. We will come to the technological aspects later, but first we will put the whole design process into its business context. Note that the term 'business' is not intended to refer exclusively to commercial organisations, but rather to any organisation performing network design or otherwise involved in the process. This can include government departments, military planners, police forces and any other defined groups. The common factor is that each of these businesses serve some kind of customer, whether on a paying basis or not.

## 1.2 Network Stakeholders

Radio networks never exist in isolation. In fact, a wide variety of people and groups will be involved in some manner during the design and operation of a radio network. This is illustrated in Figure 1.1, which shows the stakeholders that may be involved in a specific project. The figure also illustrates the interactions between stakeholders and demonstrates the principle that one stakeholder may not be able to directly influence another; they may have to go through other stakeholders or indeed there may be no direct influence path.

The importance of examining stakeholders and being aware of their presence is to help ensure that each important stakeholder is addressed during the project; the importance or otherwise of a particular stakeholder can only be determined by identifying their existence and considering them in turn. One crucial factor is not to confuse stakeholders with shareholders; a stakeholder need not be dedicated to ensuring the success or a project – they may indeed be dedicated to destroying it. The design team may therefore have a task to undermine these stakeholders by providing technical information to help refute their claims, for example. This important area is examined further in Chapter 6.

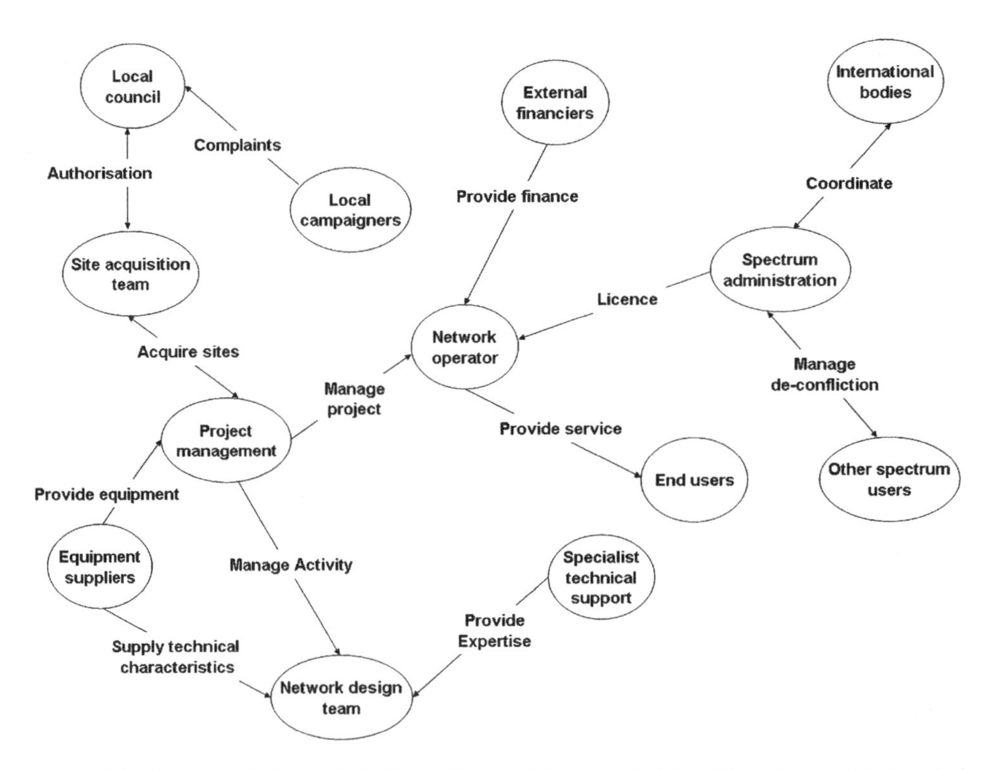


**Figure 1.1.** Network design stakeholders. Note that the list of stakeholders does not just include those dedicated to the success of the network.

## 1.3 Spectrum Coexistence

One of the most important stakeholder groupings concerns the coexistence of the network with other spectrum users. In most cases, the use of spectrum is managed by a national regulator, appointed to ensure that spectrum is used to the maximum benefit of the country concerned. This should be achieved within the framework laid out in the Radio Regulations of the ITU, which includes allocations for ITU regions and the countries within them, and also provides mechanisms for international coordination of spectrum. Within a particular country, some aspects of national regulation may also be devolved down to other organisations, either on a regional basis or on the basis of a particular type of use.

A typical situation is shown in Figure 1.2. This will vary according to country and application.

Although it may not be the network designer's role to obtain spectrum for the network (this will probably be handled by another department), the design team may be involved to provide technical details of the service to be offered, to support the application (when spectrum is allotted on the basis of a beauty contest, for example), so it is appropriate to have some understanding of what goes on at the regulatory level. It may also be beneficial to understand the regulator's role and commitments with regard to international coordination, to ensure that the network design will not be blocked on this basis. Additionally, regulators have responsibility for identifying and resolving interference issues, so it may be useful to

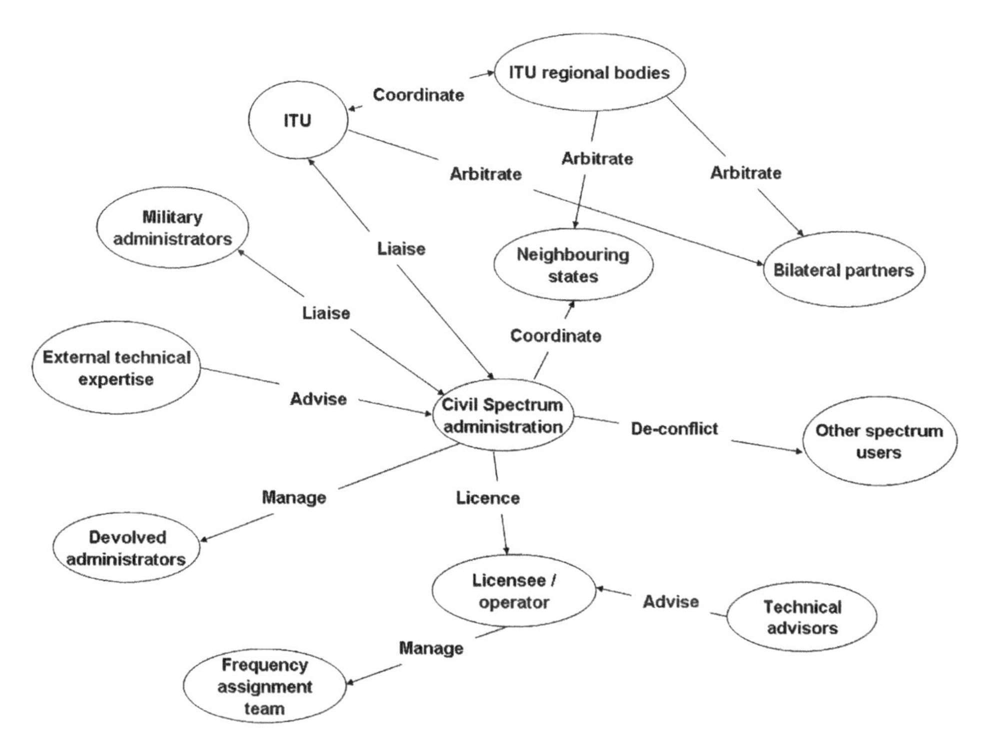


**Figure 1.2.** Spectrum-related stakeholders. Only some of the potential links are shown in the diagram.

understand how this process works in case it ever needs to be invoked if problems are encountered. This is described more fully in Chapter 2.

## 1.4 The Network Design Activity

The network design activity should not be regarded as an isolated activity within the business ('business' in this context means the business of the organisation, whether it is commercial or noncommercial in nature). The best situation is where internal stakeholders work together to meet the business needs; the worst is when internal politics and interdepartmental rivalry prevents them from gelling properly. In a good design environment, the design activity should be a process that involves the relevant departments providing required information across defined interfaces in a timely manner. Again, this applies to both commercial and noncommercial organisations, as illustrated by two different scenarios in Figures 1.3 and 1.4.

In the commercial environment, as illustrated in Figure 1.3, the aim of the network will be to generate revenue for the business. This is most easily achieved in circumstances where there is a culture of information flow between the business entities throughout the duration of the project. This often requires better management techniques than traditional methods, but there is a measurable improvement in project success.

Figure 1.4 shows the corollary in a noncommercial environment; specifically a military one. In this case, the aim is to meet the needs of the end users (deployed forces) in the best

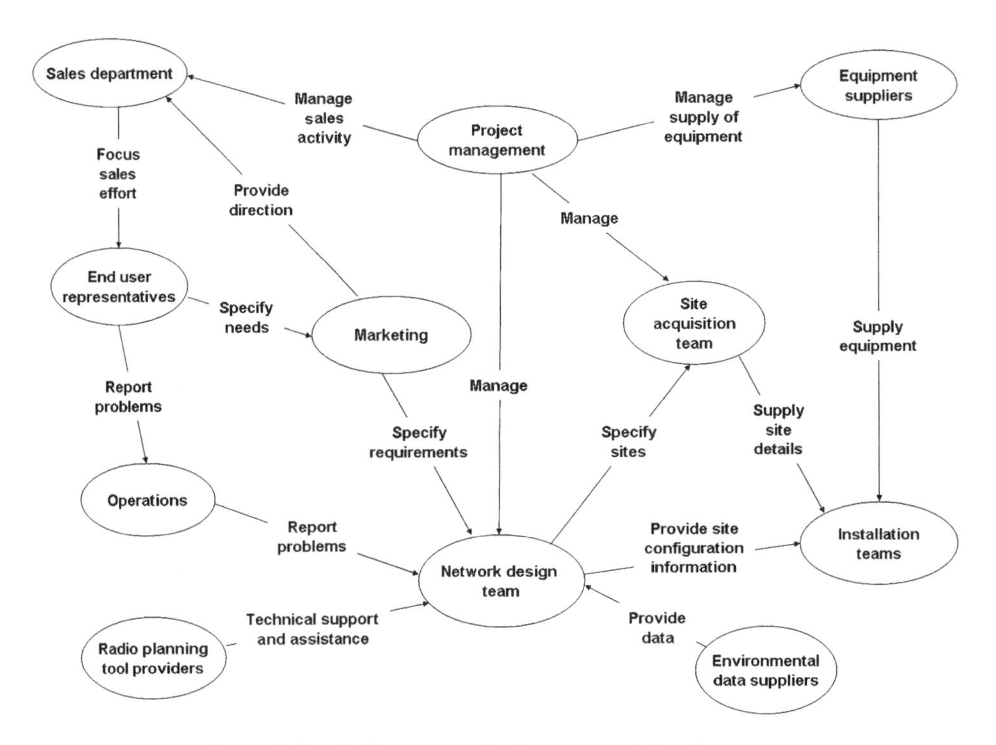


**Figure 1.3.** Design process in the commercial environment.

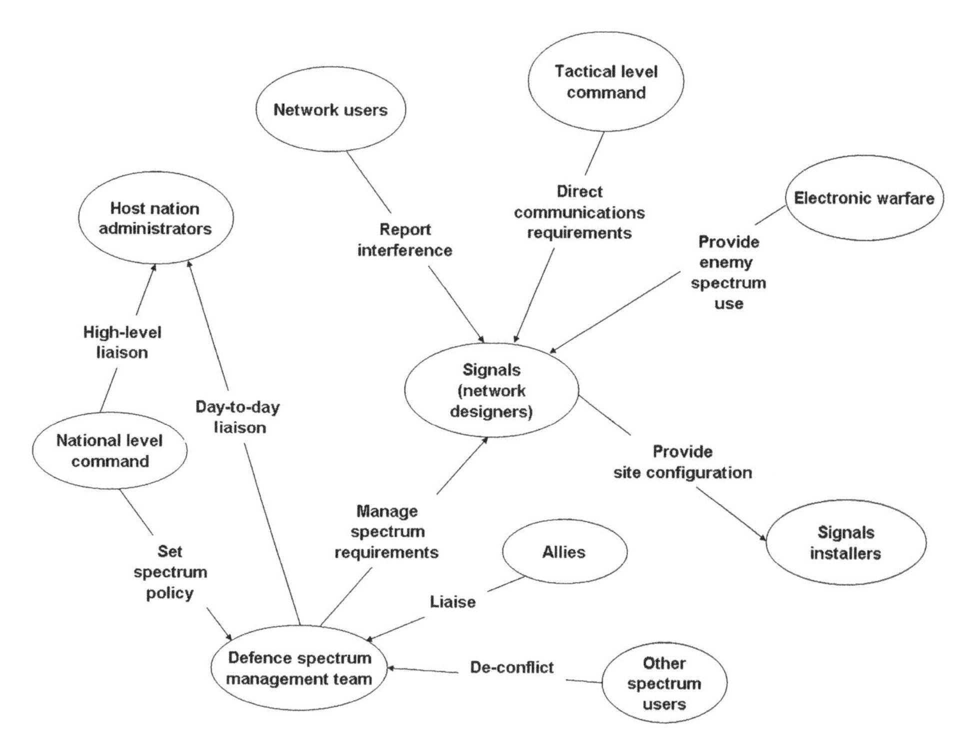


**Figure 1.4.** Design process in the military environment.

possible way so that they can best achieve their aims. The network design is still designed for customers – in this case, internal customers – and it also suffers from external constraints. Again, representatives of each of these stakeholders must be involved in the design process to ensure that the needs of each are addressed. Indeed, it may be appropriate to consider the enemy as stakeholders; they have an influence on the network in terms of design for resilience against their actions. As stated before, stakeholders need not necessarily wish to see the project work.

Understanding stakeholders and the potential risks (or benefits) they represent is important, and we will be discussing it in more depth in Chapter 6. Methods of designing and managing the design process across the whole business are also illustrated throughout Part Two. Some may argue that this is not part of classic engineering, but in terms of ensuring that a project progresses efficiently from inception to final delivery and use, designing the processes and environment to achieve success through business integration is as important as designing the purely technical aspects; otherwise the delivered system is likely to be late, over budget or it may not meet the customer needs at all.

## 1.5 Project Resources

Besides a project structure that promotes focus on the customer, successful projects will also require necessary resources for the design team. These will consist of trained personnel,

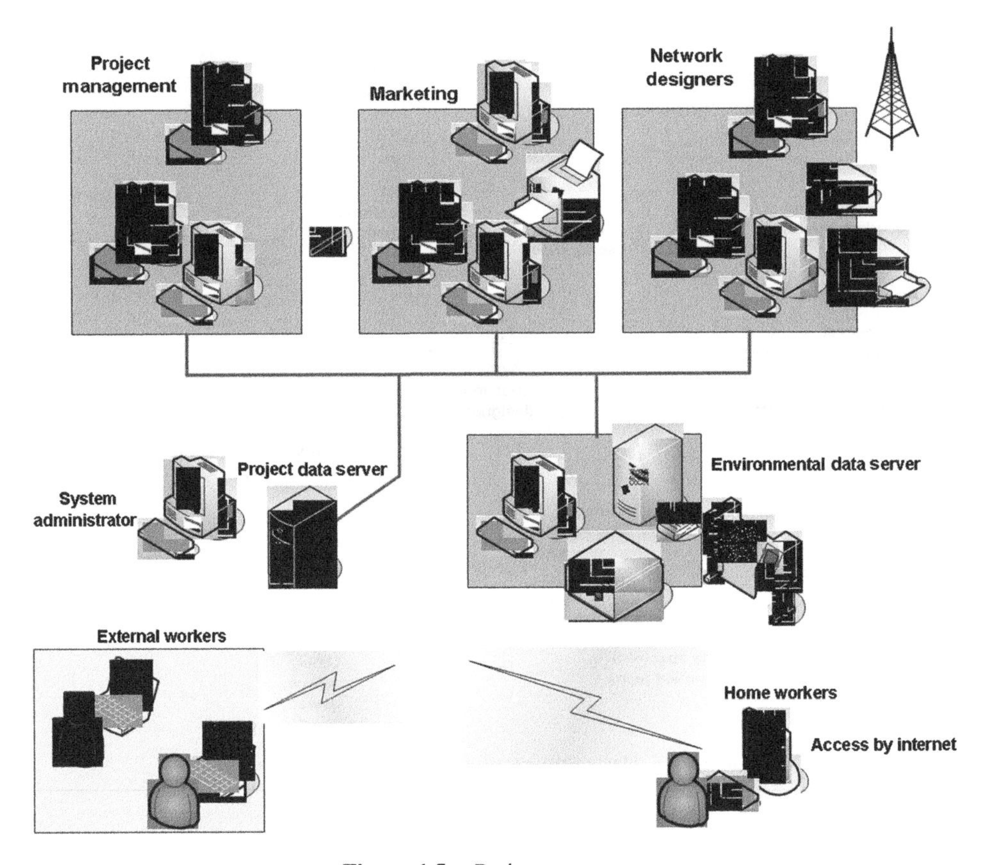


**Figure 1.5.** Project resources.

design tools and supporting data, and a budget to complete the project. Small projects may only require a single designer working on a single, non-networked PC, whereas large projects may require a networked system with shared data available to project management staff, perhaps the marketing division and network designers. It might be that to improve communication between the customer and others involved in the project that staff are separated geographically, and there may be home-workers who will access the shared resources over the Internet. Different users will also require different views on the available data, different access rights and different tools, ranging from viewers with no data input facility to full-scale design systems. An illustration of the type of system required is shown in Figure 1.5. We discuss the requirements for tools and data further in Chapter 7, and the ways in which they are used to perform various tasks are discussed throughout the rest of Part Two.

In terms of personnel for the network design process, there may be a requirement for senior consultants, who will be responsible for major engineering decisions and may work closely with customers and other stakeholders to define project parameters and help guide the customer to the required solution. There will also be the need for senior engineers capable of managing and supervising more junior staff, and for planning tool operators. In Chapter 6, we discuss our view on the skills, quality and experience we would expect of the engineers engaged in radio network design. Of course, such things are open to interpretation and others may disagree with our definitions, but our approach is based on many years of network design projects covering a large range of project requirements.

## 1.6 Validation and Verification

Besides existing within organisational and stakeholder environments, the design process also exists within the real world. The designed system must accurately predict the end result before the network is actually built, and also it must be possible to prove early on that the customer requirements will be met by the final system. Robust validation and verification methods help to achieve this. We can define validation as the process of ensuring that the engineering specification is a true reflection of the customer's intent (often posed as 'are we building the right network?'). Verification is the process of proving whether the design meets the specification ('are we building the network right?'). Without sufficient attention to these twin aspects, it is difficult to ensure that the design activity is on track to provide a good solution to the customer's needs. We raise this as a separate issue because we have seen many instances when insufficient attention has been made to this aspect of the project, often leading to major problems later in the project. However, by factoring this in as early as possible in the project, these problems can be avoided. This theme will emerge frequently throughout Part Two.

## 1.7 Evolving Needs

The final consideration in this short introduction is how the network design fits into the network life cycle. Although a network design activity can often be delimited to a specific, containerised project for which there is a finite timescale and a finite budget, it is also important to realise that networks do evolve over time – even when there was no intention of this happening at the time the network was implemented. This can often happen when the end user behaviour changes; the network becomes constrained by the number of users or when the network's life is extended beyond the initial plan, and additional work needs to be carried out to ensure it continues to function before retirement. Thus, the best design methods build-in the ability to allow the network to develop, be extended and enhanced throughout its life and to allow a graceful transition to its successor. Often, this is of low priority during the initial design phase and it is often overlooked or suppressed by management, keen to save every penny. This can rebound in the future by making changes to meet evolving needs far more expensive. We will be showing methods of working with partial legacy systems throughout Part Two.

## 1.8 A Practical Approach, Not *the* Practical Approach

As a final comment in this section, the authors would like to point out that we are presenting an approach to network design, and we are not trying to imply that the methods described are the only ones that can be used. There are many approaches to specific tasks in particular that will be at variance to the ones we describe, but as with most things in engineering, it is not usually possible to state that one method has substantial advantage over another in all circumstances. We can state, however, that all of the methods discussed in the book have been used for a large variety of projects covering commercial and noncommercial networks and their management in the civil, public safety, military, commercial aviation, maritime and a variety of governmental agency spheres and we have every confidence in them. We hope that by the end of the book, the reader will agree with us.

# 2

# Spectrum and Standards

## 2.1 Introduction

Radio networks do not exist in isolation. They share radio spectrum with other users, and it is important to ensure that licensed services do not cause undue interference with one another. This calls for coordination within the country the services operate in and also between countries. This section looks at the organisations involved in de-conflicting spectrum users and how they interact with one another.

Although the radio engineer or network designer will not come across these organisations very often, it is worth knowing that they exist and that it may be necessary to interact with some of them on some occasions. Figure 2.1 illustrates some of the elements involved in radio network design and operation, when the international dimension is also taken into account. Network designers may not be directly involved in all of these tasks, but this should highlight the way design fits into the overall picture. The design results in the selection, configuration and installation of radio sites, stations, backhaul links and other infrastructure. The spectrum to be used for the network will be drawn from the national allocation plan, managed by the national regulator, who will also be responsible for international tasks. Organisations running networks will have to manage billing and other operational tasks such as maintaining databases of the system and its subscribers.

This illustration highlights the fact that the radio network designer is actually a relatively small part of a massive machine. The fact that the designer need not necessarily consider all of this on a day-to-day basis does not mean that it disappears!

We will start by looking at international spectrum management and then look at the situation within the country of service. As we discuss these organisations, we will also look at the materials they produce that can be of use to the engineer. We will also look at bodies that are not involved in spectrum management but that are of international importance, particularly in the development of internationally agreed standards and also for important radio network design methods and algorithms.

Mobile Radio Network Design in the VHF and UHF Bands: A Practical Approach
*Adrian W. Graham, Nicholas C. Kirkman and Peter M. Paul* 

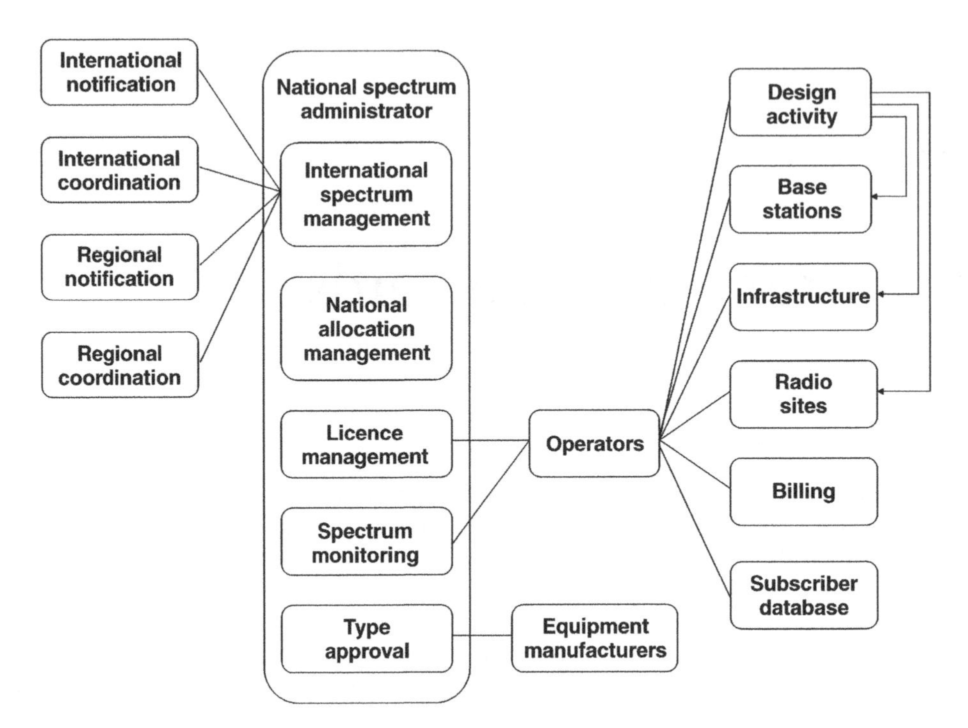


**Figure 2.1.** Elements of spectrum management and planning.

## 2.2 International Spectrum Management

### *2.2.1 The International Telecommunications Union*

The International Telecommunications Union (ITU) is an agency of the United Nations and represents the highest authority for spectrum management in the world.

The roots of the ITU grew out of the International Telegraph Union, established in 1865 to overcome the problems of nonstandardisation of telegraph transmissions at the time, which meant that transmissions crossing borders had to be re-transcribed by hand for onward transmission. As wireless telegraphy was developed, the original telegraph union came to take on radio transmissions as well as wire and in 1906 the forerunner of the radio regulations was first produced. As new technologies emerged, rules for these were added and by 1947 the ITU became a specialised agency of the United Nations and set up its headquarters in Geneva, where it still is. The ITU has been reorganised into its current form to meet emerging needs, which continue to evolve ever faster.

The purpose of the ITU is set out in its constitution. Broadly, these are to maintain and extend international cooperation, to foster fruitful partnerships and to promote the adoption of a broader approach to the issues of telecommunications.

The ITU is split into three sectors. These are the radiocommunications, standardisation and development sectors. The radiocommunications sector (ITU-R) contains the Radiocommunications Bureau (BR), which is the executive arm of the ITU-R.

The ITU-R BR develops the radio regulations (including the table of allocations) and manages the master international frequency register (MIFR). The MIFR is the master record of all systems to be protected under international spectrum management rules. It contains more than 1 250 000 terrestrial and 325 000 assignments for about 1400 satellite networks. The importance of this register is that *only* those assignments in the MIFR have to be protected – anything not in the list does not gain automatic protection from potential interferers in other countries. However, it is not mandatory to place assignments on the list if protection is either not needed or wanted (e.g. the system is too far away from any transmissions that may cause interference to it). This means that the MIFR cannot be regarded as being the definitive list of all radio systems in the world since many systems have simply not been added to it. The MIFR is maintained by use of the International Frequency Information Circular (BR IFIC), which is circulated periodically to issue notifications and the results of coordination activity.

Besides providing an agreed mechanism for international spectrum management, the ITU-R BR also produces technical information that can be of enormous benefit for the radio-planning engineer. The ITU has produced more than 4000 published titles relating to radio technology and regulation, and this is too vast and important a resource for any radio engineer or designer to ignore. These documents are produced through various study groups. The current study groups (note that study groups 2 and 5 are not listed as being active at present according to the ITU web site) are as follows:

- Study group 1 (spectrum management).
- Study group 3 (propagation).
- Study group 4 (fixed-satellite services).
- Study group 6 (broadcasting services).
- Study group 7 (science services).
- Study group 8 (mobile, radiodetermination, amateur and related satellite services).
- Study group 9 (fixed services).

These groups produce useful information for the radio engineer. Some of these are as follows:

Study Group 1
Handbook on National Spectrum Management
Computer-aided Techniques for Spectrum Management (CAT)
Handbook on Spectrum Monitoring
Development of the ITU-R SM series

Study Group 3
Terrestrial Land Mobile Radio wave Propagation in the VHF/UHF Bands
Development of the ITU-R P Series

Study Group 8
Land Mobile Handbook (including Wireless Access) Volume 2: Principles and Approaches on Evolution to IMT-2000/FPLMTS
Land Mobile Handbook (under development) Volume 3: Dispatch and Advanced Messaging Systems

Study groups 1 and 3 have also developed software programs and databases that the network designer can take advantage of to assist in the planning process. More information on what is available can be found on the ITU web site, listed at the end of this chapter. Some of the more relevant propagation recommendations for mobile radio network design are shown in Table 2.1. Each recommendation is normally referred to as ITU-R P.*x-y* (for propagation) or ITU-R SM.*x-y* (for spectrum management), where *x* is the number shown in the table and *y* is the version number of the recommendation. In the table below, the version numbers are omitted because these will change with time.

**Table 2.1.** List of some ITU recommendations appropriate to modelling of mobile networks.

| Number | Title |
|---|---|
| 310 | Definition of terms relating to propagation in non-ionized media |
| 341 | The concept of transmission loss for radio links |
| 370 | VHF and UHF propagation curves for the frequency range 30–1000 MHz (suppressed; now replaced by 1546, but still used as the basis for some coordination models) |
| 372 | Radio noise |
| 452 | Prediction procedure for the evaluation of microwave interference between stations on the surface of the Earth at frequencies above 0.7 GHz (see Chapter 4 for relevance to mobile networks) |
| 453 | The radio refractive index: its formula and refractivity data |
| 525 | Calculation of free-space attenuation |
| 526 | Propagation by diffraction |
| 528 | Propagation curves for aeronautical mobile and radionavigation services using the VHF, UHF and SHF bands |
| 529 | Prediction methods for the terrestrial land mobile service in the VHF and UHF bands (suppressed, now incorporated into 1546) |
| 616 | Propagation data for terrestrial maritime mobile services operating at frequencies above 30 MHz (suppressed) |
| 620 | Propagation data required for evaluation of coordination distances in the frequency range 100 MHz to 105 GHz |
| 678 | Characterisation of the natural variability of propagation phenomena |
| 833 | Attenuation in vegetation |
| 834 | Effects of tropospheric refraction on radiowave propagation |
| 1057 | Probability distributions relevant to radiowave propagation modelling |
| 1058 | Digital topographic databases for propagation studies |
| 1144 | Guide to the application of the propagation methods of Study group 3 |
| 1145 | Propagation data for the terrestrial land mobile service in the VHF and UHF bands (suppressed) |
| 1238 | Propagation data and prediction methods for the planning of indoor radiocommunication services and radio local area networks in the frequency range 90 MHz to 100 GHz |
| 1406 | Propagation effects relating to terrestrial land mobile service in the VHF and UHF bands |
| 1546 | Method for point-to-area predictions for terrestrial services in the frequency range 30–3000 MHz |

The ITU also organises World Radio Conferences (WRC) every 2 or 3 years. The WRC reviews and if necessary revises the radio regulations. The WRC also addresses any matter of international importance and reviews the activities of the Radio Regulation Board and the Radio Bureau, and instructs them on new tasks. The ITU also publishes the results of these meetings.

Additional information about the ITU can be found on their web site at www.itu.int, and this is well worth visiting for any radio engineer, designer or spectrum manager.

### *2.2.2 ICAO*

Civil aviation is an activity that has to be coordinated throughout the entire world. Because of this, the allotment of spectrum and standardisation cannot be left up to individual nations, or to regional organisations. Therefore, radio issues relating to commercial aviation are handled by a single organisation.

The International Civil Aviation Authority (ICAO) is, like the ITU, an agency of the United Nations. It has been working on standards for international aviation since the end of the Second World War. ICAO actively works with the ITU in respect of aircraft communications, radionavigation and radar systems, both fitted to aircraft and deployed on the ground. As such, it has a position of defending spectrum allocated to aviation against other potential uses that may compromise safety anywhere in the world.

Because of the massive rise in commercial aviation, ICAO is developing standards for new radio systems to cope the expected demands. Because aviation is present all around the world and in every country, and because aviation radio systems involve safety of life, adoption of new technologies must be phased in very slowly to allow operators to update their existing equipment, and also to ensure that the new technologies have been thoroughly tested and validated.

## 2.3 Regional Bodies

The ITU is split into three regions, and each of these holds regional radio conferences. Although these cannot modify the radio regulations, their activities feed into the activities of the WRCs. Also under the remit of the ITU, other bodies within a region can be created to deal with coordination issues, on any level down to bilateral Memoranda of Understanding (MoU) that establish cross-border rules to be applied for particular services across their adjacent borders.

Some examples of regional bodies set within the ITU framework are briefly described below.

### *2.3.1 CEPT*

CEPT (European Conference of Postal and Telecommunications Administrations, in its English form) is a European-wide body that was established in 1959 to promote cooperation on commercial, operational, regulatory and technical standardisation issues [standardisation has now passed to ETSI; see below, but other issues are also addressed by the European Radiocommunications Office (ERO), which is a part of CEPT] within the overall framework of the European Community (EC). Since 1992, CEPT essentially became a body of policy-makers and regulators covering the expanded post cold-war Europe, with 45 countries taking part. CEPT is dedicated to increasing harmonised use of the spectrum with the emphasis on practical cooperation.

### *2.3.2 CITEL*

CITEL (Organisation of American States for the Inter-American Telecommunications Commission in its English form) is a pan-American organisation covering 35 countries in the Americas, and based in Washington DC. It is the oldest of the regional organisation, and serves the Caribbean as well as North and South America. It is the main advisory body to the Organisation of American States (OAS) and exists to promote flexible regulatory frameworks that can allow new services to be introduced harmoniously in a timely manner, while seeking efficient and fair use of the radio spectrum. It also provides a mechanism for technical standardisation via its standardisation committee to promote interoperability for networks and services within the region.

### *2.3.3 Regional Commonwealth in the Field of Communications*

The Regional Commonwealth in the field of Communications (RCC) coordinates development of networks, technical standards and spectrum management throughout the countries of the CIS (Commonwealth of Independent States). It is also active in joint research and development programs and training as well as interacting with other international bodies. It has 12 member states.

### *2.3.4 Asia-Pacific Telecommunity*

The Asia-Pacific Telecommunity (APT) was established in 1979 and now has 33 members. It acts as means of allowing member states to prepare for global ITU conferences, and is dedicated to the expansion of telecommunications throughout the Asia-Pacific region, the development of regional cooperation, the exchange of information and the development of common standards.

### *2.3.5 Gulf Cooperation Council*

The Gulf Cooperation Council (GCC) exists to allow Persian Gulf states to coordinate telecommunications tariffs, allows the adoption of the GSM standard throughout the region and as a regional input into the ITU.

### *2.3.6 African Telecommunications Union*

The African Telecommunications Union (ATU) coordinates the development of African telecommunications by serving as a regional discussion forum.

### *2.3.7 National Bodies*

The ITU provides a framework for international regulation, but most of the actual work is carried out by each member state's national administration. To illustrate national example, we can use United Kingdom as an example.

In the United Kingdom, OFCOM is responsible for civilian spectrum and the Ministry of Defence (MOD) for military spectrum. OFCOM reports to parliament, and the MOD reports to the Cabinet Office.

National regulators perform a number of roles. These include participating in the ITU, providing advance notification that certain systems such as satellite systems are going to be put in service, coordinating with adjoining administrations, often via bi- or multilateral agreements of MoU and spectrum management within their region of responsibility.

National spectrum administration consists of adjudicating licence applications, managing those licences and collecting the appropriate fees as well as performing the roles of allotment and assignment of bands of frequencies or individual channels, and dealing with interference issues. These allotments and assignments will be within the scheme laid out in the international Table of Allocations. It is regarded as acceptable for national administrations to devolve responsibility for specific parts of the spectrum to other organisations and this has been done in the United Kingdom, for example. This is illustrated by the devolution of broadcast management to the BBC and IBA, short-term outside broadcast to JFMG, civil aviation management to the CAA and NATS and amateur services to the RSGB. Major allotments (such as those for GSM and UMTS) are largely managed by the operators themselves, who will determine the actual assignments to be used. The situation is likely to change with the advent of spectrum trading (currently being introduced in the United Kingdom), but it is likely to see more rather than less devolution of responsibility.

Other countries will differ significantly in their methods of managing the spectrum from an administrative (rather than technical) standpoint, and to find out how this is done in each country; it is usually a good idea to start by visiting the Internet web site of the relevant authority.

## 2.4 Other Useful Bodies

### *2.4.1 Introduction*

In addition to the bodies described previously, there are a variety of other organisations that generate standards and produce technical material that can greatly assist the radio engineer. This section provides a brief description of some of these organisations. It is recommended that engineers familiarise themselves with the type of materials available from these organisations and routinely check for new information. In most cases, a visit to the relevant web site is the first step. The web site addresses at the time of writing are found at the end of the chapter. In this list we have not included generic academic search engines, which are also an invaluable source of technical papers, as most engineers should be familiar with.

### *2.4.2 ETSI*

The European Technical Standards Institute (ETSI) is an independent, not-for-profit organisation that hosts the Information and Communication Technologies (ICT) standardisation body. This produces standards for fixed services, TV broadcast and information services as well as mobile technologies. In the mobile environment, ETSI has produced some of the most influential standards in modern mobile communications, such as GSM and

TETRA, and they are founding members of the 3GPP initiative. ETSI works in partnership with other bodies such as the ITU, CEPT, international fora and national bodies. All of the ETSI-produced standards are available for free, and can be downloaded from their web site, which at the time of writing is www.etsi.org. The standards obtained from ETSI represent the most authoritative texts on that technology and thus should be used as a reference for any engineer working in those areas. The specific sources mentioned have been included, as they have proved beneficial to the authors in the past.

### *2.4.3 COST*

The council for European Cooperation in the field of Scientific and Technical Research (COST) is a long-running European body supporting cooperation among scientists and engineers across Europe, which is supported by the EU framework Programme. Some 34-member states and others from beyond Europe cooperate to COST in some form. COST covers a very wide range of domains across human scientific endeavour. The Telecommunications and Information Science and Technology (TIST) domain is the most relevant to mobile radio network engineering. TIST research sub-domains include 'radio propagation and interference' and 'wireless and mobile communications' as two separate domains. One of the most relevant recent research projects (termed 'actions') to the mobile radio engineer is COST Action *231 'Evolution of Land Mobile Radio (including personal) Communications'* (also often referred to as COST 231), which summarises relevant research in mobile radio.

### *2.4.4 IEEE*

The Institute of Electrical and Electronic Engineers (IEEE) is a nonprofit organisation and is a professional association for the advancement of technology covering a wide range of applications, from medical to aerospace systems, and including telecommunications. The IEEE acts as a major source of technical and professional information to engineers, and it produces 128 transactions, journals and magazines and holds more than 300 conferences a year. It also creates standards, of which there are about 900 active and more than 400 in development, including the currently vogue IEEE 802 series.

### *2.4.5 IET*

The Institute of Engineering and Technology (IET) has replaced the Institute of Electrical Engineers (IEE) following a merger with the Institute of Incorporated Engineers (IIE). The IET provides engineers with valuable information contained in its books and conference proceedings, IEE proceedings, Electronic Letters and also via its library and archive service. The library is based in central London, UK.

### *2.4.6 NTIS*

The National Technical Information Service (NTIS) is an agency of the US Department of Commerce. It has acted as a central resource for access to government funded scientific, technical, engineering and business research for more than 60 years. The service has access to more than three million publications covering over 350 subject areas.

### 2.4.7 *NTIA and ITS*

The US National Telecommunications and Information Administration (NTIA) is responsible for providing technical information to the President on telecommunications. As well as managing federal use of the spectrum and other tasks, it sponsors the Institute for Telecommunications Sciences (ITS), which performs radio-related research work and publishes some of the results. Some of these are of benefit to the mobile radio engineer.

For further information:

Much of the material for this chapter has been created from material on the web sites of the organisations discussed. At the time of writing these web sites are as follows:

1. The ITU: www.itu.int
2. CEPT: www.cept.org
3. ETSI: www.etsi.org
4. CITEL: www.citel.oas.org
5. RCC: www.rcc.org.ru
6. Telecommunications Regulation Handbook
   Module 1: Overview of Telecommunications Regulation
   Ed. Hank Intven, McCarthy Tetrault
   The World Bank
   ISBN 0-9697178-7-3
   http://www.infodev.org/files/1080_file_module1.pdf
7. APT: www.aptsec.org
8. COST: www.cost.esf.org
9. IEEE: www.ieee.org

# 3

# Mobile Radio Technologies

## 3.1 Introduction

Mobile radio is used for a wide variety of applications over land, sea and in the air, and these networks are based on a range of technologies. Some of these are cutting edge, but many are legacy systems based on older designs. Such systems are used for voice communication, a variety of data applications and can be used to assist navigation and localisation. In this chapter, we have chosen to include aeronautical and maritime mobile broadcast – not including commercial broadcast such as TV and radio – and we have also included DF and localisation systems, because it is important to examine the variety of systems that fulfil the need for mobile users.

It is worth reviewing these technologies in order to gain a good overview of all types of mobile radio networks, and not just the ubiquitous mobile phone technology market. In this section, we will look at some radio networks and the technologies used to implement them. This will include older system types because many operators still use them; indeed these networks are continually being modified and in some cases still rolled out. Often this is because of the time taken to agree and introduce new systems for worldwide application, for example the new systems for civil aviation currently undergoing agreement and specification by EUROCONTROL/ICAO and the FAA are due to be introduced some time after 2015. Other reasons for using older technologies include the length of procurement cycles, particularly in the military sector, cost issues and indeed cost effectiveness; in some cases an older, mature technology may meet the requirements of a network without the need to resort to newer, more expensive and often more technically risky technologies.

We will not attempt to fully describe each technology but rather those aspects that are relevant to actual network design; there are many other books dedicated to describing each of these technologies in great depth, and these should be used in order to understand the aspects that do not influence the physical design of the network. As has been noted in other parts of the book, most of the design aspects of a radio technology are fixed and are not under the control of the radio network designer, and indeed an in-depth knowledge of how

Mobile Radio Network Design in the VHF and UHF Bands: A Practical Approach
*Adrian W. Graham, Nicholas C. Kirkman and Peter M. Paul* © 2007 John Wiley & Sons, Ltd

they work is not always necessary (but is always desirable). Those who wish to find out more about other aspects of a particular technology should refer to the organisation responsible for its specification.

## 3.2 Mobile Radio Network Users and Networks

One of the reasons for the range of radio technologies available is that each technology offers a slightly different service, which offers different benefits for different user groups. The suitability of a technology to a specific network requirement will depend on many things, which will include such considerations as follows:

- Cost-effectiveness.
- The technology being mandated for a particular application.
- Support for authorised frequency bands.
- Functionality.
- Security.
- Ability to meet service levels (quality of service, service type, coverage area).
- Acceptable risk levels (ability of technology to deliver, reliability of radio infrastructure suppliers).
- Availability (particularly for new radio technologies).
- Availability of accessories (hand-portable radios, other system components).
- Interoperability with legacy systems or other systems within range of the service area.

Each of these factors will have a different level of importance for a particular project. For emergency service networks, security and ability to meet the service levels may be of the highest importance. For commercial networks, the design will have to be balanced against the budget and cost may be the critical factor, or in a highly competed market, ability to meet service levels and functionality may be the crucial aspects. In most cases, there will be a trade-off between several of these factors. Sometimes the decision will be very straightforward, and it may well be completely constrained, for example when UMTS has been specified as the technology to be used for a particular application, but on other occasions it may be necessary to prepare complex studies into all aspects of the network before a technology can be selected. In such cases, it is vital to make the right decision, since it is one that will have to be lived with for the life of the network and in commercial systems, it may mean the difference between the success and failure for the entire business. We will now look at a few typical groups of mobile radio technologies.

## 3.3 Types of Mobile Network

In this chapter, we will split radio network types into their main mode of operation. This affects both their applicability and the planning methods used to design them:

- Direct mode.
- Single site.
- Simulcast (quasi-sync).
- Trunk.

- Cellular.
- Composite.
- Other approaches.

We now look at the characteristics of these types of networks and when they might be applied.

## 3.4 Direct Mode

The direct mode of operation is where one mobile station talks directly to another mobile, or group of mobiles, without going through a fixed base station. This is still used within some networks as an alternative to the mobile-fixed-mobile method, for example by police subscribers working near an incident, but it is more typically used by military forces. This is illustrated in Figure 3.1.

Calls may be made to individual users or groups of users, depending on which are tuned to the same channel, but there is no distinction between different subscribers and thus the call goes out to all (an 'all-informed' net). Calls are typically single-channel simplex in nature (only one subscriber can talk at a time).

Direct mode networks are mostly used for military radio tactical networks over short distances and normally limited to individual squads. They are also typically used by other organisations requiring temporary communications over relatively small areas, such as mountain rescue teams, maritime search and rescue, inter-ship communications, marshals for sporting and other public events (e.g. car, motorcycle and bicycle racing, marathons, protests and so on) and for businesses where internal communication is required to staff on the move throughout the business such as security and warehouse personnel.

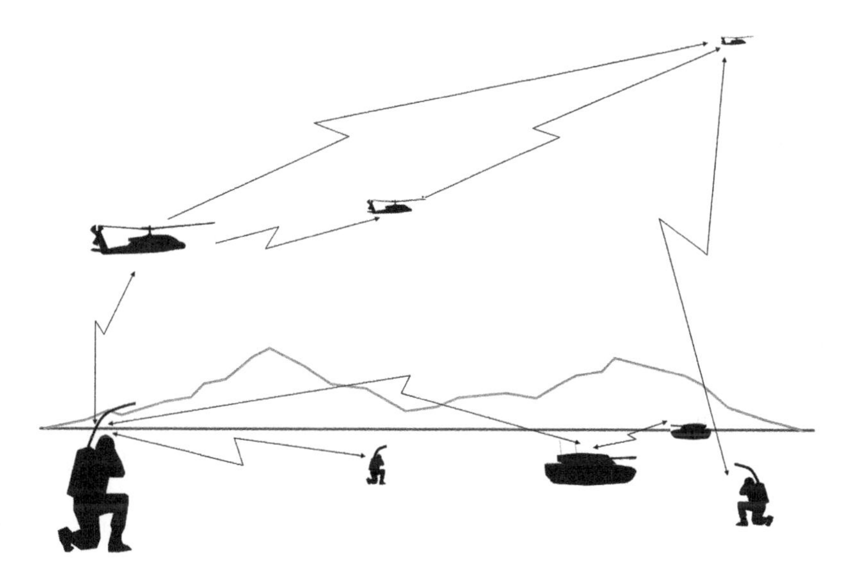

**Figure 3.1.** Direct mode.

Direct mode radio networks have a number of advantages for these operators, including:

- No costs for infrastructure.
- Very fast deployment time (only time for units with mobile radios to deploy).
- Not limited to a set service area (can be deployed anywhere).
- Simple to operate, even for nontechnical operators.
- No permanent footprint (nothing left once subscribers depart).
- Short communications range among subscribers is useful to prevent enemy detection, intercept and direction finding.

There are also a number of major constraints with direct mode:

- Difficult to plan in advance without deploying mobile personnel as essentially 'fixed' elements over the duration of the activity (i.e. some subscribers would have to go to a specific location and stay there as essentially fixed stations). Otherwise, can only be planned statistically, and performance ensured by operating procedures rather than direct planning.
- Short range of links will limit maximum coverage, and there will be no guarantee that every subscriber can access all of the other subscribers.
- Only suitable for small number of subscribers, otherwise traffic will limit the performance of the system.
- Ability to communicate depends on radio discipline. A single rogue user can block other calls by poor procedure or intent.
- Highly vulnerable to jamming.

Despite these limitations, direct mode operation networks will continue to exist for a long time to come, even if this is only one mode of operation with other modes available. Their use for temporary operations will continue to be required.

## 3.5 Single Site

The simplest type of mobile radio network with fixed infrastructure features a single site, providing coverage over a relatively small service area. These systems often use simplex operation, and each subscriber in the network can join in a conversation, but only one person at a time can talk. Typically, there would be a dispatcher at a central location who will provide a link to the rest of the world, such as organisational controllers. This is illustrated in Figure 3.2, which shows a dispatcher making a call heard by all mobile subscribers in the network. Only those mobiles within the direct mode range of the calling subscriber will hear the uplink call, which often leads to the situation where only the dispatcher's side of the call can be heard.

Although the single site mobile radio network is one of the simplest and oldest architectures, it is still in widespread use today for many applications, including:

- Aeronautical applications, for airport ground movements.
- Maritime, for port control and related activities.

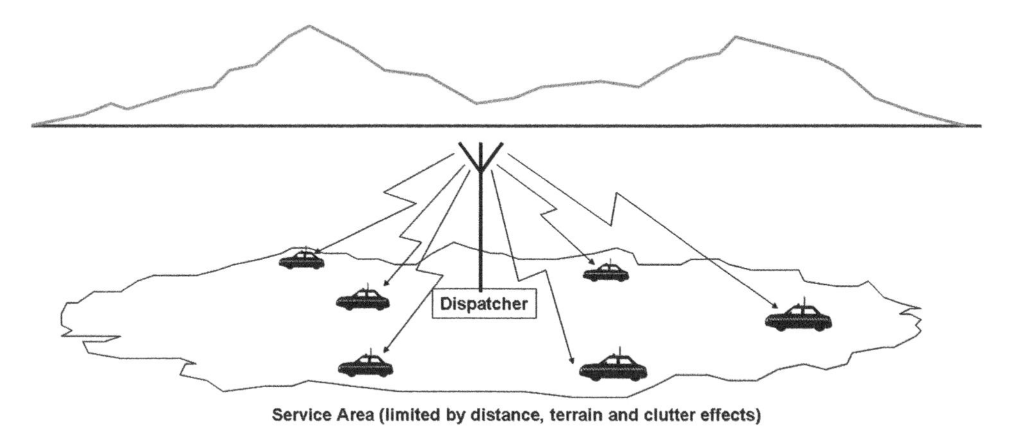


**Figure 3.2.** Single site mobile radio network, with dispatcher transmitting.

- Small businesses operating over a limited geographical area, such as taxi companies.
- Emergency services still using old technology.

The single-site architecture has a number of limitations that can make it unsuitable for many applications, including:

- Limited coverage to that achievable from a single site.
- Limited capacity to that supported by the number of subscribers and their radio discipline.
- Proneness of the system to abuse, where one caller (usually not an authorised subscriber) can block the system by continually transmitting.

However, the system still has some advantages that make it appropriate for use in certain circumstances, such as:

- It is of low-cost, both for the base station and mobile units.
- It provides longer range operation than mobile–mobile direct mode operation.
- It allows spectrum sharing between many localised users.
- It provides a simple standard for use all over the world (important for aeronautical and maritime operations).
- For combined operations where individual subscribers can hear both the downlink and uplink directions, it allows subscribers to be all-informed and thus gain an understanding of how the operation is unfolding.

Single-site networks are typically very easy to design. The only variables are antenna location, height above ground and type. Usually, COTS (Commercial-Off-The-Shelf or 'standard product') equipment is purchased for both base station and mobile. The power of both is likely to be fixed to the maximum licensed for the application. The antenna location can also often be prescribed (for example it must be at the taxi company call centre).

Frequencies will be externally assigned by a regulator or through delegation to an SMO (Spectrum Management Operator); the licensee will simply apply for a licence in the band available.

The single-site network demonstrates that radio communications (and its design) do not always depend on the latest technology and the latest planning needs, and thus there is a requirement to retain ability to design older styles of network.

## 3.6 Simulcast

Where there is a requirement for extensive geographic coverage, but low traffic demand, the simulcast or 'quasi-synchronous' (often abbreviated to 'quasi-sync') architecture can be used. In this case, instead of there being one base station, there are a number of them. Each station transmits the same information on same frequency (small offsets in frequency or launch delay may be applied to improve system performance) and synchronised to minimise interference or at least move it to less important parts of the network. The structure of a simulcast network is shown in Figure 3.3.

Mobile subscribers in areas served by more than one base station will be received at each of the serving base stations. The signals received at each base station are combined, with the system 'voting' to the system with the strongest received signal. Other signals present with a lower strength and different phase will then be ignored. This can improve uplink performance at the edge of range. For the downlink, when two base station signals arrive at nearly the same level but different phases, the mobile system can suffer from 'beating' where audio noise is produced which makes it more difficult to understand the voice message.

Simulcast has long been used for applications such as emergency services, maritime services covering large areas such as the Thames Estuary and aeronautical communications systems between the ground and aircraft.

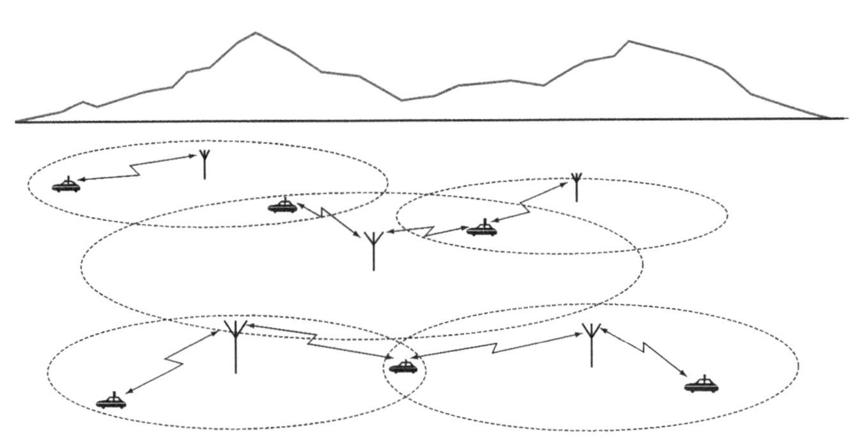


**Figure 3.3.** Simulcast (quasi-synchronous) network: note that mobile subscribers in the boundary area are served by more than one base station.

The principal benefits of simulcast are as follows:

- Its lower cost to implement than cellular systems.
- It can allow coverage of large areas using minimal infrastructure.
- It is a well-established technology and thus lowers risk to implement than new technologies.
- Because it is well-established, it is suitable for application anywhere in the world (particularly for aeronautical and maritime applications).
- Because a single frequency (or single pair) is used, all the subscribers will hear the dispatcher and may hear the uplink, thus are kept informed of activities occurring.

The main limitations are as follows:

- The mobile units must be able to work with long propagation delays and are more complex (and thus more expensive) than some other systems.
- It cannot accommodate high traffic demand.

Simulcast networks are more complex to design that single-site networks, because of the following reasons:

- The design of the network to achieve a given coverage service area is more complex because there are more variables. In general, it is desirable to choose sites that are situated in high locations with a good take-off angle with respect to obstructions within the horizon. In some cases, the aim will be to provide extra coverage for each new site added, with minimum overlap with existing sites. In other cases, such as aeronautical applications, there may be a requirement to have overlapping coverage to meet safety requirements and to provide higher availability (see Chapter 5).
- The design will have to include planning for frequency offsets, launch delay and tweaks to coverage design to adjust interference away from operationally important areas.
- It is also necessary to link the base stations together via landlines or backhaul (point-to-point) links to enable calls to be routed through the network.

Simulcast will continue to be around for some years, even though in most cases simulcast networks have been replaced with more modern technologies (such as TETRA in the case of emergency service networks). Although this is generally the case, networks operating in the aeronautical and maritime arena will persist for a long time to come.

## 3.7 Trunked Radio Systems

In general, providing more capacity to mobile subscribers requires an increased number of circuits to be available. A circuit is an individual communication channel that can accommodate up to one call at a time. Depending on the technology employed, one circuit can take up one frequency (or two frequencies in duplex operation), one time-slot or one code. However, mobile subscribers will not need to occupy a circuit all the time, and thus a single circuit can be shared by many subscribers (as described in Chapter 10). This gives rise to the concept of trunked radio networks, as illustrated in Figure 3.4. This shows a plan view

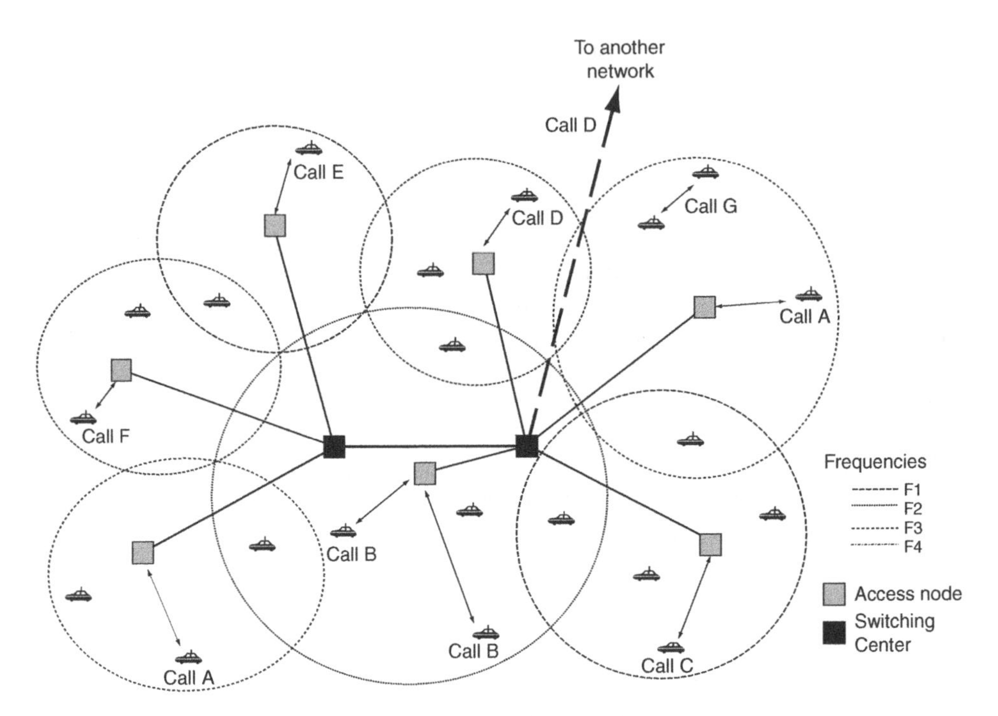


**Figure 3.4.** Trunked radio network concept.

schematic diagram of a trunked radio system with a number of calls in progress. In the following description, we will assume that there is only one channel available to mobile subscribers in the service area of each access node at a time. Also note that unlike the previous examples, this network features a number of frequencies used to allow concurrent use in areas where access node coverage is overlapping. Note also that frequencies can be reused in areas far enough apart to prevent interference (see Chapter 13).

The access nodes are the fixed base stations, providing access to the rest of the wired network. Switching centres will route calls as appropriate. This means that for Call A, which is to another mobile in a different access area, the call is only broadcast on the relevant access node and not to every access node in the whole network. This means that concurrent to Call A, the other calls shown can be carried at the same time. Call B does not cross access nodes, but is to another mobile in the same area. In this case, the call is handled locally by the access node without recourse to the switching centres. Calls C, E and F are all made to a central control room and thus do not reappear in the mobile part of the network. Call D is made to a completely different network which could be another radio network or a landline system, via one of the switching centres. Finally, while all this is going on, Call G is made between two mobile units on an entirely different frequency and they make no use of the fixed infrastructure. Thus, in total, there are seven concurrent calls being handled. Each access node has one call in progress, and the backhaul infrastructure is carrying five calls simultaneously.

Note that there are other users in the system who are not concurrently making calls, but who are still network subscribers. Twenty-one subscribers are shown, and they will make calls at different times. Thus in this simple example used for illustration, 21 subscribers are served by seven access nodes and four frequencies. In practice, real systems will be larger in scope and more efficient in terms of numbers of subscribers served by access nodes and frequencies.

In essence, the trunked radio architecture provides a pot of shared channels available for subscribers.

Trunked radio systems have several advantages over the network architectures discussed so far, which are as follows:

- The service area can extend over any size of geographic area, including very large areas.
- It is spectrally efficient, through frequency reuse.
- It has many signalling features such as group calling, emergency alert and selective mobile inhibit (to remove stolen or lost radios from the list of subscribers).
- It has a roaming capability, so subscribers can easily move through the whole network area.
- It is more cost-effective than cellular systems for systems that do not need their high traffic capability.
- The infrastructure requirements are lower than for cellular systems, so they can be deployed more quickly and cheaply.

Trunked radio systems have been used widely for public and private mobile radio, emergency services and military networks. Modern technologies such as TETRA and TETRAPOL can be deployed either as trunked radio network or as cellular systems and thus provide the best of both worlds.

The main disadvantages of the trunked radio network architecture are that it is more expensive than the simpler methods and can handle less traffic than cellular infrastructures.

In terms of network design, the crucial aspects of trunked radio planning include the following:

- Network coverage design to ensure network coverage matches the wanted service area. For low traffic density systems, this will involve selecting sites based primarily on their ability to cover wide areas; for high traffic density systems, sites that provide extensive coverage are not desirable.
- Providing capacity to meet traffic demands across the network.
- Adjusting coverage of individual sites to meet traffic capacity.
- Frequency assignment to minimise interference.
- Call handover.
- Backhaul design.

Trunked radio systems are more of an approach to network architecture rather than a specific technology. Examples include early military systems such as Ptarmigan, right through to many TETRA networks currently being deployed.

## 3.8 Cellular Systems

Most people are aware of mobile radio through cellular systems, which support mobile phone operation. Cellular systems started with the analogue systems such as TACS. Currently, GSM is the most popular system deployed around the world, and 3G systems such as UMTS and CDMA 2000 are currently being rolled out in various countries. The basic cellular concept is illustrated in Figure 3.5. In the figure, the designation inside each cell is representative of the frequency used, so 'F1' is frequency one and so forth; subscribers falling within one cell will be served by the fixed infrastructure present in that cell. The repeating pattern is representative of the approach taken to reuse frequencies, albeit usually with greater separation using more frequencies and also modified to adjust for real world condition. Also, in practice the cell sizes will be constrained by terrain, clutter and traffic demands and thus will vary in size and shape across the network. For high traffic demand areas, each cell will be split into smaller cells, often by using directional antennas to split the cell into three or four sectors, each capable of handling part of the traffic demand (see Chapter 14).

The strengths of the cellular approach include the following:

- The ability to handle very large traffic demands, and thus many subscribers.
- High volume for mobile production means low cost per unit.
- Good spectral efficiency.
- Ease-of-use for the subscriber; all connections, handovers and so forth are handled automatically by the system so the subscriber need not have any technical expertise at all.
- Supports roaming, including international roaming.
- Because of the high traffic handling capability, traffic-intensive services can be provided (e.g. video, music).

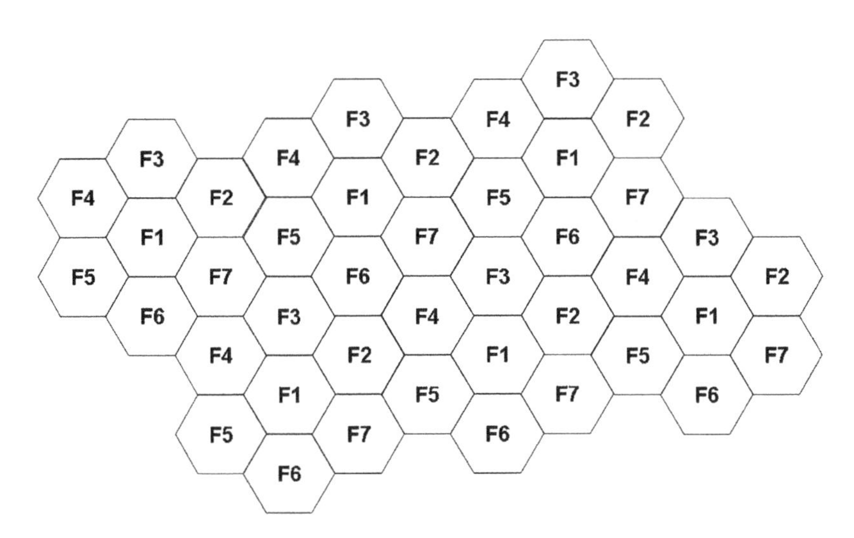


**Figure 3.5.** Basic cellular concept.

The limitations of cellular systems include the following:

- High cost to implement and maintain.
- Very complex to design.
- Need for a large number of fixed masts; this can be very unpopular with the public.

The radio aspects of the design of cellular systems are more complex than the other systems described, and this complexity has increased with the 3G designs. In general, the design must balance coverage and traffic capacity at the same time, keeping within the constraints of the frequency allocation and avoiding interference. This will often mean overlaying cell layouts on top of one another and providing repeaters and micro-cells to cover troublesome areas. The system must also be capable of providing regions for mobile subscribers to be 'handed over' between fixed base stations as they move around the customer. In addition, the network must be able to route and handle traffic through the backhaul network, all the while maintaining a high level of availability and reliability for subscribers.

3G systems in particular pose a significant design challenge. This is because there is only one channel that is shared by all users. Because of the way the technology works, each subscriber appears as noise to the other; the more the subscribers that are using the service at a time, the higher the noise and thus the higher the signal required to operate in the presence of that noise (see Chapter 13). Dynamic power control is used to manage this situation as far as possible, and thus the link budgets used (see Chapter 5) are variable. Also variable during design are the locations of subscribers as well as potential base station locations and antenna heights, patterns, azimuth and tilt. Thus there are many interrelated variables that make the design difficult to produce, even though each element of the design involves the same principles involved in design of previous systems. In practice, the location of potential base stations may be constrained for legacy reasons (for example using the fixed infrastructure locations of a GSM system as the base for a 3G network), and once subscriber behaviour has been adequately quantified to the necessary resolution (see Chapter 10), then some of the design constraints become fixed.

Cellular systems are predominant in terms of revenue and subscribers by a long way in the mobile market and will continue to be so for the foreseeable future.

## 3.9 Composite Systems

In practice, many systems will be a composite of the approaches discussed. For example a TETRA system may be designed on a cellular basis in the urban areas, but as a trunked radio network in the rural areas. The system may also allow individual subscribers to talk to one another without requiring fixed infrastructure via direct mode. Where the technology allows, the designer should determine which approach is likely to be most appropriate for specific parts of the network and not just for the network as a whole, because it may be appropriate to vary it.

## 3.10 Other Approaches

The approaches discussed so far cover the majority of the mobile networks in use, but others will require slightly different design approaches. These may be broadly characterised as

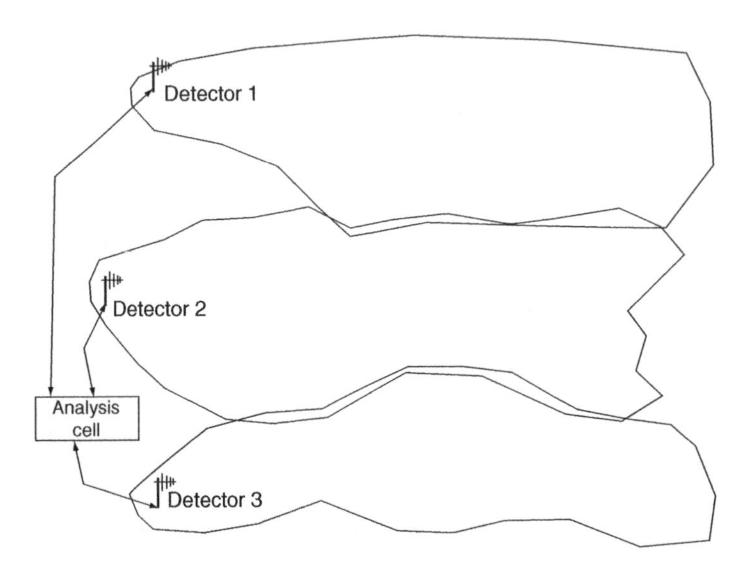


**Figure 3.6.** Maximum coverage for intercept network; the irregular areas are the coverage area for each station. Each station is linked to an analysis cell by a backhaul link.

coverage maximisation with minimum equipment, and concurrent coverage by a set number of fixed stations.

In the first case, the idea is to ensure no overlap of coverage as far as possible and the maximum amount of coverage per station, so that the entire network covers the largest possible area. In this situation, it is acceptable if only one station detects a transmission, for example. This is illustrated in Figure 3.6, which shows a network set up to detect transmissions over a particular area.

Each detector is shown with its coverage area. The coverage areas are irregular due to terrain and clutter limiting coverage. In particular, for this type of display, it is likely that the horizontal boundaries would be caused by hill ridges, so that the area of overlap is small. This is not a problem since all detections are passed back to an analysis cell that collates the data for further processing. In this type of network, the aim is to maximise coverage with the minimum amount of assets. This type of architecture is also common for intercept systems, monitoring systems and distance measuring equipment.

Often, however, having coverage from only one base station is not enough; it is necessary to have two or more stations in range concurrently. This calls instead for the greatest possible area served by the requisite minimum number of sites. This is a common requirement for direction finding (DF) systems such as military systems, VHF omni-directional ranging (VOR), digital VOR (DVOR), digitally resolved direction finding (DRDF) and other aeronautical and maritime systems.

An illustration of the requirements is shown in Figure 3.7, which shows the total coverage area for a three-element DF system and also the total effective coverage, given that the system needs at least three concurrent DF hits in order to automatically localise the transmission.

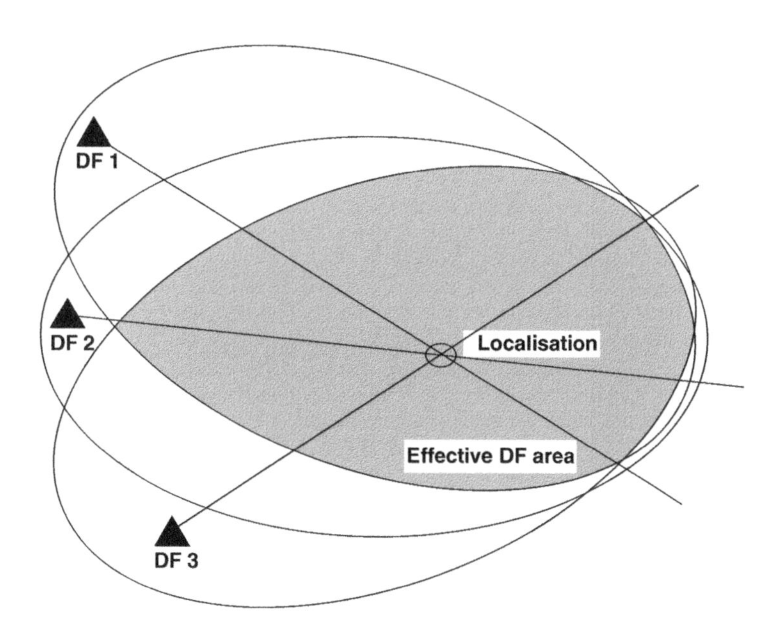


**Figure 3.7.** DF system that requires three simultaneous DF hits for automatic localization.

Figure 3.7 shows that the effective service area is restricted to the shaded area where all three elements provide coverage, not the composite of the three individual coverage areas. An example of a localisation following detection on all three DF stations is also shown, to illustrate the point.

These other types of networks have slightly different design criteria than the other described. It is always important to understand the key requirements for any network under consideration, and not to assume that, because other networks have been designed in a specific way, the approach is acceptable for the next network design.

## 3.11 Fixed and Mobile Convergence

Currently, much business effort is going into a convergence of service provision, so that a single service is provided by a range of methods. For example a sports headline ticker tape service may be provided simultaneously by broadband wireless, fixed lines, GSM, GPRS, UMTS, WiMax and point-to-multipoint services, so that subscribers can pick it up by the most appropriate means wherever they are. This means that planning tools for such services must have the ability to design each service provision method and collate the whole design into a single system. It does not, however, mean that individual design processes need to be changed, but rather that each method has to recognise the design requirements for each part; thus mobile wireless provision must be made for those situations where the customer may not be able to access the system via another method. Although this complicates the mobile network design process, it does not fundamentally change the individual processes within it.

# 4

# The Mobile Environment Part 1: Propagation Mechanisms and Modelling

## 4.1 Introduction

In this chapter, we will be looking at the mobile radio environment and methods of modelling radio propagation in the VHF and UHF bands. This is a large subject and is perfectly capable of taking up an entire book by itself, but we will endeavour to restrict the scope of this chapter to the necessary minimum. We will not cover the basics of electromagnetic theory and the mathematics underlying the models described, but for those who seek a deeper understanding there is a recommended reading list at the end of the chapter. We will also only look at a representative sample of the models available; there are many others in use.

A radio link consists of a transmitting element, the communications channel and the receiving element. In this chapter, we will be examining the communications path from a practical perspective and introducing some radio propagation models that have found practical application in the design of mobile radio networks. Figure 4.1 shows a typical land mobile path from a fixed base station to a mobile element. The antenna on the mobile element typically receives the signal via a number of paths with varying strength and phase and seldom has direct line of site with the base station antenna.

All mobile radio network engineering depends on the models used to predict the behaviour of the radiowaves transmitted in the network, and therefore the correct selection and configuration of propagation models is of critical importance for any project. It is essential that anyone calling themselves a mobile radio planning engineer or designer understands the subject in great depth. The contents of this chapter therefore act as a starting point for further study and, since the modelling of radio propagation is continually evolving,

Mobile Radio Network Design in the VHF and UHF Bands: A Practical Approach
*Adrian W. Graham, Nicholas C. Kirkman and Peter M. Paul* 

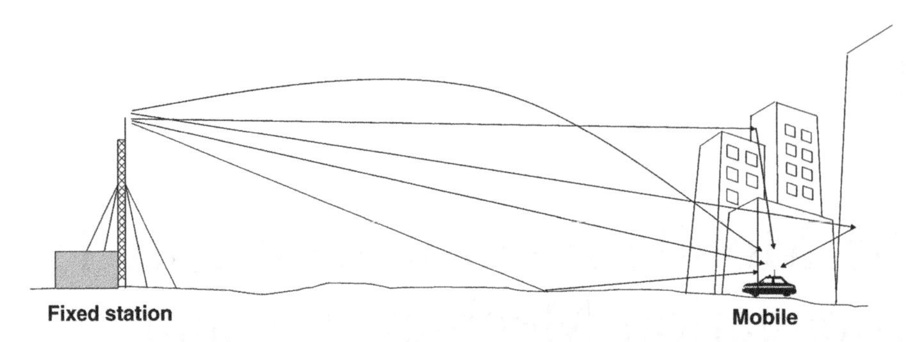


**Figure 4.1.** Radio energy transmitted from a base station to a mobile subscriber typically arrives via a range of paths. The radio signal can be reflected from buildings, refracted through the atmosphere and diffracted over and around objects.

it is the start of what should be a career-long study. We start by a short refresher on the electromagnetic spectrum.

## 4.2 The Electromagnetic Spectrum

If we restrict ourselves to the radio part of the electromagnetic spectrum, we have the naming conventions shown in Table 4.1.

Table 4.1 shows that mobile services exist in most of these bands, but we restrict ourselves to those in the VHF and UHF bands. This is because most current development is occurring

**Table 4.1.** The radio part of the electromagnetic spectrum.

| Band | Abbreviation | Frequency range | Typical applications |
|---|---|---|---|
| Extremely low frequency | ELF | Below 3 kHz | Worldwide submarine and underground communications, remote sensing |
| Very low frequency | VLF | 3–30 kHz | Worldwide ship telegraphy, long range fixed services, navigational aids |
| Low frequency | LF | 30–300 kHz | Long range for ship communications, broadcasting, navigation aids |
| Medium frequency | MF | 0.3–3 MHz | Land, sea and air mobile, broadcasting, navigation |
| High frequency | HF | 3–30 MHz | Land, sea and air mobile, broadcasting, fixed point to point |
| Very high frequency | VHF | 30–300 MHz | Land, sea and air mobile, broadcasting, navigation |
| Ultra high frequency | UHF | 300–3 GHz | Land, sea and air mobile, broadcasting, navigation, fixed, radar, satellite, telemetry |
| Super high frequency | SHF | 3–30 GHz | Fixed point to point, fixed satellite, radar |
| Extra high frequency | EHF | 30–300 GHz | Fixed and mobile point to point (line of sight), satellite |

in these bands and also because at lower frequencies the propagation mechanisms and hence planning methods differ from the situation at VHF and above. At HF and below, terrain obstacles form a very limited constraint on propagation, with atmospheric conditions, environmental noise and ground conductivity having more of an influence. Planning tools and methods for these frequencies must therefore be more concerned with atmospheric conditions, ground conductivity and noise influences such as sunspot numbers. There are dedicated tools and methods available to deal with this, such as the VOACAP, ICECAP and GRWAVE prediction systems, and International Telecommunications Union (ITU) recommendations such as ITU-RP.368 (groundwave propagation), ITU-R P.533 (skywave propagation) and supporting recommendations covering the inputs to such models such as ground conductivity maps (ITU-R P.832). We will instead focus on VHF and UHF.

## 4.3 Propagation Mechanisms at VHF and UHF

At frequencies above HF, the wavelength of the signal is small as compared to the size of terrain features, building obstructions and other clutter structures, and hence these features have a considerable effect on the propagation of RF energy. The wavelength can be easily computed from the simple formula;

$$\lambda = \frac{c}{f} \qquad [1]$$

where
$\lambda$ is the wavelength in m
$c$ is the speed of light in air (approximately $3 \times 10^8$ m/s)
$f$ is frequency in Hz (cycles/s)

This is shown in Table 4.2 for specific frequencies, to illustrate the relationship.

As a result of this, propagation in the VHF and UHF bands is typically dominated by the following propagation mechanisms:

- Diffusion between transmitter and receiver due to distance.
- Reflection from 'flat' surfaces.

**Table 4.2.** Relationship between frequency and wavelength.

| Frequency (MHz) | Wavelength (m) |
|---|---|
| 3 | 100.00 |
| 30 | 10.00 |
| 100 | 3.00 |
| 300 | 1.00 |
| 500 | 0.60 |
| 800 | 0.38 |
| 1000 | 0.30 |
| 2000 | 0.15 |
| 3000 | 0.10 |

- Scattering from irregular surfaces.
- Refraction due to changes in the atmosphere (especially with altitude).
- Diffraction due to solid obstructions.
- Absorption due to objects that attenuate RF energy.

For VHF and UHF, atmospheric effects such as precipitation (rain, fog, snow, hail) do not affect transmission unless they have a marked effect on the refractive index (see Section 4.3.4). This differentiates these bands from higher bands above approximately 6 GHz, where precipitation has a marked and indeed limiting effect on radio links.

Since almost all mobile networks are vertically polarised, we shall not be examining issues relating to horizontally polarised transmissions, except for backhaul, which we shall be looking at in Chapter 12.

Before going on to examine some popular propagation models for VHF and UHF, we will first look at these mechanisms and how they affect radio propagation in general terms.

### *4.3.1 Distance*

Any radiated signal will reduce in power the further away it is from its source. In most conditions where no other factors are present, this will take the form of the well-known free space loss equation for an isotropic antenna (shown in its logarithmic form);

$$L = 32.44 + 20\log f + 20\log d \qquad [2]$$

where
$L$ is path loss in dB
$f$ is frequency in MHz
$d$ is distance in km

This equation shows that not only does loss increase with distance, but also with frequency.

Although this is the most fundamental equation associated with radio propagation, it can rarely be used alone with confidence. This is due to the other mechanisms described below, which tend to ensure that, especially for land mobile applications, conditions approximating the free space scenario very seldom arise. In practical terms the only conditions where the free space condition is realised in practice is between two highly directional antennas with clear line of sight and no obstructions anywhere near the link. Otherwise, the free space loss equation should not be used for practical work, except on occasions as a method of determining the limiting case for radio link range (except under abnormal conditions).

The main reference for free space loss is ITU-R P.525, and other models will also incorporate a variation on this mechanism based on mathematical treatment or empirical data.

### *4.3.2 Reflection*

Reflections of radiowaves occur when the wave encounters large flat surfaces. The degree of flatness must be compared to the wavelength of the radio energy, rather than to the optical case. In many cases, the surface will not be perfectly flat, but may be undulating, yet

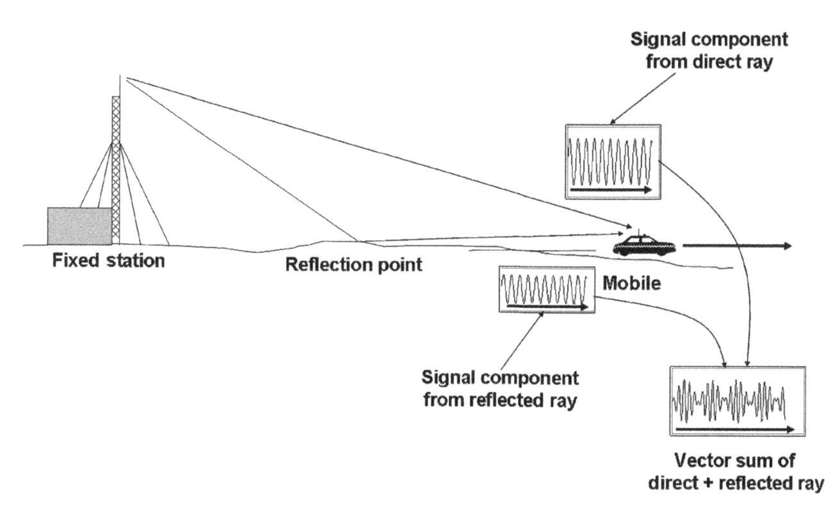


**Figure 4.2.** A mobile path showing a reflection point. The signal varies as the mobile moves, and it will vary according to the vector sum of all components received. This is typically a complex interaction, resulting in a complex received waveform.

reflection will still occur. Reflections cause changes in signal phase, possibly absorption or refraction of some of the energy (into the reflecting material), and reflection of the rest. Although the mathematics of such interactions is well understood, the conditions found in most practical situations will only approximate to the mathematical ideal. The closest approximation will occur when a signal is transmitted over a flat calm lake, but it also happens in a variety of other situations.

A reflection will provide a secondary path between transmitter and receiver, as illustrated in Figure 4.2. The two waves will interfere at the receiving antenna location, since the difference in path length (and thus arrival time) will typically be small. There will be some occasions on which the signals constructively interfere, and therefore a higher signal level will be experienced at the receiving antenna, as shown in the figure. On other occasions the signals will interfere destructively, and thus the signal received will be less than that would otherwise be received. This can be modelled explicitly either by using a mathematical construction to represent the physical scenario (see Section 4.6 on point-to-point models) or by using empirical data based on measurements (see Section 4.5 on point-to-area models).

The vector sum of the received signals will vary in amplitude as the magnitude and phase relationship between the direct and reflected wave change due to movement of the receiver. This is illustrated in Figure 4.3, which shows the vector sum in both linear and logarithmic units. The conversion between the two is simply obtained by converting the linear data to its logarithmic equivalent, while referencing the values to a positive number so that none of the values become negative. If measuring real signal variations, the display on any signal analyser will normally be similar to the logarithmic version.

The variations in the graph are occurring on the same scale as individual wavelengths, and the regions where the signal level is significantly lower than the mean are known as fades. These are discussed in greater detail in Chapter 5, which looks at statistical methods of

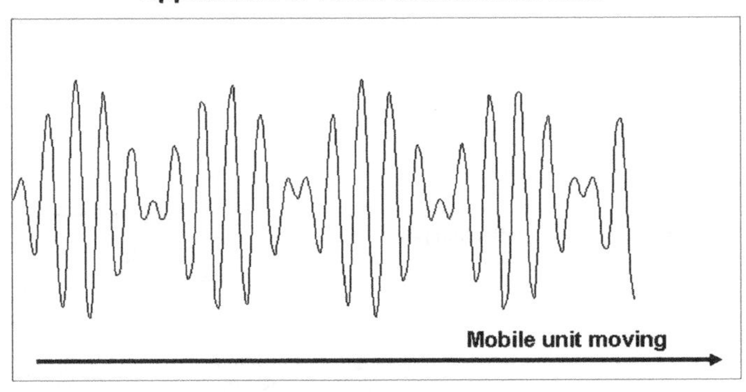


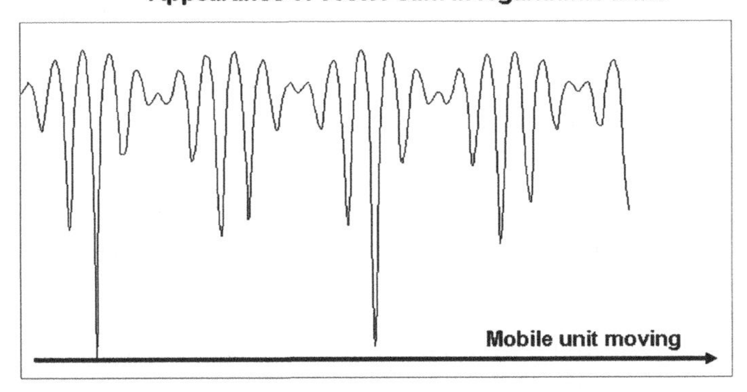


**Figure 4.3.** Vector sum of two signals in linear and logarithmic terms. Again, this is the situation for a mobile subscriber on the move.

accounting for this type of variation. The next section describes another mechanism also described in the same manner, which is scattering.

### *4.3.3 Scattering*

Scattering is similar to reflection, but is the condition that arises when the surface is not smooth and thus the reflections are not coherent or organised in any manner. It is therefore more typical of the situation encountered in practice. The difference between scattering and reflection can be illustrated by an optical example. If one were to look at a sunset over a calm lake on a clear evening, one may see a reflection of the sun on the water. By contrast, other details will also be visible; the lake itself, the shoreline, clouds, etc. These features are still illuminated by the same source (the sun), but the mechanism in this case is scattering rather than reflection, and it is the mechanism by which our eyes see them. This illustration is shown graphically in Figure 4.4. Also shown is a reflection point from a nearby island, caused by the angle of the ground at that point relative to the angle of the incident ray and the position of the receiving eye.

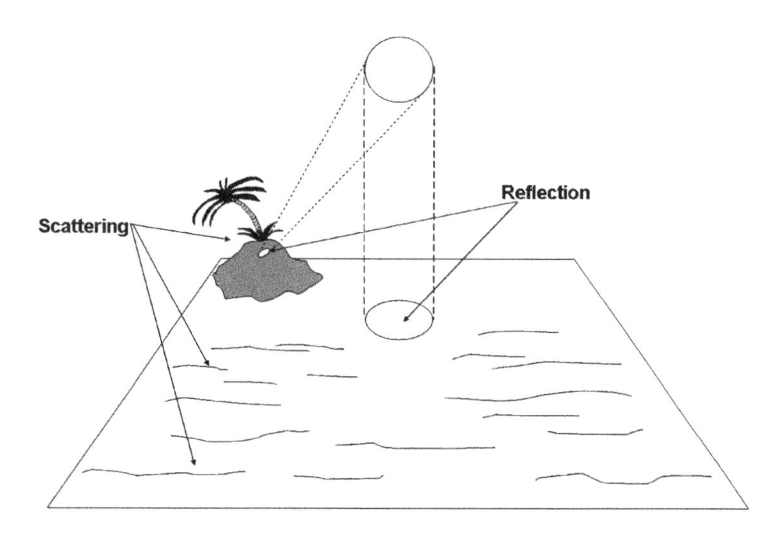


**Figure 4.4.** This illustration uses an optical example to show the principles behind radio scattering and reflection. From the vantage point of an observer looking at this scene, the energy from the Sun arrives directly (allowing us to see the Sun itself) by a reflection in the water (highlighted) and also from a point on the island. The principle of light scattering allows the observer to see the island and the surface of the sea even though they are not actually producing visible energy themselves. If scattering did not exist, we would be unable to use our sense of vision as we do. For radio frequencies, the same principles as illustrated in this example still apply.

Scattering is far more complex than reflection to model accurately, and even small changes in receiver position result in large changes of received signal level. This will occur over distances of $1/2\lambda$ (see Table 4.2), and the mechanism is known as fast fading. It works in a similar way to reflections in that the received signal is a vector summation of contributions from different paths. There are, however, likely to be a larger number of scattering paths each of which will have a lower strength than the direct reflection case. The form of signals with a scattering component is of the same type as that shown in Figure 4.3, but the signal variation is likely to be most complex.

In most cases, sufficient detail to accurately model the mobile environment will not be available, and in any case it will change rapidly as elements move. This will not only depend on the movement of the receiver but also on any other sizeable moving object in the environment that will scatter radio energy. This means that for most mobile applications, discrete modelling is neither possible nor desirable and instead, account of this is taken by applying statistics to the level of the received signal (see Chapter 5). In practice, most propagation models will provide a median received signal level based on macro effects of distance, terrain, clutter and atmospheric effects, which together account for the mechanisms termed 'slow fading'. Other model types based on ray-tracing do attempt to take these factors directly into account. They are computationally intensive and, as discussed, require that the environment be modelled to a high degree of precision. If the actual situation varies at some time from that of the model, then this high degree of precision does not lead to a high degree of accuracy.

Scattering from refractive irregularities in the troposphere (the region of atmosphere from ground level up to about 8 km at the poles and some 18 km at the equator) can be beneficial and allow long-range communications to occur over several hundred kilometres, well beyond the radio horizon. These are known as troposcatter links. Because of the scattering losses involved, it is usually necessary to use high power and directional antennas to make such links work. However, the same mechanism can also allow propagation beyond the desired path length in nontroposcatter links and often lead to interference to other networks at some distance.

### 4.3.4 Refraction

Refraction is caused by variations of the atmospheric radio refractive index at different levels in the atmosphere. For terrestrial mobile networks, the most important areas are those relatively close to the surface of the Earth (typically up to about the first kilometre). In general, the refractive index reduces with height due to the lower atmospheric pressure, but there will be variations due to weather effects.

Typically, the effect of the variation in radio refractive index will cause radiowaves to 'bend' downwards, allowing reception of radio signals over the apparent horizon, even without relying on diffraction effects (see next section). This means that the radio horizon is further away than the optical horizon, as illustrated in Figure 4.5a.

It is possible to model this situation by making a correction to the apparent radius of the Earth, which in real life is approximately 6370 km. By making the apparent radius larger to match the median refracted path of radiowaves, we can alter the geometry such that the bent

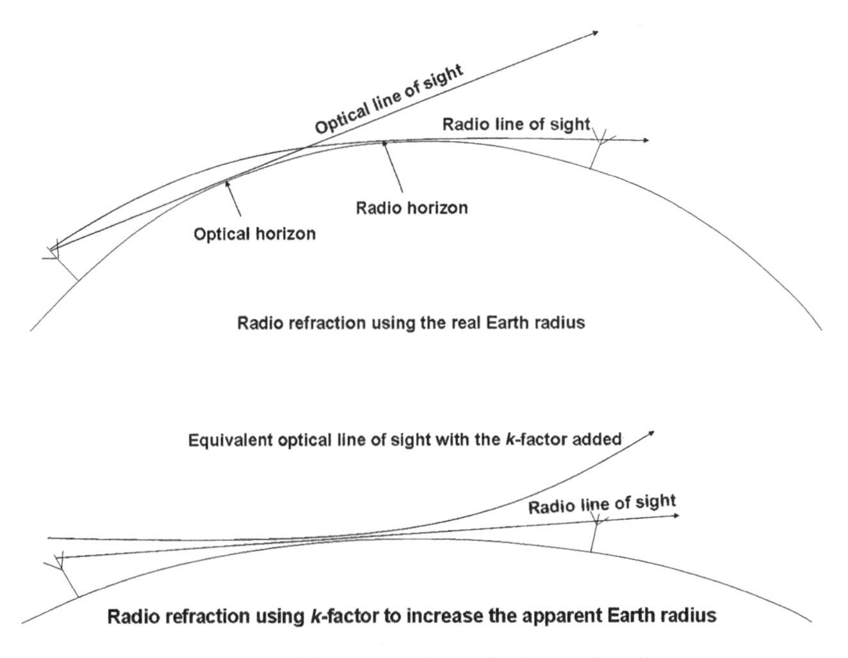


**Figure 4.5.** Illustration of the effective Earth radius.

radiowaves appear straight, as shown in Figure 4.5b. This revised radius is known as the 'effective Earth radius' and for North-western Europe is typically 4/3 of the actual size of the Earth, with a figure of 8500 km commonly used. The multiplier of 4/3 is referred to as the *k*-factor. It is important to note, however, that this is based on atmospheric conditions, which will vary in actual value continuously, and this can lead to variations in the mean over extended periods. The ITU produces maps of median refractive index that can be used for planning in specific parts of the world for different times of the year (ITU-R P.453). The variation of $k$ can cause the effective Earth radius to increase, even to the point where the Earth appears flat (causing the phenomenon known as ducting). There can also be occasions when the radio energy is bent into the ground before reaching the real horizon, which is known as super-refraction.

The radio refractive index can be computed from measurements of atmospheric pressure, water vapour pressure and absolute temperature by:

$$N = N_{dry} + N_{wet} = 77.6\frac{P}{T} + 3.732 \times 10^5 \frac{e}{T^2} \qquad [3]$$

where

$P$ is atmospheric pressure (hPa)
$e$ is water vapour pressure (hPa)
$T$ is absolute temperature (K)

The long-term height dependence on the refractive index is given by:

$$N_s = N_o \exp\left(\frac{-h_s}{h_o}\right) \qquad [4]$$

where

$N_o = 315$
$h_o = 7.35$ km
$h_s$ is the height in km

The gradient of the refractive index is important and is provided in ITU-R P.453 for the first 1 km of the atmosphere. This value, $\Delta N$, is used to compute the *k*-factor to determine the effective Earth radius to be used in specific circumstances when modelling radio propagation. A value of $-40$ is typical, and approximates the value $k = 4/3$.

The value of $k$ can be derived from:

$$k = \frac{157}{157 + \Delta N} \qquad [5]$$

Figure 4.6 shows a graph of the standard deviation of time variability around the mean signal strength received at a particular distance, derived from ITU-R P.1406. It demonstrates how the median signal strength received from a transmitter varies due to changes in the value of $k$ over time. It shows that the variability is highest for UHF frequencies in the maritime environment, where the $k$ factor can often be large. This means that when planning maritime networks or links, it is important to consider how much further than the intended link the energy will travel. This is important from an interference point of view in the civil world, and for unintended signal detection by the enemy in the military context.

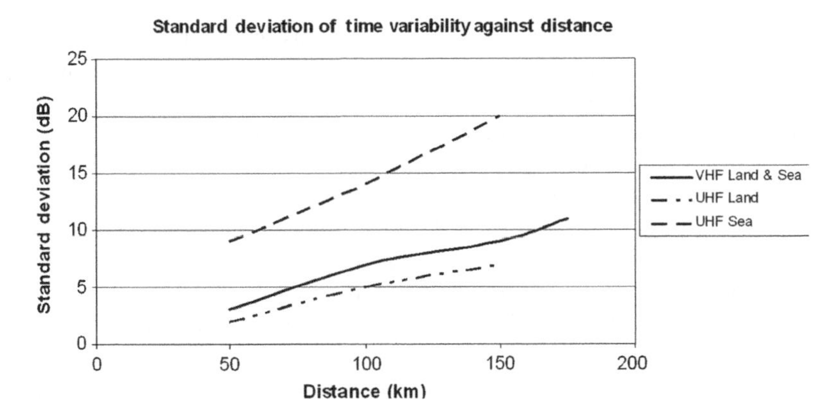


**Figure 4.6.** Standard deviation of time variability against distance. This shows by how much a signal received will vary in strength over time, simply due to changes in the atmosphere. The highest amount of variability is seen for UHF frequencies over the sea.

As can be seen in Figure 4.6, the behaviour of radiowaves due to refraction are more pronounced at longer ranges than they are for shorter ones. This is expected as, the longer the path, the more effect atmospheric changes are bound to have. This can often mean that when link-planning ranges are short, the variations due to changes in the refractive index are less important than other mechanisms. The same is not true for interference analysis, however, when it may be necessary to determine signal levels at long range when the propagation conditions support extended ranges, usually for short but significant periods of time.

As has already been stated, the refractive path can vary when atmospheric conditions are different. This is shown in Figure 4.7, which shows four main conditions: the typical condition when $k$ is at its mean value; super-refraction, where the $k$-factor is negative and the signal does not travel far before encountering the ground and either reflecting upwards or being absorbed; sub-refraction, where $k$ is less than 1 and the signal radiates into space and the condition where $k = \infty$, and the signal travels in a path at a constant altitude above the Earth's surface. This occurs where $\Delta N = -157$, as can be determined from Equation [5].

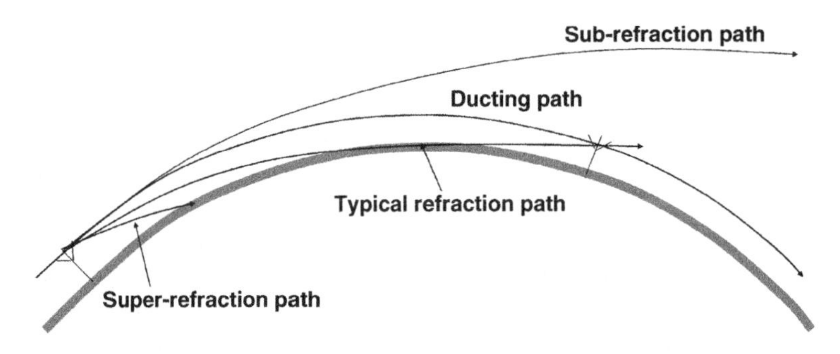


**Figure 4.7.** Types of path due to atmospheric changes.

In the case where the signal path exactly matches the curvature of the Earth, the waves are neither bent downwards towards the Earth's surface nor upwards towards space and there is effectively no horizon. This will typically only occur within a given vertical region of the atmosphere and such regions are known as 'ducts'. These ducts can be at the surface of the Earth (surface ducts) or at a higher altitude (elevated ducts). Not only will these ducts lead to the direction of energy parallel with the ground, but the different refractive index of the neighbouring regions of the atmosphere may tend to channel energy into the duct and not allow it to escape. In this case, energy becomes concentrated in the duct and can travel great distances; far further than it would be able to travel if the energy is radiated equally in all directions. This can lead to exceptionally long propagation paths. Such paths are generally unwelcome since they can result in interference but occur too infrequently to be useful. However, they can be valuable for signal detection and interception, and this is a mechanism sought after for electronic support (ES) in military and intelligence operations.

### 4.3.5 *Diffraction*

In many instances, the path between a fixed base station and mobile element will be obstructed by terrain and ground clutter. In this case, there is no direct line of sight path, and the principal mechanism may be diffraction of the signal over or around the obstructions, as shown in Figure 4.8. There may be more than one obstruction in a path, in which case the signal may be diffracted over each obstacle. The loss in signal level compared to the free space condition can be very significant for diffraction paths, and the maximum range that a link can be established over will be much reduced. The path shown in Figure 4.8 is heavily obstructed and has multiple diffraction features. These can be identified by the contact points of the polygon draped over the path as though it were a piece of string drawn tight between transmitter and receiver (in fact we will see later in the chapter that this is exactly the mechanism used by some models to determine diffraction points).

In some cases, losses due to diffraction will be such that energy arriving from directions other than that of the transmitting element will dominate or compete with the direct path. In such cases, there may be paths due to reflection and scattering, leading to multipath

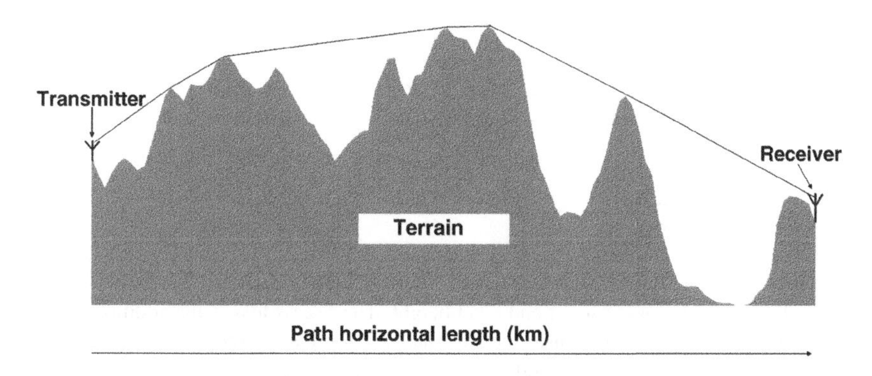


**Figure 4.8.** Example of a diffraction path.

propagation, troposcatter or there may be no coverage at all. The path shown in Figure 4.8 is highly obstructed, but there can also be conditions where diffraction across an obstruction is slight, or is caused by the curvature of the Earth rather than specific terrain features. These conditions must also be catered for by propagation models. This is often referred to as diffraction loss.

### 4.3.6 *Absorption*

Radio energy will also be absorbed by media such as building materials, vegetation, vehicles and body mass for personal systems. This will particularly be true when the mobile antenna is embedded within the radio clutter, for example when walking down a street surrounded by buildings, within a forest or inside a building. Since the materials that may absorb energy will vary in properties and in thickness, it will normally be possible only to derive typical figures for level of loss in most cases. Figure 4.9 shows a graph of typical body loss for both waist- and head-height receivers derived from ITU-R P.1406. It can be seen that the loss is significantly higher when the antenna is worn at waist height.

Of course, when considering the variations between the height and bulk of human beings, it is clear to see that these mean values will vary significantly between individuals and thus they can only be a guide. The same principles are true for buildings which vary much the same way in terms of height, number of floors, construction materials, number of internal walls, window size and so on.

These additional loss factors will be discussed in greater depth in Section 2, where practical examples will be shown. We will now look at how the propagation mechanisms so far discussed can be modelled for radio network design purposes.

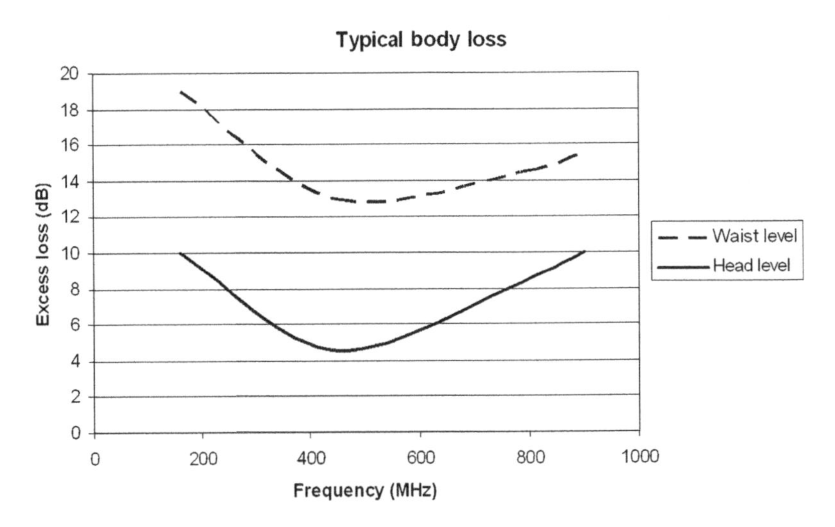


**Figure 4.9.** Typical body loss at waist and head height. The excess loss is the additional attenuation purely due to the presence of the body near the antenna, and thus these figures are referenced against an antenna without a person nearby. The body loss figures vary significantly based on body mass and shape of individual people.

## 4.4 Introduction to Propagation Modelling

Propagation modelling is the fundamental building brick of mobile radio network design. Without the ability to accurately model the performance of a network before it is built, it is necessary to build the network, test the actual performance and then use trial-and-error to optimise the design. This can be time consuming, expensive, risky and the ultimate design is unlikely to be anything like optimal. With accurate propagation prediction and the simulation methods that can be built around it, network design can be fully tested before any hardware is deployed – or even acquired. We can also use the same methods to help choose between potential technologies and architectures, optimise the design by trying different solutions (either manually or via automatic planning methods) and determine detailed metrics about the expected performance of the network, all without expending unnecessary expenditure on hardware and surveys.

In terms of mobile radio propagation prediction, the types of model available broadly fall into two categories. These are point-to-area (also referred to as site-general) and point-to-point (also referred to as site-specific) models. In this regard, a point-to-point model does not refer to a fixed link, but rather that it is possible to explicitly describe the path between transmitter and receiver in some detail, when the assumption is made that a receiver is at a particular location. By repeating this process for all possible locations of the receiving antenna, it is then possible to determine the network behaviour for a mobile subscriber. By contrast, point-to-area models use less precise methods of describing the path itself but rather characterise the path according to a number of parameters that are expected to be present. The decision of whether to use a point-to-point or point-to-area model will depend on a number of factors which are discussed in Section 4.10. At this stage it should be noted that there is no one clear winner in the search for the 'best' model – different models have different applicability for different situations, and it will be necessary to determine which is best for a specific task. It should also be noted that the distinction between the two model types can be blurred, with semi-empirical models (based on generalities derived from measured data) combining aspects of deterministic models (based on the physical characteristics of the link between transmitter and receiver) with correction coefficients or additional terms derived from measurement. The next two sections outline models that fall into these camps, albeit with the caveat noted above.

## 4.5 Point-to-Area Models

### 4.5.1 General Properties of Point-to-Area Models

Point-to-area, otherwise known as site-general models, are intended to provide a general estimate of radio propagation based on nominal characteristics rather than specific path data, such as that shown in Figure 4.8. The general form of such a model [1] is:

$$E_r = -\gamma . \log(d) + K(P_{BS}, f, h_{BS}, h_{MS}) \qquad [6]$$

where

$E_r$ is the received field strength in dBμ/m
$d$ is the path length in km
$\gamma$ is the inverse exponent used with path length

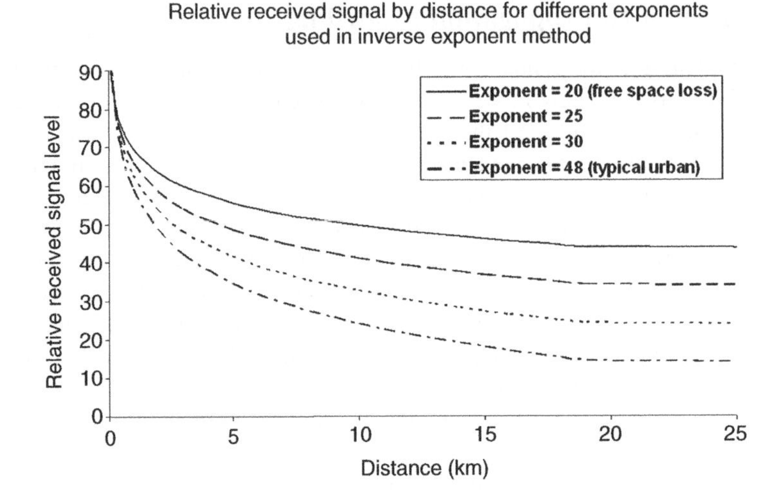


**Figure 4.10.** An illustration of how much predicted signal level will change with different exponents used in a simple inverse-exponent prediction. The absolute levels are not listed, but the relative differences are in dB units (e.g. dBm or dBuV/m, depending on the reference). The difference between the free space exponent and a typical urban value is pronounced.

$K$ is an offset based on power ($P_{BS}$, normally in dBW or dBm), frequency ($f$, normally in MHz), height of base station antenna ($h_{BS}$) and height of the mobile station antenna ($h_{MS}$), both of which are normally expressed in metres above ground.

The typical graphical form of such an equation is shown in Figure 4.10, where four different exponents are shown. Note that the free space loss equation:

$$L = 32.44 + 20\log f + 20\log d \qquad [7]$$

is a specific implementation of this if one assumes that the received field strength is:

$$E_r = P_{BS} - L \qquad [8]$$

And thus:

$$E_r = P_{BS} - 20\log d - 20\log f - 32.44 \qquad [9]$$

The term $20\log f$ is a frequency-dependent example of the function $K$, and the constant 32.44 occurs as a result of converting units to their most commonly used form. Figure 4.10 shows a graphical representation of the inverse exponent equation for different values of $\gamma$, using relative units for the received signal level.

The exponent of 20 is the equivalent of free space loss, which shows the highest received field strength values. The exponent of 48 is fairly typical for land mobile applications in urban or mixed urban/suburban environments and the excess attenuation above free space loss is apparent. The other exponents of 25 and 30 lie in between and would represent different mixed of rural and rural/suburban environments, and possibly higher antennas for the mobile elements.

The simple inverse exponent method is suitable for determining typical cell radius in mobile systems, but is less suitable for any kind of detailed planning. The basic method can, however, be modified to account for typical terrain variations in the coverage area. This introduces the concept of effective base station height and terrain undulation height ($\Delta h$), and modifications to the inverse exponent based on local clutter type.

Although point-to-area models will in general show poorer correlation with measured results compared to well-calibrated point-to-point models, this does not mean that they are not very useful. They can be very useful for dimensioning of proposed networks at an early stage, before specific base station site locations have been identified, they can also be used when terrain and clutter data of good quality is not available, and they are also more appropriate for analysis of very long paths for interference assessment. We will now look at some models in widespread use.

### *4.5.2 ITU-R P.370 and ITU-R P.1546*

ITU-R P.370 (which will now be referred to as the '370 model' for the remainder of this section) has been used as a recommended model by the ITU for many years. It has now been replaced by ITU-R P.1546 (the '1546 model'), which is an extended and enhanced version, but it is still important because it forms the basis of many interference methods for domestic and international use. The 1546 model has been derived from the 370 model combined with the Okumura-Hata model described in the next section.

The 370 model provides a series of prediction curves for the frequency range 30–1000 MHz, over ranges of 10–1000 km, for transmitting antennas at effective heights between 37.5 and 1200 m, and receiving antennas at 10 m above local ground. Correction factors are provided to allow prediction between the curve values and for lower antenna heights, and also for other factors such as transmitting power. The curves are provided for a variety of environments such as land and warm/cold seas. The concept of the effective height is not the same as the antenna physical height above the ground on a mast; instead it is the height compared to the average terrain height within 3–15 km of the antenna location.

The empirically derived field strength values are provided for 50 % availability (see Chapter 5) within a 500 by 500 $m^2$, and for time percentages of 50, 10 and 1 % for VHF and UHF bands, with an additional 5 % time percentage for VHF. The concept of percentage time is useful for examining occasional interference due to variations in the atmospheric refractive index.

The curves shown in the report may be used directly by the radio-planning engineer, but more often they will be accessed by a computer implementation of the recommendation that will incorporate all of the aspects of the model. The computer tool operator will have to specify the parameters to be used as illustrated in Figure 4.11.

In this example, the operator has to select the required percentage of locations (availability) and the percentage of time. The path profile influence identifies some options provided for accounting for terrain effects in the model, and the delta $N$ value is the change in the refractive index between 1 and 15 km above the ground to be used to set the effective Earth radius. The 0–>10 km options are used to provide estimated field strength values over short ranges, since the 370 model does not compute values for this. The values within the

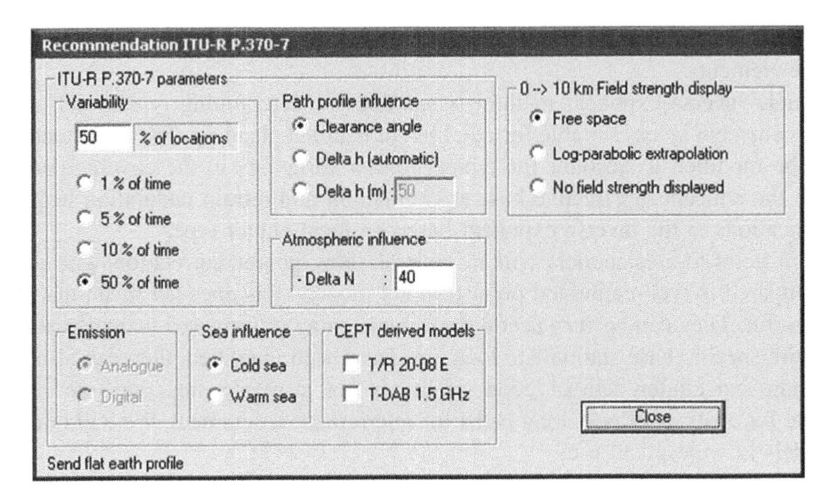


**Figure 4.11.** Typical input data required to specify configuration of the ITU-R 370 model. Reproduced by permission of P & D Missud.

0–10 km range are not likely to be very accurate and should not really be used for any kind of planning. The 370 model has most application for long ranges, particularly for examining coverage for different percentages of time. This is illustrated in Figure 4.12, which shows an example plot for a transmitter at 50, 10 and 1 % of the time. The 50 % time contour is the

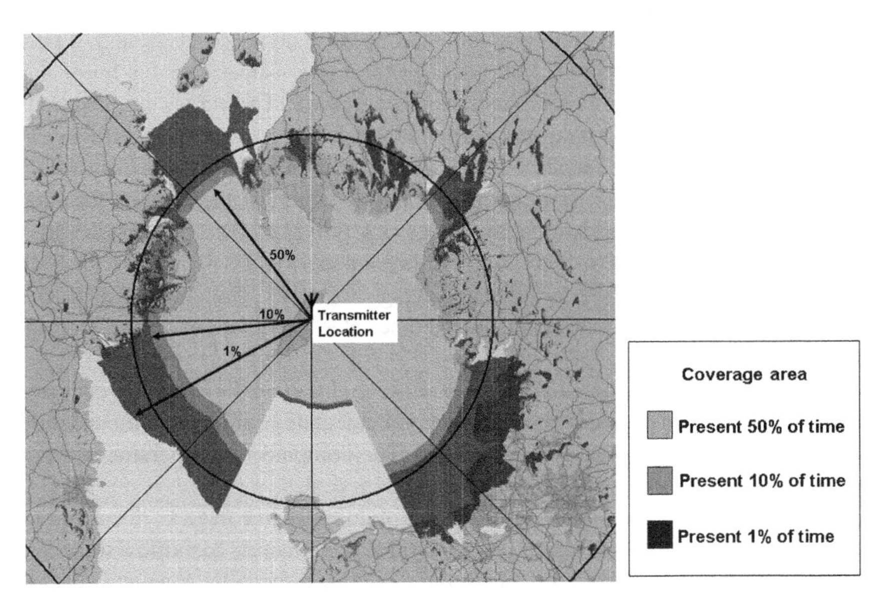


**Figure 4.12.** Prediction of field strength for different percentages of time, showing how for a small percentage of time the energy radiates a far longer distance than usual. Under these circumstances, other networks can experience interference not normally present. Reproduced by permission of ATDI Ltd.

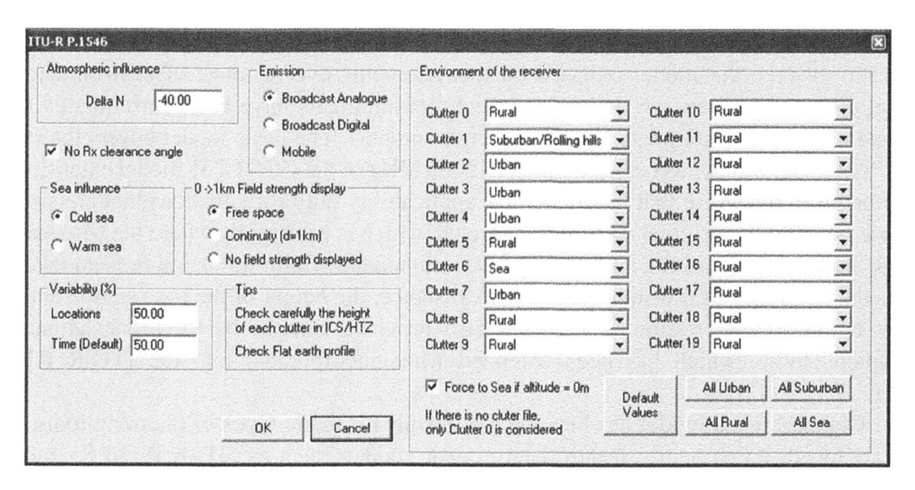


**Figure 4.13.** Input data requirements for ITU-R P. 1546 model. Reproduced by permission of P & D Missud.

one closest to the station, the 10 % contour is the next and the 1 % is the furthest out. The range circles are at 100 km intervals.

The figure shows that for small percentages of the time the transmitted energy travels substantially farther than the median condition. In this example, the 1 % condition easily travels an extra 50 km from the median value, and in act there are individual locations of potential interference out to beyond 250 km. This shows the potential for this transmitter to cause interference to other radio users for short periods of time well beyond what would be expected for most of the time. Note that all of the models described in this section will produce similar types of display if set up in the same way (although the actual values and shape of the coverage pattern will be different). For this reason, a coverage plot is not shown for the other models discussed.

ITU R-P.370 has been replaced by ITU-R P.1546 ('the 1546 model'), which has been derived from a composite of ITU-R P.370 and Okumura Hata. The 1546 model covers a wider frequency range than the 370 model; 30–3000 MHz. The 1546 model also takes account of radio ground clutter, which can lead to more realistic results. This makes the configuration of the 1546 model more complex than that of the 370 model, as shown in Figure 4.13

Radio engineers will need to familiarise themselves with the data entry methods of the specific tool being used in a particular project. As always with sophisticated software systems, it is vital that the data used in the input part of the process is accurate, otherwise the output data may well be wrong and, worse, may look right. Obviously, such errors propagated through the design process can have major implications later on.

### *4.5.3 Okumura-Hata, COST 231 Hata and Other Point-to-Area Models*

The Okumura Hata model has been used extensively to model land mobile services, particularly for urban environments, even though it does not necessarily provide the best correlation with measured data when compared to point-to-point models. As with the 370 and 1546 models, the model is based on measurements over a range of environments at several different frequencies, and these initial measurements have been corroborated by a large number of measurements in differing environments. The basic model covers the range 100–1500 MHz, which has been extended to 2000 MHz in the COST 231 model extension. It is important to recognise that the model has applicability only for situations that are similar to those in which the original data were collected, which is principally where the base station is high above clutter and the mobile element is embedded within it. Okumura Hata takes as inputs the frequency of transmission, the link distance, the height of the base station and the height of the mobile station. The model also provides corrections for urbanisation and vegetation clutter, which has been extended when incorporated into the ITU-R P.1546 model.

The Okumura Hata model has become the default for many types of planning tools, and properly tuned, it can nearly match deterministic models such as ITU-R P.526 for mobile applications, for short distances. It is not suitable for point-to-point link analysis, and it is important to apply the model only to scenarios close to those of the original measurements; it is definitely not recommended to use the model for any other application unless verification studies and model tuning is carried out for the actual circumstances of the network being planned.

Okumura Hata gained popularity partly due to the fact that it is not computationally intensive and that the data precision required to model the environment is relatively low. This lent itself to computer systems that were, at the time, slow and with limited memory. Modern personal computers (PCs) are typically far more powerful and have the ability to process large amounts of data. Thus, it increasingly becomes more justified to use more computationally intensive algorithms using higher precision data (where available).

Besides Okumura Hata, a host of other propagation models have been developed to more accurately model specific applications, and these continue to appear and evolve. It is worthwhile for the radio-planning engineer to learn of the existence and applications of these models so that when new projects appear, an objective assessment of the best model to use can be made. Although there is insufficient space to discuss these models in depth here, it is worth mentioning that different models will not only be applicable to different frequency ranges but also to different scenarios. Besides reviewing the model reference, ideally back to the original paper but at least to the degree where the limits and boundary conditions can be established, an examination of the input characteristics will help the experienced radio-planning engineer to determine the applicability of a model. Thus, the input for the ITM model, shown in Figure 4.14 in a specific implementation, shows that the model deals with atmospheric variability for a variety of climate types and for different generic types of ground. It is thus a long-range model suitable for modelling performance over different percentages of time.

There are two point-to-area models that require specific mention, since they are among the few suitable for aeronautical applications. The prediction of propagation for high altitude depends on different factors than those that are dominant at lower altitudes. Thus it is important not to simply assume that a terrestrial model will be suitable for aeronautical applications, unless the aircraft are operating at very low altitude (which can of course happen in military and emergency services applications).

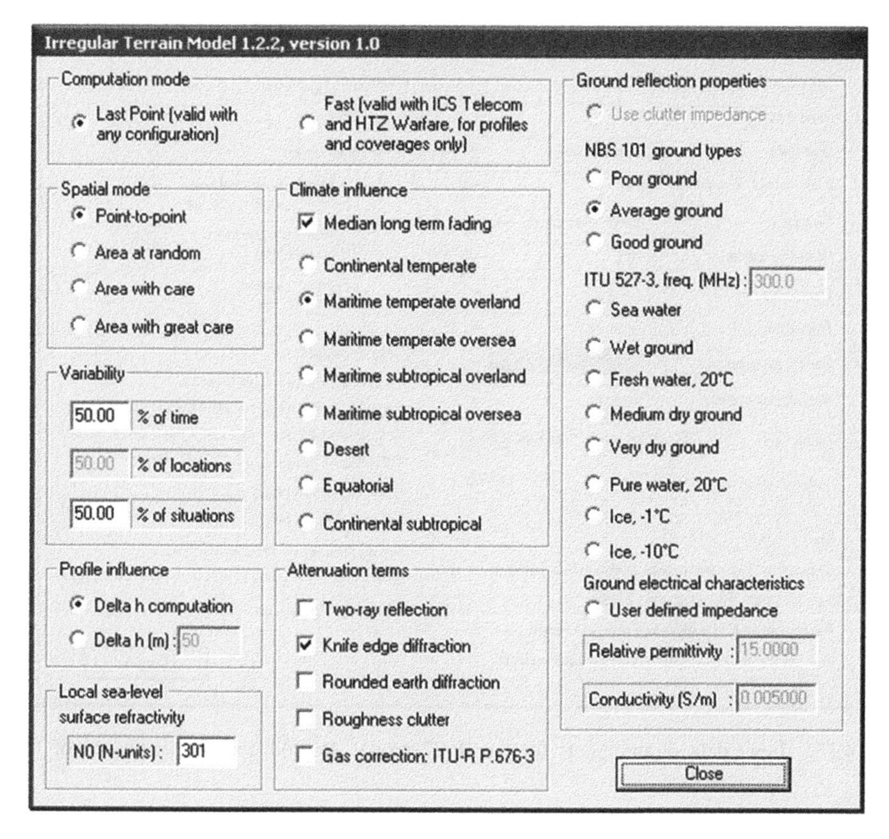


**Figure 4.14.** Input data requirements for the ITM model. Reproduced by permission of P & D Missud.

### *4.5.4 IF-77 and ITU-R P.528 Models*

The 370, 1546 and Okumura-Hata models have most application for terrestrial applications and should not really be used for aeronautical work. Two point-to-area models have been developed specifically for aeronautical applications. These are the IF-77 (also known as the Johnson-Gerhardt model) and ITU-R P.528 (the '528 model') models.

The IF-77 model provides for long range modelling over a range of environments, including effects such as sea surface roughness, climactic variations and the presence of storm cells (a storm cell is a region of heavy rain). An illustration of the input requirements is shown in Figure 4.15, as implemented in a specific planning tool. The limits of validity are also shown, which is a useful guide to have displayed as an operator reminder. The values for sea state and reflection material are included to allow the algorithm to determine reflection points for the reflected path, which is normally present in aeronautical applications. Often the reflection point is near the transmitter and the reflected signal is of almost the same strength as the direct ray as it arrives at the receiver. This results in very deep fades, known as 'lobing', due to the typical pattern seen by drawing a graph of constant signal strength

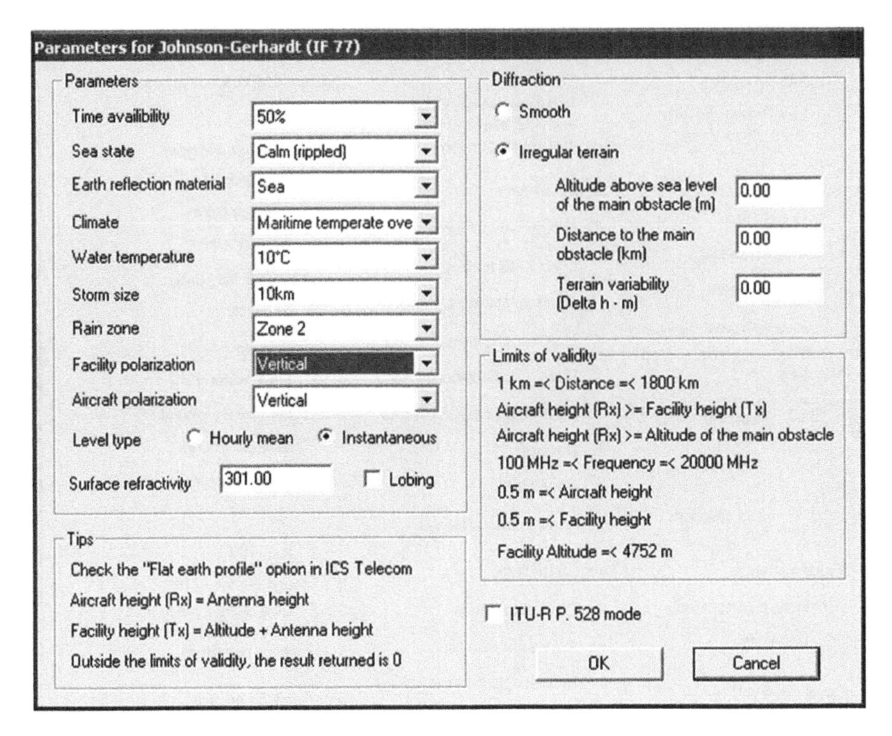


**Figure 4.15.** Input data requirements for the IF-77 model. Reproduced by permission of P & D Missud.

against altitude around a transmitter on the ground. Such a pattern shows lobes where the signal interferers constructively and near-nulls where the signals are 180 degrees out of phase.

The 528 model is a subset of the original IF-77 model, but it omits the inclusion of lobing effects because this needs detailed input information and if this is not available, the results may be misleading.

The 528 model allows for antennas between 0.5 and 20 000 m altitude, with the mobile element always being higher than the fixed installation. The model covers a wide frequency range, 100 MHz to 20 GHz, and ranges of up to 1800 km. The form of the graphs is similar to that of the 370 and 1546 models when viewed on a graph of field strength versus distance. An example of a tool to display the curves is shown in Figure 4.16. This shows a family of curves showing different altitudes of aircraft, with the lowest altitude corresponding to the left-most curve and the highest altitude corresponding to the right-most curve. Free space loss is also displayed for comparison purposes. The graph shows the loss experienced on the $Y$-axis, increasing towards the bottom of the display (not the top), and distance in kilometres on the $X$-axis. Note that in aeronautical terms, the requirement for model applicability extends beyond 1500 km.

Figure 4.16 has been annotated to highlight three important regimes that can occur for any long-range propagation. The first is the free space loss regime, where the aircraft and base station are in radio line of sight and losses fall off at the same rate as free space loss, but

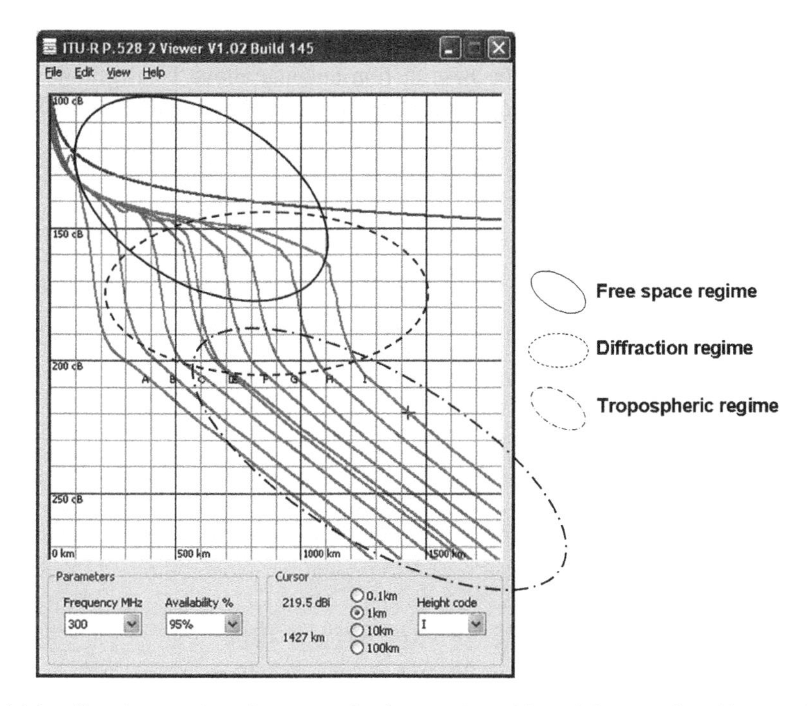


**Figure 4.16.** Signal strength against range for the ITU-R P.528 model. Reproduced by permission of ATDI Ltd.

usually with a larger loss. The second is where radio line of sight is lost and the signal propagates by diffraction over the Earth curve. The third occurs where tropospheric effects (for the given percentage of time) lead to lower losses than diffraction at that distance. This is not just applicable to aeronautical applications.

A radio-planning tool for aeronautical applications should include a suitable aeronautical model, because it is not appropriate to use terrestrial models for conditions they were not designed for. It is also important to note that the 528 model was generated using data predominantly in a continental environment, and so it should be used with care for maritime paths.

### 4.5.5 Other Point-to-Area Models

Besides the models already described, a variety of other models have also been developed. Some others that engineers may come across include the Lee model, which is based on measurements taken at 900 MHz. The equation of the Lee model is:

$$L = 10n \log d - 20 \log h_{\mathrm{b(eff)}} - P_0 - 10 \log h_{\mathrm{m}} + 29 \qquad [10]$$

where

$L$ = loss (dB)

$n$ is a distance correction factor that varies between 4.35 for open areas, 3.84 for suburban areas and differing values for urban environments.

*d* is distance in km.

$H_{b(eff)}$ is the effective height of the base station antenna above local ground, based on extending the average terrain slope of the area around the base station to the location of the antenna.

$P_0$ is an additional term derived from measurements that is −49 in open areas, −61.7 in suburban areas and differing values for urban environments.

$h_m$ is the height of the mobile.

Another model quoted in literature is the Egli model, which has been expressed by Delisle in the approximate form:

$$L = 40 \log d + 20 \log f_c - 20 \log h_b + L_m \qquad [11]$$

where

$L$ = loss (dB)

$f_c$ = frequency

$h_b$ = height of base station

And $L_m$ is defined by

$$L_m = 76.3 - 10 \log h_m \text{ for } h_m < 10\,\text{m} \qquad [12]$$

$$L_m = 76.3 - 20 \log h_m \text{ for } h_m >= 10\,\text{m} \qquad [13]$$

This is effectively a model that adds a clutter factor to account for the difference between free space loss and the measurements actually obtained.

Other models exist and continue to be developed. For long-range interference models, point-to-area models will probably continue to be the most appropriate methods for evaluating propagation. For the shorter range situation, certainly less than about 50 km, it is likely that point-to-area models will gradually be replaced by point-to-point models that show a better prediction performance. This change is principally due to the increased availability of accurate environmental data and to the improvements in software planning and simulation tools and the computers used to run them on.

## 4.6 Point-to-Point Models

### *4.6.1 General Properties of Point-to-Point Models*

Whereas point-to-area models derive general propagation rules from specific measurements, point-to-area models, which are also called site-specific or deterministic models, attempt to derive propagation predictions based on the physics and mathematics of the link. The intention is to provide computationally efficient and fast simplifications of the link scenario while retaining the salient features of the link. The most common method of achieving this is to use the concept of the Fresnel ellipse and variations on knife-edge diffraction.

The Fresnel ellipse concept is shown in Figure 4.17. A Fresnel ellipse is the locus of points where, if there is a reflecting surface, the reflected signal will arrive at the receiving point with the same phase as the original signal. This occurs when the delay due to the additional distance of the reflected path is exactly equal to *n* wavelengths, where *n* can be

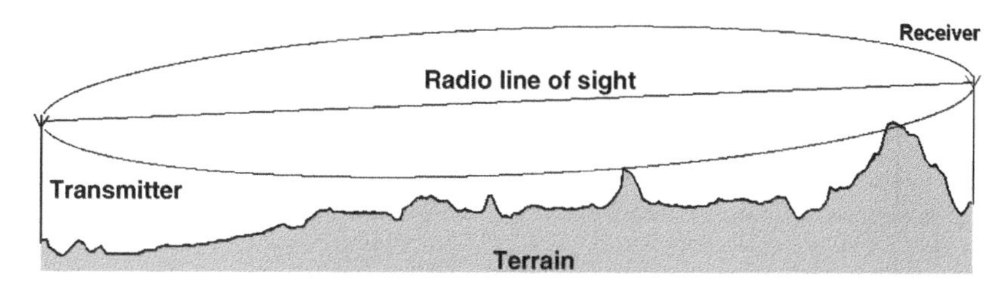


**Figure 4.17.** Fresnel ellipse.

any integer larger than one. In other cases, the reflecting signal will arrive out of phase with the original and will thus cause a reduction in received signal strength.

In the Figure, the transmitting antenna is at the end of the straight line at the left of the diagram, and the receiver is at the right hand end. The line is the direct radio path between the transmitter and receiver. Around the direct ray is a cigar-shaped mathematical construction which is a Fresnel ellipse. This is a 3-dimensional zone through which the radio wave substantially passes. The radius of the ellipse in the vertical plane is expressed by:

$$R_n = 550\left[\frac{nd_1d_2}{(d_1+d_2)f}\right]^{\frac{1}{2}} \qquad [14]$$

where

$R_n$ is the radius of the Fresnel ellipse
$n$ is number of the Fresnel ellipse
$d_1$ is the distance between the transmitter and the point being calculated
$d_2$ is the distance between the received and the point being calculated
$f$ is frequency in MHz

For an ideal radio link, there should be no obstructions within the zone where $n = 1$ (also known as the first Fresnel zone). In this case, the link behaves broadly as though it were in free space (although other mechanisms may prevent this from being true in practice). The link in Figure 4.17 shows a link for which this is broadly true, except there is a small intrusion towards the end of the link. If we can determine the extent to which the terrain protrudes into the Fresnel ellipse, it is possible to determine an estimate of the loss over that of the free space condition, and this is what point-to-point propagation models endeavour to do. Before focusing on individual models, we will look at the simplest mathematical construction to approximate real paths, which is known as a knife-edge.

A knife-edge is considered to be a single obstruction of negligible thickness that protrudes into the ellipsoid. There are three basic conditions that may arise, as shown in Figure 4.18. Condition (a) is where the knife-edge does not intrude into the first Fresnel ellipse. In this case, there is no excess loss due to the intrusion. In condition (b), the intrusion does enter into the ellipse, and some of the energy will be blocked by the intrusion, and there will be a reflection from the point, as shown by the dotted line. This may interfere constructively or destructively with the direct path. The third condition (c) is where the intrusion blocks the

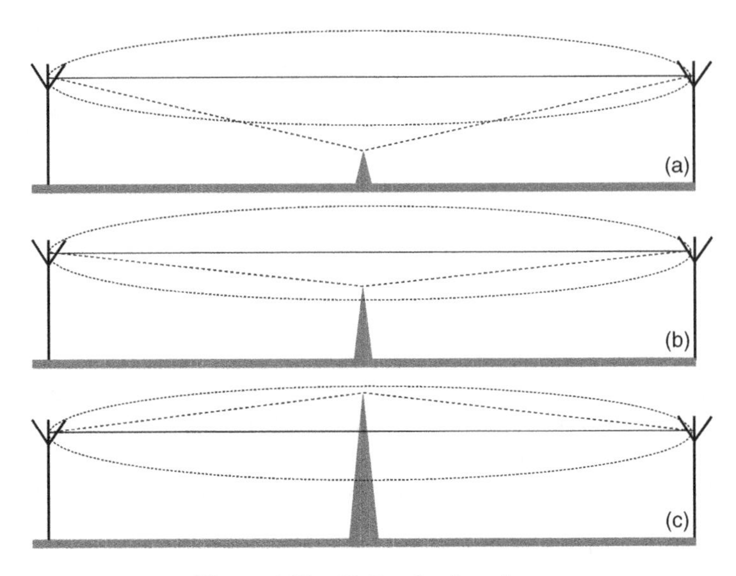


**Figure 4.18.** Knife edge intrusions.

line of sight and most if not all of the ellipse. In this case, the energy arriving at the receiver is from the diffraction paths over the obstruction.

It is possible to determine the degree of excess loss due to obstructions in all of these cases by considering a mathematical construction:

$$v = h\sqrt{\frac{2(d_1 + d_2)}{\lambda d_1 d_2}} \qquad [15]$$

where

$v$ is a frequency (wavelength) dependent dimensionless measure of intrusion, known as the Fresnel–Kirchhoff diffraction parameter

$h$ is the height of the intrusion above or below the straight line joining the transmitter and receiver; $h$ may be positive or negative

$d_1$ is the distance from the transmitter to the intrusion

$d_2$ is the distance from the intrusion to the receiver

$\lambda$ is the wavelength of the transmission in m

This construction is based on the mathematical relationship between a direct path and a path passing the point of the obstruction. A more detailed derivation is contained in reference/further reading item [7]. The benefit of the Fresnel–Kirchhoff diffraction parameter is that diffraction loss can be determined from it by using a complex integral that can be graphed as shown approximately in Figure 4.19.

The *Y*-axis shows how many dB are lost due to the obstruction (e.g. $-8 = 8$ dB less than would be present just due to free space loss).

A value of 0 for $v$ indicates that the intrusion is directly in line with the transmitter and receiver. A negative value indicates that the intrusion is below the direct line, so line of sight

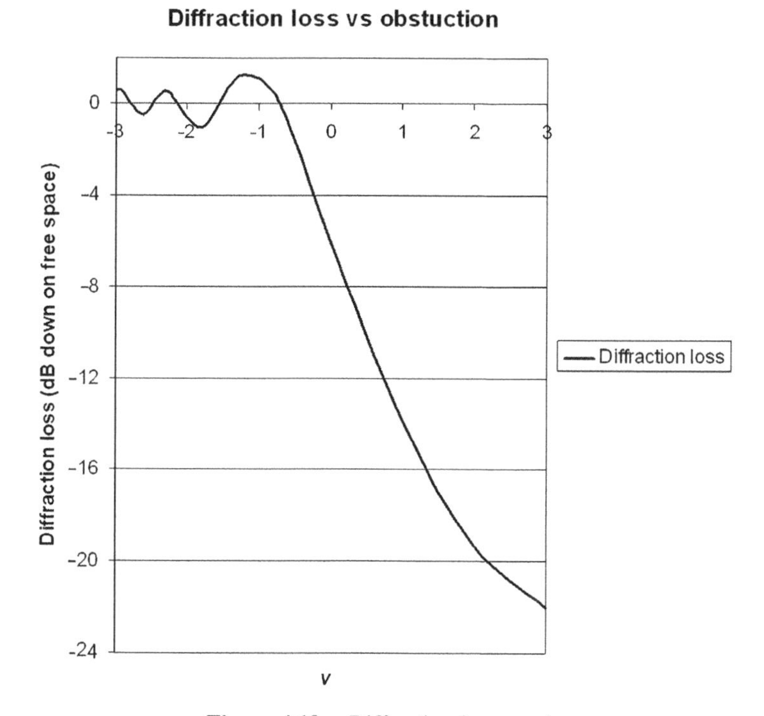


**Figure 4.19.** Diffraction loss graph.

is maintained. A positive value indicates that the line of sight is blocked. Note the sinusoidal region below $v = 0$. This is where constructive and destructive interference occurs, leading in some cases to a higher value than would be expected in free space, because of the reinforcement of the additional reflected component.

In practice, radio paths for mobile networks will almost always be more complex than the single knife-edge model. The need therefore is of a model that expands the basic theory to cope with more than a single diffraction object for the generic situation (as shown in Figure 4.8, in which there are five diffraction objects). Some of the more well-known models are discussed in outline next. The discussion focuses on the way the path is represented, since this is the principal difference between the models. This is illustrated in Figure 4.20.

### *4.6.2 Bullington Method*

The Bullington model is the simplest of these models. It replaces all terrain obstructions with a single equivalent knife-edge which is located at the intersection of the optical horizon of both the transmitting and receiving ends of the link, as shown in Figure 4.20a. The model is rather coarse and ignores other obstructions that may be present, so it is mostly important from a historical perspective rather than being a model for modern radio network planning. The RPO 3 model that was extensively used by the UK military for many years was based on the Bullington model.

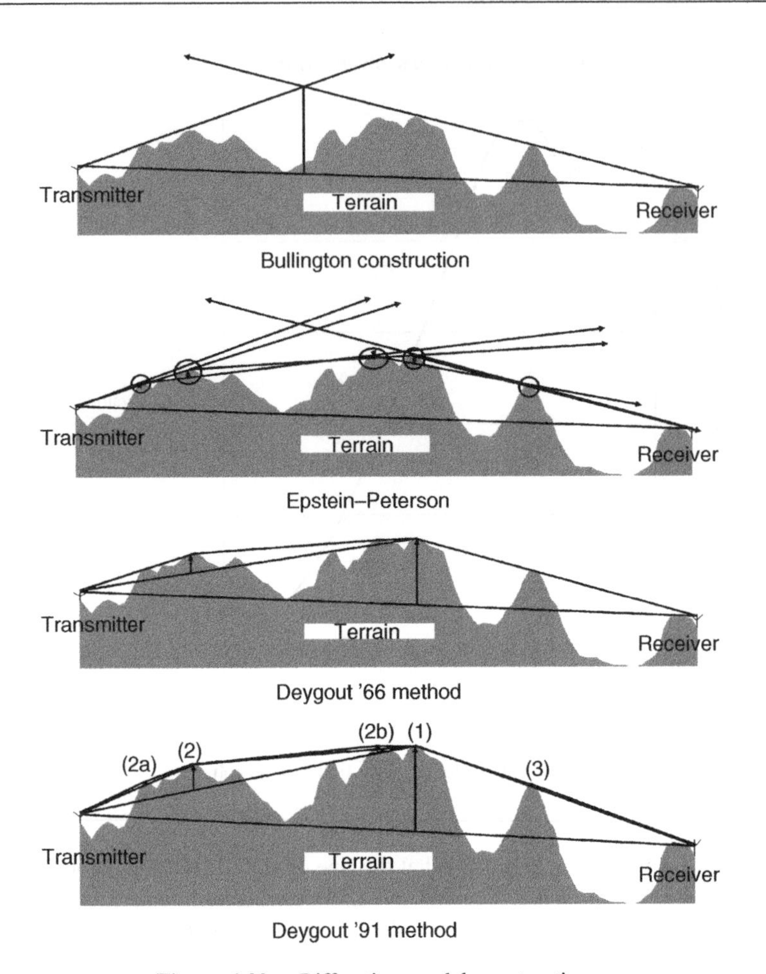


**Figure 4.20.** Diffraction model constructions.

### *4.6.3 Epstein–Peterson Method*

The Epstein–Peterson model uses a multiple knife-edge construction to represent each obstruction in turn. This is more accurate than the Bullington but it is difficult to apply when there are two or more obstructions close together, as is shown in Figure 4.20b.

### *4.6.4 Edwards and Durkin Method*

This is identical to the Epstein–Peterson method for up to three obstacles. For more obstacles, a Bullington-like construction is made between the first and last obstruction. The excess loss is calculated for the first obstacle, the last obstacle and the middle Bullington construction of the remaining obstructions. This gets round the problem of the situation when diffraction objects are close to each other.

### 4.6.5 Deygout Method

The Deygout model introduced the concept of a main diffraction edge, which is found when the value of $v$ is largest. Then, the path between the primary obstruction and the transmitter, and between the primary obstruction and the receiver are examined to look for further diffraction objects on these two sub-paths. If a diffraction edge is found, the loss due to this is added to that of the main diffraction edge. This is done up to twice, if there is a diffraction object on both the transmitter to main diffraction object path and the main diffraction object to receiver path. This construction was found to be an improvement over the Bullington method, but still not ideal. This is shown in Figure 4.20c.

In 1991, Deygout produced a refined method to account for more complex paths. In this case, the model still searches for the main diffraction edge, but then recursively searches for additional diffraction paths until all diffraction objects have been discovered. This is illustrated in Figure 4.20d, where all five diffraction objects are shown to be accounted for. This is an improved model, but it is computationally far more complex than the other models; in theory there could be lengthy recursion. This is worst over sea paths or flat terrain, where the curvature of the Earth forms the diffraction objects, and there will be a diffraction object for each point in the path. However, when implemented in a practical manner to avoid such problems, the Deygout '91 model shows very good correlation with measured results. In fact, Deygout was working on another refinement of the model in 1994, which in the event remained unpublished. However, an implementation of the model is found in the ATDI planning tool ICS Telecom, which was used as the test bed for the new model. It typically shows the best correlation with measured results of all the diffraction models described.

### 4.6.6 ITU-R P.526 Model

The ITU-R P.526 model is a fully referenced deterministic model based on the principles already discussed for point-to-point models. The recommendation offers nomograms for noncomputer evaluation of spherical Earth diffraction for 30 MHz and above, and for irregular terrain uses the Deygout '91 method limited to a maximum of three diffraction edges. It differs from the Deygout construction in that two secondary diffraction edges are still considered for line-of-sight paths. The model also extends the knife-edge principle to the concept of the finitely conducting wedge, which has thickness. It is assumed that no energy is transmitted through the wedge. The calculation is more computationally intense than the Deygout '94 model, but does not include the potentially more costly recursive method.

Since the 526 model is based on the link physics, it is generically applicable to most types of radio link and therefore it is safer than using an empirical method that has not been tested against the application. For macro cell planning, the 526 model used with clutter corrections can show very good performance. It has also been extensively peer reviewed as part of the recommendation approval process and has been compared with extensive measurement data.

## 4.7 Hybrid Models

Hybrid models share the characteristics of both empirical models (point-to-area) and deterministic models (point-to-point). This provides a potentially computationally less costly implementation that does account for gross terrain and obstruction features. As an example of this approach, we can look at the Allsebrook and Parsons model.

This model combined terms derived from measurements, combined with a single diffraction object intended to represent the final building over which the signal propagates. The general form of the model is:

$$L_T = L_p + L_B + \gamma \qquad [17]$$

where

$L_T$ is the total loss

$L_P$ is the plane-Earth loss

$$L_B = 20 \log \left( \frac{h_0 - h_m}{548\sqrt{(d_m \times 10^{-3})/f_c}} \right) \qquad [18]$$

$\gamma$ is approximated by

$$\gamma = -2.03 - 6.67 f_c + 8.1 \times 10^{-5} f_c^2$$

The basis of the model is illustrated in Figure 4.21. Compared to other more modern models, Allsebrook and Parsons should be regarded as relatively simple and will not show as good results as other, more modern models.

Just as with the point-to-area models, the configuration of propagation settings in modern planning tools must be carried out before launching a simulation. A typical example of the requirements is shown in Figure 4.22. In this case, two separate basic model types, the Fresnel and ITU 525/526 model, are provided, along with a group of different ways of calculating diffraction geometry and also the method of sub-path calculation, which is used to predict propagation when terrain features are present within the first Fresnel ellipse but do not break radio line of sight. Besides the options shown, the engineer may also have to configure other factors, of which the main one may be the effective Earth radius, or $k$-factor to be used, because this will influence the geometry of the link.

The method of configuring propagation models in practical terms will vary from tool to tool, but the important point is that the engineer is responsible not only for selecting a

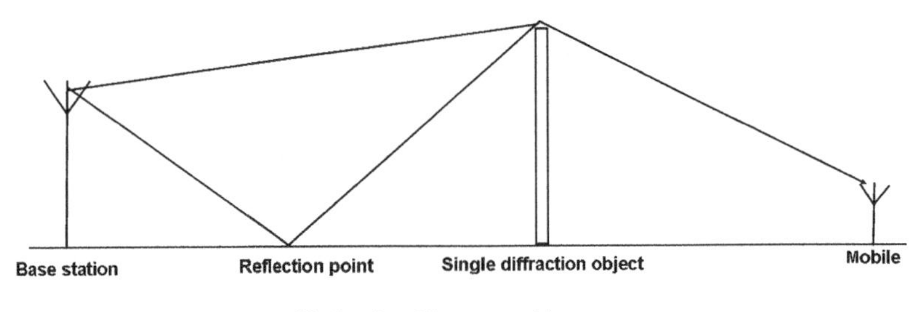


**Figure 4.21.** Allsebrook and Parsons model.

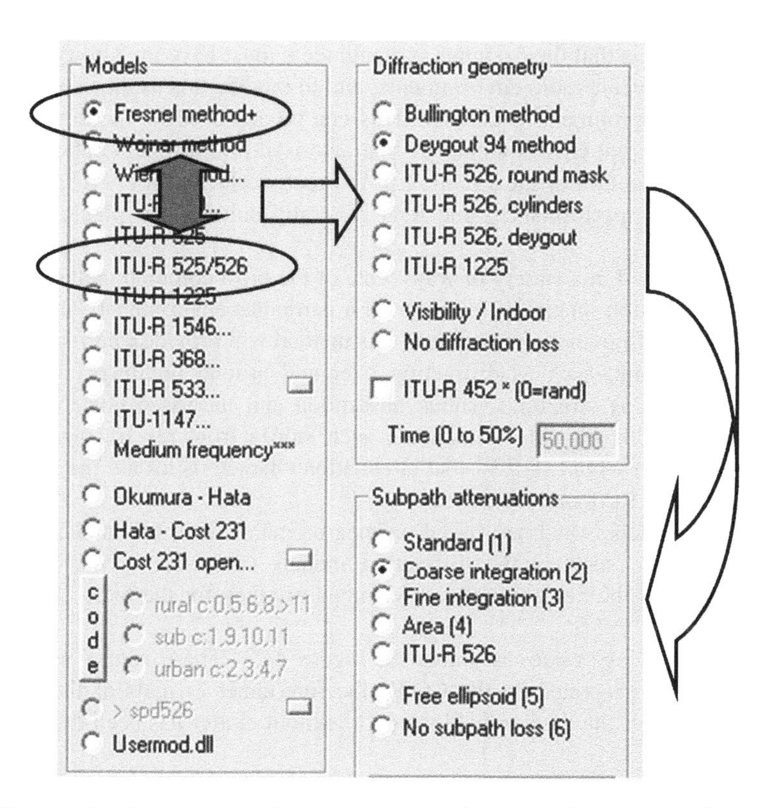


**Figure 4.22.** Configuring point-to-point propagation models. Reproduced by permission of P & D Missud.

suitable propagation model to perform a particular simulation, they must also take responsibility for ensuring that the model is configured properly. Such a configuration exercise requires not only knowledge of the available propagation models, but which configurations are valid and how they should be configured in the tool to be used.

## 4.8 Radio Clutter in Propagation Models

Radio clutter is a representation of the mobile station local environment (it is described in greater detail in Chapter 7). Base station locations will normally be planned during the design process to be as optimal as possible from a radio propagation perspective, free from nearby obstructions. Mobile elements will not generally position themselves for radio link advantage, but rather due to other, nontechnical, reasons. For public users, this will be predicated on where they need to be as a result of their lifestyle; at work, travelling to and from work, shopping centres or perhaps the local bar. For private users such as the emergency services, this will be predicated on the demands of their jobs; they will travel to crime scenes, city centres and recreational centres such as football grounds or city centres at night. Military radio users will go where the operation requires as the situation evolves. In each case, the radio system must support the subscriber, not the other way round. This also means that the designer and engineers must have the ability to model the mobile element in adverse radio environments, and to modify this as the subscriber moves from one type of environment to another. This can be achieved by the concept of radio clutter, in which different environments are characterised into different categories. This is necessarily a simplification of the situation in the real world, but it provides a pragmatic method of predicting performance, and it is generally suitable for planning real-world systems.

Clutter can be treated in a variety of ways. One of the most common methods is to apply an additional attenuation factor for receivers in a particular environment. This attenuation factor can be derived from measurement, and this method will provide a correction factor for all mechanisms present, such as diffraction over and around buildings, reflections and scattering from building structures, signal absorption and increased noise over the rural baseline. Although it is possible to estimate such values from ray tracing methods, the current state-of-the-art would suggest that attenuation characteristics are the most practical approach to planning for mobile networks.

The empirical models, which are based on measurements, tend to include these effects. Deterministic models, such as the diffraction models discussed so far, need to have corrections applied to the field strengths determined to allow for the radio clutter environment.

An example of excess values is shown in Figure 4.23. These values were originally derived from figures generated by the CCIR (the forerunner to parts of the ITU). These values are applied after the median field strength without clutter has been derived from the diffraction model.

Although these results are based on fairly old measurements, they still appear valid when used in modern propagation planning models. Further values have been derived for higher frequency systems but in any case, wherever possible, it is best to tune the model by using a calibration survey. This will allow the model to be made more accurate for the specific application being planned. The concept of clutter is further described in Chapter 5.

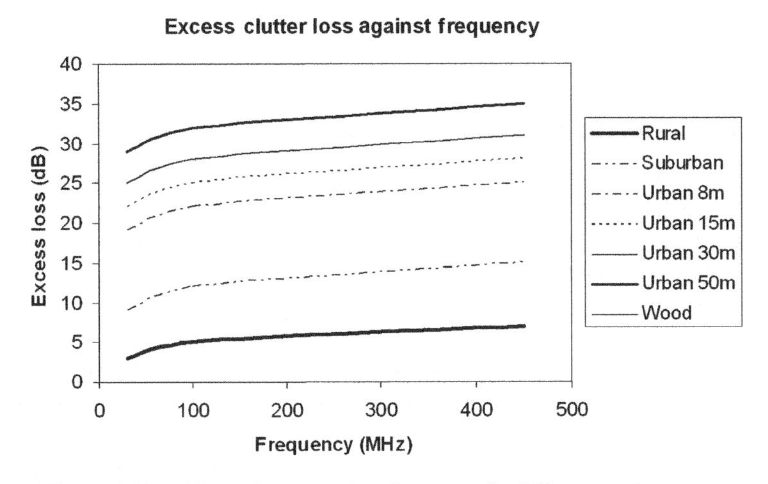


**Figure 4.23.** Clutter losses against frequency in different environments.

## 4.9 Tuning Propagation Models

The process of tuning propagation models is very important, and can have major impact on their accuracy.

To illustrate this, we can look at a specific example, based on a test transmitter at 450 MHz. A mobile survey was carried out using a vehicle with a survey system, out to ranges of approximately 30 km, covering a range of rural, suburban and urban environments. The field strengths recorded were compared to predictions using the 526 model and the clutter attenuation values shown in Section 4.8. The following results, taken over 6987 samples, were obtained:

$$\text{Mean error}: \quad -1.4\,\text{dB}$$
$$\text{Standard deviation of error}: \quad 8.2\,\text{dB}$$

The mean error is fairly good but the standard deviation of 8.2 dB could do with improvement, given that the 526 model is a point-to-point model and not a point-to-area model. Further analysis of the comparison between measured and predicted results were performed to split the mean error and standard deviation per clutter category, which is shown in Table 4.3.

The attenuation values used for these clutter categories were then modified and the prediction run again. With the tuning, the new figures became:

$$\text{Mean error}: \quad 0.2\,\text{dB}$$
$$\text{Standard deviation}: \quad 7.2\,\text{dB}$$

These figures are improved for this particular project. This demonstrates the benefits of performing a calibration survey for each specific project, even when the model and clutter coefficients are well established for the frequency band and application. This is because the overall accuracy depends not only on the model itself, but also the environmental data available and the way in which mobile users will use their terminals. The environmental data is important in terms of its precision (how many square metres on the ground does a single data point represent?) and its accuracy (how well does the ground clutter data correlate with what is actually on the ground?).

Measurement data can also be used to tune the clutter data itself, since if the conditions present at areas where the measured data differs from that predicted are analysed by physical survey, then the clutter at that location can be changed to reflect what is actually there. This can often occur, for example where a coarse propagation model does not include such

**Table 4.3.** Clutter tuning.

| Clutter category | Mean error (dB) |
|---|---|
| Rural | 0.4 |
| Suburban | −3.75 |
| Urban 8m | −8.73 |
| Urban 15m | −5.27 |
| Wood | −0.96 |

salients as wide roads within urban environments. Additionally, road and railway cuttings and embankments may not be represented.

## 4.10 Factors in Model Selection

### 4.10.1 Introduction

It is important that propagation models are selected appropriately for the application and the task being conducted. This will require considering a number of issues before settling on the best model and configuration. In this section, we will look at some of the key considerations. This analysis should be done each time a model is to be used. Some of these factors can overlap, but it is important to consider the issue from a number of perspectives.

### 4.10.2 Frequency Range

*Check that the proposed simulation is within the valid frequency range.*

Each propagation model is valid only for a specific frequency range. In general, the deterministic, or site-specific, models have the widest supported frequency range. For example the 526 model is appropriate for the entire VHF and UHF bands, whereas empirical models such as Okumura Hata are valid only between 150 and 1500 MHz unless extended by other work. Other models, specifically when tuned for a specific application, will have a narrower frequency band. In general, the more tuned a model is, the narrower the valid applicability at the derived accuracy figures.

### 4.10.3 Link Length

*Check that the model is appropriate for the link lengths to be simulated.*

When considering radio links in the VHF and UHF bands for relatively short ranges out to the horizon and a little way beyond, the dominant additional attenuation factors will be due to terrain and clutter features. Where the terrain and clutter can be explicitly described, a deterministic model such as the 526 model will typically be best. Where the terrain and clutter can only be generalised, a model such as Okumura Hata may be suitable. For longer ranges, atmospheric effects will have more importance, and terrain and clutter will in contrast have less effect. It is also worth noting that Okumura Hata, even with the corrections in the 1546 model, is only usable out to 100 km, and the 526 model performs poorly for large numbers of diffraction edges or for smooth Earth. In this case, the curves of the 370 or 1546 model may be more appropriate (the 1546 model will switch to the longer range empirical curves automatically as the range increases). Likewise, for aeronautical applications, the 528 model can be used for long-range simulation.

### 4.10.4 Radio Environment

*Check that the model is appropriate for the environment.*

Not all models are suitable for all environments. This is particularly true for point-to-area models which are based on measurements taken in specific circumstances. For example the Okumura Hata model is suitable for urban and suburban models where building clutter surrounds the mobile element and forms the dominant excess loss mechanism, so long as the environment is close to that of the original measurements which were performed in Japan. It is less suitable for environments where terrain is the dominant mechanism, where a point-to-point model would show greater accuracy. It is therefore important to check that a proposed propagation model is applicable to the environment in which the network is to be used. If it is not explicitly known that the model is appropriate for a specific environment, it will be necessary to choose a model, perform a calibration survey and analyse to ensure its applicability and also to tune the model if necessary. In general, a calibration survey is always a good idea for any sizeable planning project in any case.

### *4.10.5 Antenna Height*

*Check that the model is valid for **both** the base station and mobile station antenna heights.*

Antenna height at both the fixed and mobile end of the link is important. Some propagation models are appropriate only for particular link configurations (for example the 528 aeronautical model is only valid when the mobile element is higher than the fixed site), and for terrestrial systems it is important to know whether the mobile element is likely to be embedded in the clutter or above it. This will affect the clutter loss that needs to be applied.

### *4.10.6 The Application*

*Check that the model is applicable to the application being planned.*

A crucial consideration is the application the simulation activity is to be put to. For example, if the purpose of the activity is to dimension a network by determining a rough estimate of the number of base stations needed, then a point-to-area model, possibly working with low-resolution terrain and clutter data, will probably be entirely appropriate and will give the wanted approximate answers. In this case, adding to computation time by using point-to-point models will not add any benefits. Conversely, using too coarse a model in detailed planning is dangerous. One clear distinction that must be made is the contrast between a planning model and an interference model; the two are not the same and should never be confused. The 526 model is an example of a planning model, and the 1546 model is an example of a model suitable for interference prediction for long ranges and for small percentages of the time.

### *4.10.7 Available Data*

*Check that the model matches the data available.*

Finally, another important issue is the consideration of the environmental data that will be used with the propagation model. If the only data available is coarse or of dubious accuracy – which sometimes does happen – then it is pointless to insist on using a highly accurate, computationally intensive propagation model. Not only will this take longer, it may produce results that display salients that do not exist in real life (in the same way that quoting the results of a calculation to too many decimal places may allow the reader to infer that the calculation is more accurate than it actually is). This may unduly influence the planning activity. When considering the available data and the propagation models to be used with it, it is best to take a pragmatic view.

## 4.11 Abnormal Propagation Conditions

Most propagation models will provide answers for median conditions when used with their default values. Atmospheric conditions will vary, however, and it is important to be able to determine propagation under less usual conditions. Models such as 1546 and 528 incorporate conditions to be expected for 50, 10, 5 and 1 % of the time. This will include most anomalous conditions to be encountered. For other models, it may be necessary to change the $k$-factor or apply margins to account for adverse conditions (this is discussed in Chapter 5).

## 4.12 Propagation Model Summary

Many different propagation models have been developed over the years, and there is no single model that meets all modelling needs, when considering the frequency of operation, the distribution of mobile elements throughout service area, the length of links, environmental data available to use with the model and computational cost, among others. When selecting the right model to use, it is best to consider all of the factors discussed in Section 4.10, each of which will affect the suitability of a specific model.

Calibration surveys are a vital tool for any major new project, particularly if it differs in any way from previous projects. This allows the models available to be tested with the data available in order to determine which shows the best available for that project. To do this, it is important that the measurement system mimics the way that real subscribers will use the system.

Table 4.4 shows a comparison of some models available in a specific radio-planning tool (ICS Telecom). It includes some models not described, but which are alternatives to those

**Table 4.4.** Comparison of models with measured data.

| | No clutter | | Clutter | |
|---|---|---|---|---|
| UHF Survey (0–30 km) | ME | SD | ME | SD |
| Bullington, standard | 19.2 | 7.0 | 8.8 | 8.8 |
| Deygout, no sub | 16.5 | 7.0 | 6.5 | 8.2 |
| Deygout, sub (standard) | 9.3 | 7.4 | 0.1 | 7.6 |
| Deygout, sub (coarse) | 11.4 | 8.0 | 2.0 | 8.1 |
| Deygout, sub (fine) | 6.7 | 6.7 | −1.9 | 7.4 |
| Deygout, sub (area) | 10.4 | 6.7 | 1.0 | 7.8 |
| ITU 526, deygout | 6.7 | 8.4 | −1.4 | 8.2 |
| ITU 526, round | 14.4 | 8.7 | 5.3 | 8.3 |
| ITU 526, cylinder | 14.7 | 7.3 | 5.3 | 7.4 |
| Okumura Hata (no diff) | 14.1 | 12.3 | 14.1 | 12.3 |
| Hybrid-Ok Hata (526) | 4.8 | 8.5 | 4.8 | 8.5 |
| FCC 98 | 6.2 | 11.4 | −2.2 | 10.3 |
| ITM 122 | 7.1 | 11.2 | −1.4 | 10.8 |
| Millington | 19.0 | 7.1 | 8.6 | 8.9 |
| 452 | 7.0 | 8.5 | −1.2 | 8.6 |
| 370 | 13.3 | 17.4 | 3.5 | 16.8 |
| 1546 | 14.3 | 14.9 | 3.5 | 16.8 |

described. The mean error (ME) and standard deviation (SD) of error are shown in dB for untuned models both with clutter attenuation and without it. The most obvious factor in this table is that the incorporation of clutter loss is important for all models other than those based on Okumura Hata (which already incorporates clutter terms). The Deygout models are all based on different implementations of the Deygout '91 model, and in this case they show the best results, with the smallest mean error and standard deviation of error. None of the models have been tuned and far better results would be expected after tuning.

Once a model has been selected from this list, it is then possible to further tune the model for the specific application. This process changes the configuration from a generally applicable model to one optimised for the particular project; it should not be used on any other project unless it can be proved that the two projects share similar characteristics. If this cannot be done, then a calibration survey should be carried out and the whole process repeated from scratch.

For the ITU models, ITU-R P.1144 gives a list of the models produced by Study group 3 and their applicability to different applications.

As a final point, it is worth remembering that it is not likely that a single model will be appropriate for all tasks in a project; the best model should be selected for individual tasks, not whole projects.

## References and Further Reading

All ITU publications are available from the ITU web site, which at the time of writing was www.itu.int.

[1] ITU Terrestrial Land Mobile Radiowave Propagation in the VHF/UHF Bands ITU-R Working Party 3K of Study Group 3

[2] Propagation of Radiowaves, Ed. M. P. M. Hall, L. W. Barclay and M. T. Hewitt, Institute of Electrical Engineers

[3] ITU-R P.370-7: VHF and UHF propagation curves for the frequency range from 30 to 1000 MHz

[4] ITU-R P.453-6: The Radio Refractive Index

[5] Digital Line-of-Sight Radio Links, A. A. R. Townsend, Prentice Hall

[6] COST 231 (available from the COST organisation web site)

[7] The Mobile Radio Propagation Channel 2nd Edition, J. D. Parsons, 2000, John Wiley & Sons Ltd, Chichester, ISBN 0-471-98857-X

[8] Antennas and Propagation for Wireless Communication Systems, Simon R. Saunders, 1999, John Wiley & Sons Ltd, Chichester, ISBN 0-471-98609-7

[9] ITU-R P.526-8: Propagation by Diffraction

[10] ITU-R P.1546: Method for point-to-area predictions for terrestrial services in the frequency range 30 MHz to 3000 MHz

[11] ITU-R P.452: Prediction Procedure for the Evaluation of Microwave Interference Between Stations on the Surface of the Earth at Frequencies Above About 0.7 GHz

[12] ITU-R P.1144: Guide to the application of the propagation methods of Radiocommunication Study Group 3

[13] Mobile Communications Design Fundamentals William C. Y. Lee, Howard W. Sams & Co.

# 5

# The Mobile Environment Part 2: Fading, Margins and Link Budgets

## 5.1 Introduction

In this chapter, we will return to our examination of the mobile environment and will look at fading mechanisms for both slow and fast fading. Both are crucial for mobile radio network planning because they influence the application of propagation models to determining system availability for mobile subscribers, and thus the design of the entire system. These mechanisms are not normally included in the macro-cell propagation model (such as those described in the previous chapter), but they must be accounted for when determining how the user will perceive the service availability. As such, they must be included in the link budgets used in the project. Link budgets are crucial to radio network engineering and we will be looking at these later in the chapter. The basic principles of building a path budget are relatively simple, but it can be difficult to create a realistic budget for a variety of practical reasons. For example obtaining equipment characteristics, which should be a straightforward activity, can often be difficult as the information may not be available – or the information that is available is unreliable. Another area of difficulty can be the determination of the correct values to be used for calculating link margins, based on the fading characteristics and practical loss factors to be encountered in reality. We will examine the various components of the link margin and identify their important characteristics. In addition, we will look at the practical issue of how a calculated link margin can be represented in a particular planning tool, since the different parts of a link budget will probably have to be entered into different dialogues in the radio network-planning tool.

We will start looking at fading and some relevant statistics that can be used to represent it. Where possible, formulae will simply be presented and their importance explained rather than delving into the mathematical principles behind them; these are adequately described in many other books and papers that the engineer can read at their leisure.

Mobile Radio Network Design in the VHF and UHF Bands: A Practical Approach
*Adrian W. Graham, Nicholas C. Kirkman and Peter M. Paul* 

### 5.1.1 Statistics Relevant for Fading

In this section, we will not cover the fundamental mathematics of statistics, as these can be found in many other publications. We would particularly recommend ITU-R PN.1057: probability distributions relevant to radiowave propagation modelling. Statistics are important in radio network planning because not all aspects of propagation can be modelled using the methods discussed in Chapter 4, in particular fading mechanisms because they depend on factors that can be easily modelled. Instead, we can use the properties of known statistical distributions to describe how the received signal level will vary around the median value predicted by a propagation model, so long as we know which particular distributions to apply. In this section, we will look briefly at the lognormal, Rayleigh, combined lognormal and Rayleigh, Nakagami-Rice and Ricean distribution. These are commonly used in radio modelling.

For most real processes, statistical distributions can be used to determine the variation of a variable around its normalised value. For radio prediction purposes, we are principally interested in whether a signal received at a location will be above or below a certain threshold level. This is illustrated by a generic probability distribution in Figure 5.1. In this case, the distribution is normalised on a specific value, often the lowest median operating signal level that a radio receiver can operate with to achieve a given degree of performance. In this case, the radio signal would be acceptable for 50% of samples and not acceptable for the remainder. This is a level of performance that subscribers would be unhappy with; a service that fails half of the time does not offer a good level of performance. We would therefore wish to determine the limit of acceptable performance as being some value higher than the minimum operating level. This would have the effect of shifting the curve in Figure 5.1 to the right, and setting the minimum median operating level to some point on the left hand side of the curve, such as point (A) or (B). The key thing we need to do as radio

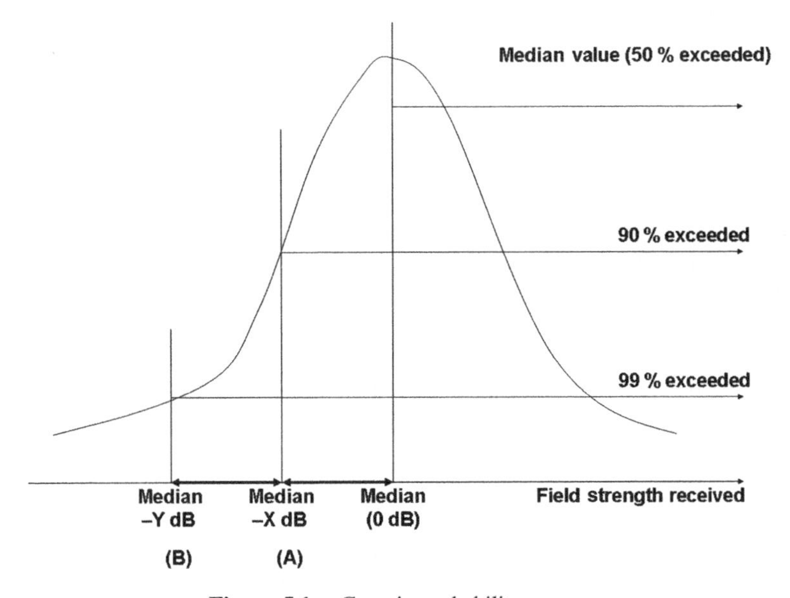


**Figure 5.1.** Generic probability curve.

engineers is to determine the number of dB we need to shift the curve to the right (margin) in order to achieve a given percentage of signals that will be above the reference value (values of 50, 90 and 99 % in the figure).

As a simple illustration, if the value $X$ is 10 dB, then shifting the reference value by 10 dB would change the percentage of signals above the reference value from 50 to 90 %. If the median value was −90 dBm, for example, and if we set a limit of −80 dB, then we can be sure that 90 % of the signals experienced would be above −90 dBm, even taking account of the variability of the signal. We will now go on to look at specific distributions. We will not cover derivation of the mathematics, so look at the list of further reading material for further information.

### *5.1.2 Lognormal Distribution*

Most people are familiar with the Normal or Gaussian distribution, which describes a wide variety of physical variations. Since in radio prediction we are normally working in logarithmic units, the logarithmic version of this is the most relevant. However, this does not behave quite in the same way as the linear version. Some of the key differences are[1]:

- The normal distribution is normalised about zero. The lognormal distribution is normalised around a positive value.
- The normal distribution is symmetrical in linear terms, and the mode, median value and most probable values are identical. In the lognormal case, this is not true; it is symmetrical in logarithmic units. If $m$ is the mean value and $\sigma$ is the standard deviation of the lognormal case, then the following apply to obtain the linear equivalent:

  The most probable value is

  $$\exp(m - \sigma^2) \qquad [1]$$

  The median value is

  $$\exp(m) \qquad [2]$$

  The mean value is

  $$\exp\left(m + \frac{\sigma^2}{2}\right) \qquad [3]$$

  The root mean square (RMS) value is

  $$\exp(m + \sigma^2) \qquad [4]$$

  The standard deviation (SD) is

  $$\exp\left(m + \frac{\sigma^2}{2}\right)\sqrt{\exp(\sigma^2) - 1} \qquad [5]$$

Figure 5.2 shows the probability density ($P(x)$ – PDF curve) and cumulative ($F(x)$ = CDF curve) distributions of the lognormal curve, with the most probable, mean and median values shown. The equations of the PDF and CDF are shown below[1].

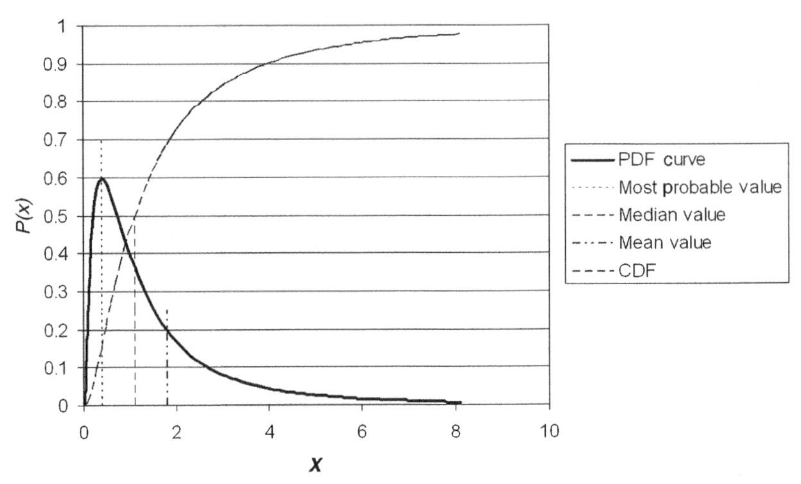


**Figure 5.2.** Log normal distribution curve.

$$P(x) = \frac{1}{\sigma\sqrt{2\pi}}\frac{1}{x}\exp\left[-\frac{1}{2}\left(\frac{\ln x - m}{\sigma}\right)^2\right] \quad [6]$$

$$F(x) = \frac{1}{2}\left[1 + erf\left(\frac{\ln x - m}{\sigma\sqrt{2}}\right)\right] \quad [7]$$

The lognormal distribution is found in many aspects of propagation, relating to signal level or field strength variation. In the same way that the normal distribution is often seen in circumstances where there are a large number of independent variables, the same is true for the lognormal distribution.

### *5.1.3 Rayleigh Distribution*

The Rayleigh distribution is applicable to non-negative numbers. Unlike the lognormal distribution, the Rayleigh distribution is not symmetrical in logarithmic units. The main characteristics can be determined by using the standard deviation $\sigma$ as shown below[1].

The most probable value is

$$\sigma$$

The median value is

$$\sigma\sqrt{2\ln 2} \quad [8]$$

The mean value is

$$\sigma\sqrt{\frac{\pi}{2}} \quad [9]$$

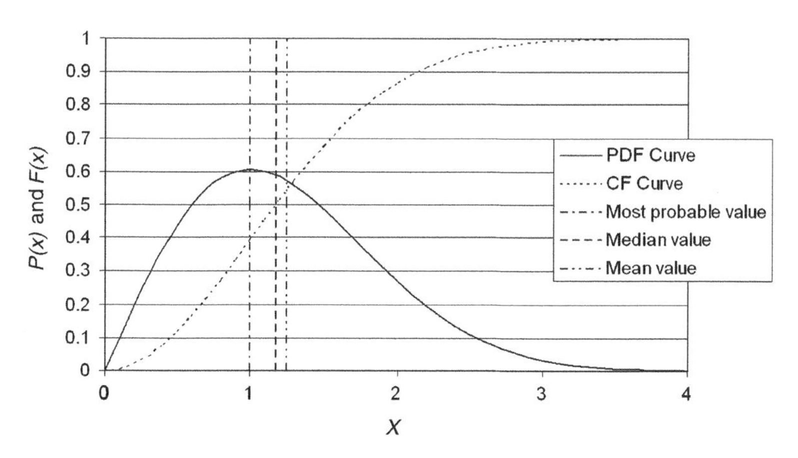


**Figure 5.3.** Rayleigh distribution curve.

The root mean square value is

$$\sigma\sqrt{2} \quad [10]$$

The standard deviation is

$$\sigma\sqrt{2-\frac{\pi}{2}} \quad [11]$$

The form of the Rayleigh distribution is shown in Figure 5.3, which also shows the most probable, median and mean values. The height-to-width aspect ratio changes with the standard deviation. The equations for $P(x)$ and $F(x)$ are shown below.

$$P(x) = \frac{x}{\sigma^2}\exp\left(-\frac{x^2}{2\sigma^2}\right) \quad [12]$$

$$F(x) = 1 - \exp\left(-\frac{x^2}{2\sigma^2}\right) \quad [13]$$

Rayleigh distribution is associated with the vector addition of numerous signals of varying magnitude and phase. This is important for the modelling for the fast fading often seen in mobile radio networks as we shall see.

### *5.1.4 Ricean Distribution*

The Ricean distribution represents the case where there is a steady value that obeys lognormal distribution (offset to give a positive value) and a varying component that is

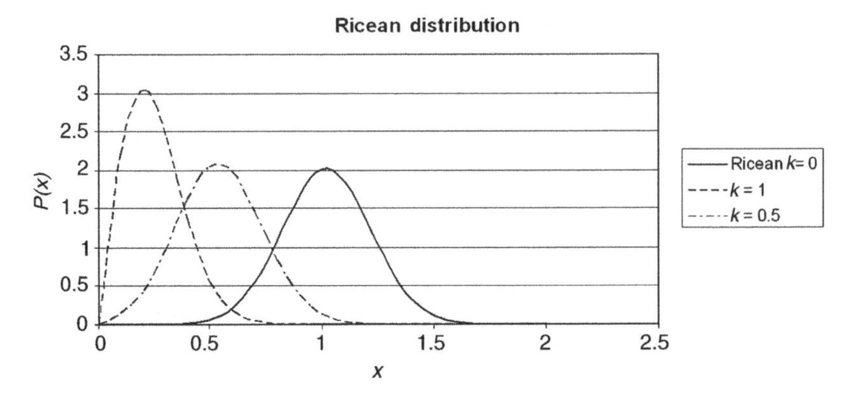


**Figure 5.4.** Ricean distribution curve.

more like the Rayleigh case in nature. The ratio of the steady-state to the varying signal level is represented by a dimensionless parameter $k$ where:

$$k = \frac{P_c}{P_r} \qquad [14]$$

where
$P_c$ is the power of the constant part
$P_r$ is the power in the varying part

Figure 5.4 shows the Ricean distributions for three different values of $k$, and it can be seen that for $k = 0$ (meaning there is no constant power component) the distribution is of the same form as the Rayleigh distribution. The equation of the curve can be expressed by[2]:

$$P(x) = \frac{x}{\sigma^2}\exp\left(-\frac{x^2}{2\sigma^2}\right)\exp(-k)I_0\left(\frac{x\sqrt{2k}}{\sigma}\right) \qquad [15]$$

where
$I_0$ is the modified Bessel function of the first kind and zero-th order.

Note that $k$ can also be expressed as

$$k = \frac{s^2}{2\sigma^2} \qquad [16]$$

where $s$ is the magnitude of the constant part relative to the varying part.

The cumulative function of the Ricean distribution is shown in Figure 5.5 for the same values of $k$. This is derived from the integral of the $P(x)$ Equation [15].

### *5.1.5 Other Statistical Distributions*

The three distributions covered above are not the only ones used in radio engineering, but they are probably the most commonly used. Engineers may encounter other distributions such as Nakagami-Rice (Nakagami-n), Nakagami-m and gamma, for example (as described

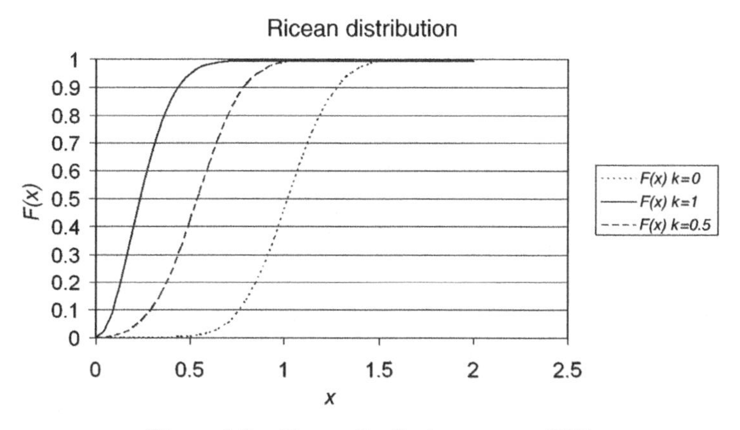


**Figure 5.5.** Ricean distribution curve – CDF.

in ITU-R P.1057), and the engineer should familiarise themselves with the properties and application of these distributions.

## 5.2 Slow Fading

### *5.2.1 Slow Fading (Shadowing) Mechanisms*

Slow fading (also known as shadowing) mechanisms are those that cause variations in the local median signal level received at a constant distance from a transmitter. Such mechanisms include terrain effects, vegetation and clutter, and they affect the median field strength received over a region of the ground. Typically, the median signal level is considered as relatively constant over a region of approximately $40\lambda$, which is the precision to which a typical 'macro' point-to-point propagation model provides a result, and also the region over which measurements are averaged for surveys (See Chapter 15 on verification which explains the importance of the $40\lambda$ term). This region is also frequently called the 'short sector'. Table 5.1 shows the value of $40\lambda$ for a range of frequencies.

An example of how shadowing will affect radio propagation is shown in Figure 5.6. The lightest irregular line shows the terrain. The curved line shows the calculated field strength by free space loss only. The mid-grey irregular line (the lower of the two) shows the field strength predicted by a model that takes some account of shadowing mechanisms. Notice

**Table 5.1.** Length of 40 wavelengths.

| Frequency (MHz) | $40\lambda$ (m) |
|---|---|
| 30 | 400.00 |
| 80 | 150.00 |
| 150 | 80.00 |
| 300 | 40.00 |
| 450 | 26.67 |
| 900 | 13.33 |
| 1800 | 6.67 |

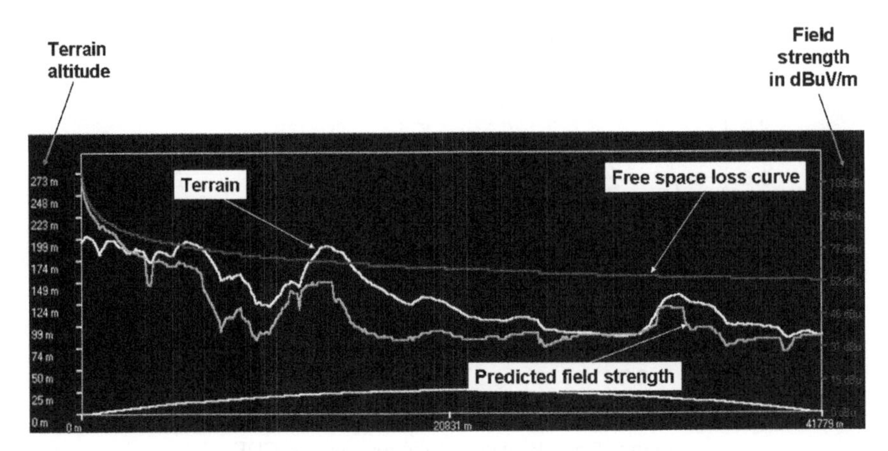


**Figure 5.6.** Slow fading shown on a sample path profile taken from a radio planning tool. Reproduced by permission of P & D Missud.

that unlike the free space loss curve, the more sophisticated model displays greater variation with distance and, although it falls off generally with distance, there are occasions when an increase in distance results in an increase in predicted field strength. This is due to the path at a more distant point suffering from less diffraction than a point at a shorter distance.

The signal variation at a given range from a transmitter will vary following a lognormal-type distribution as shown in Figure 5.2. It can be seen from Figure 5.6 that the variation of signal is not directly accounted for point-to-area models, but is for point-to-point models.

In radio network planning, we are typically interested in determining where links will work effectively. For the planning of CDMA-based systems, this will principally be predicated on intra-network interference, but for systems that are not inherently noise limited, we are interested in determining where the received field strength level will exceed a pre-defined minimum value, including losses due to distance, shadowing and, as we will see, fast fading mechanisms. We will now look at how this can be determined.

## *5.2.2 Slow Fading and Propagation Model*

### 5.2.2.1 Point-to-Area Models

Point-to-Area models that consist only of an inverse exponential term do not include the effects of shadowing. In order to use them for any kind of planning – even network dimensioning – it will be necessary to include a term in the margin (the level above the minimum signal the system can operate effectively with) to include the effects of this mechanism. Some point-to-area models do take some account of gross terrain effects and clutter, but even in this case, there will still be a variation between measured results and those predicted. It is possible to use statistics to determine the probability of coverage within a particular service area (area served by a base station, in this case assumed to be serving omni-directional) by using either the cell edge or cell area methods, as described next.

#### 5.2.2.2 Cell Edge and Cell Area Method of Cell Radius Prediction

Figure 5.7 shows a typical inverse-exponent propagation model with the median predicted field strength shown. However, shadowing will ensure that the actual value measured at a constant distance around the transmitter will vary according to the path between the transmitter and the receive location. This will normally vary according to a lognormal distribution since it is a function of many uncorrelated factors, so if we know the standard deviation of the signal, we can determine the difference in dB between the median value and some other value representing a change of a percentage of samples exceeding a given value (see Figure 5.1). If, as in Figure 5.7, the value exceeded by 90 % of samples corresponds to a difference of 10 dB, then we can draw a second exponent curve 10 dB lower than the first one, together with a straight line indicating our minimum wanted field strength value (arbitrarily set at 25 dBμV/m in this figure, for the purposes of illustration).

Examining the figure shows that if one were to use the nominal field strength figure, then the expected radius of the effective coverage from the base station would be about 18 km. In this case, only 50 % of locations at the edge of the coverage area would be covered. However, if the 10 dB margin to account for shadowing is used, so 90 % of locations at the edge of coverage are served, then the radius is reduced to nearer 9.5 km. This is the cell edge method.

It can be shown that for an inverse-exponent variation with distance, the probability of exceeding a specific value at range $r$ is[2]:

$$p_e(r) = \left[1 - erfc\left(\frac{FS(r) - FS_m}{\sigma}\right)\right] \qquad [18]$$

where

$FS(r)$ is the nominal field strength at range $r$ in dBμV/m
$FS_m$ is the minimum acceptable field strength (dBμV/m)
$\sigma$ is the standard deviation of the shadowing loss.

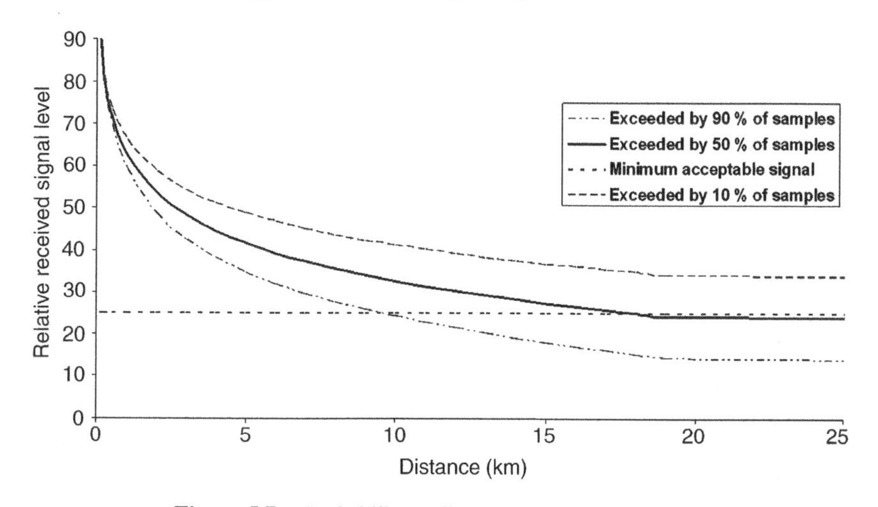


**Figure 5.7.** Probability ordinate exceeded by distance.

One problem with this method is that although the probability of exceeding a minimum field strength at the edge of the coverage area may be expressed as a given probability, mobile subscribers closer to the base station will benefit from a better service (as shown in Figure 5.7). Sometimes, an assessment is made over the whole of the service area rather than just at the edge. This can be related to the probability of coverage at the edge by the following equation[2]:

$$p_{\text{cell}} = p_e(r_{\max}) + \frac{1}{2}\exp(A) \times (1 - erf(B)) \qquad [19]$$

where

$$A = \left(\frac{\sigma\sqrt{2}}{10n\log e}\right)^2 + \frac{2M}{10n\log e} \qquad [20]$$

$$B = \frac{\sigma\sqrt{2}}{10n\log e} + \frac{M}{\sigma\sqrt{2}} \qquad [21]$$

This can be used to determine the probability of coverage throughout a whole service area, not just at the edge. In practice, it would be more normal to use a computer-based radio-planning tool to determine the predicted coverage, and then use a point-based summation technique (described in Chapter 9). If the system is configured correctly, this will give a more meaningful result, which will far better approximate to the situation found in reality.

One point of particular note is that on some occasions, engineers attempt to use the cell-edge to cell-area conversion when using point-to-point prediction models. This is entirely wrong and should be avoided because the mathematics described above would be completely wrong in this case.

#### 5.2.2.3 Point-to-Point Models

Point-to-point models will take account of most of the effects of shadowing, but because the model will simplify the path mathematically and because not all features will be included in the model, there will still be model variability between measured and predicted results. The degree of this can be determined by comparing measured results against those predicted by the model and determining mean error and standard deviation.

## 5.3 Fast Fading

### *5.3.1 Fast Fading Mechanisms*

Slow fading occurs over distances of many wavelengths, but there is also a mechanism that causes variations in signal level over ranges of $1/2\lambda$. This is known as fast fading, and it is due to scattering and reflections, mostly in the vicinity of the receiving antenna and particularly if it is near obstructions. There will be many such paths and they will change with small movements of the antenna and also through movement of the objects causing the scattering and reflection. This situation would not only be difficult to model due to its complexity, but such modelling would only be valid for the specific condition modelled; if any component within the model is moved, the results would change significantly.

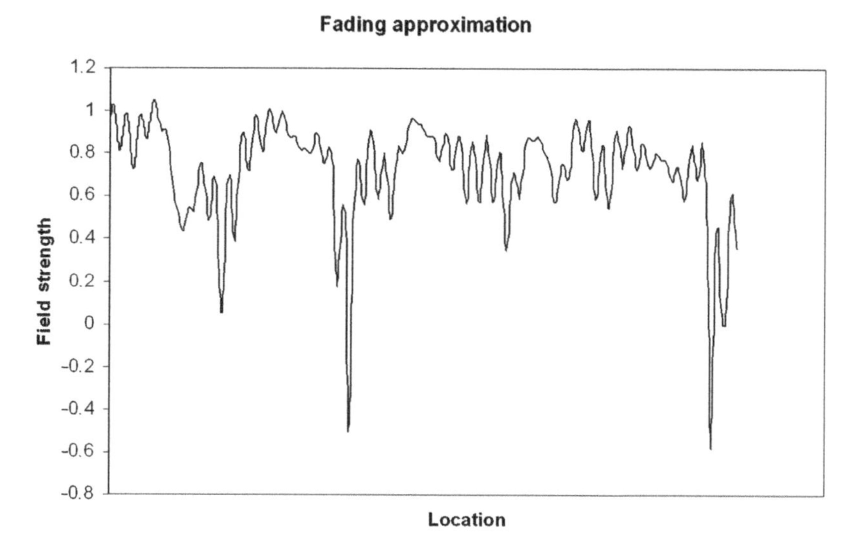


**Figure 5.8.** Illustration of signal strength variation due to fading.

Additionally, it would be next to impossible to collect sufficiently detailed information regarding the mobile environment. For these reasons, fast fading is not normally modelled, but instead is dealt with by statistical means. In fact, fast fading is familiar to most people who use a mobile phone; when coverage is poor, it is sometimes necessary to move around the immediate environment until an acceptable signal is found – the variability is due to fast fading effects.

Fast fading takes the form shown in Figure 5.8. This shows a signal of varying strength as the various components vector sum to produce the value at a specific point. The location axis represents movement within a single short sector, in any direction and not necessarily in a straight line.

Measurements and theoretical work have both confirmed that fast fading for mobile networks follows a distribution that can be approximated by Rayleigh statistics when the mobile element is in a heavily cluttered environment and there is no direct line of sight with the base station. This is frequently the case in mobile systems, particularly towards the edge of the coverage area. The physical distribution of fast fading may be imagined as shown in Figure 5.9, where some of the small areas within the short sector are above the median value and some are below. The distribution within the short sector is random. The median signal strength varies with slow fading as described in the previous section.

Fast fading is also present in aeronautical systems, but in this case, the effect is more typical due to changes in the phase of the direct path and at least one reflected path, which is common in such links.

Fast fading is not modelled using macro-cell propagation models but instead is accounted for statistically. The process is to determine the appropriate statistical representation for the receiver environment and then apply the appropriate dB correction (the 'margin') to ensure that the target signal strength is exceeded for a given percentage of locations within the short sector.

Most propagation models produce a median value where it can be expected that 50 % of locations in the short sector will be above the median value, and 50 % will be below. If the

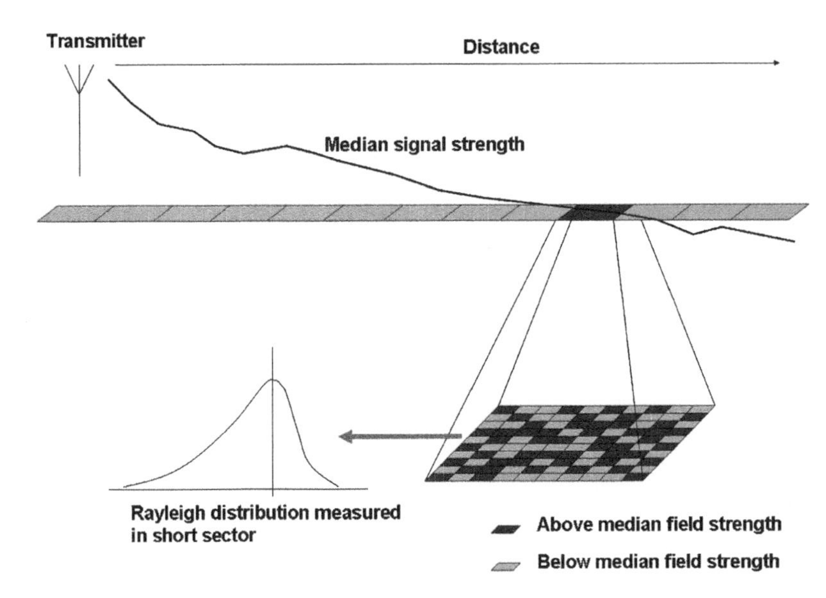


**Figure 5.9.** Slow fading and fast fading.

median value happened to correspond to the minimum operating signal level of the radio receiver, then the system will work for 50 % of the time, and not for 50 % of the time. This would not be acceptable to most subscribers, so it would be necessary to increase the percentage of the curve in excess of the minimum value. For example if it were assumed that the value (Median – X) dB were the minimum operating value for the system, then 90 % of signal would be above this value, and if it were (Median – Y) dB, then 99 % of measurements would exceed this value.

The dB corrections described are the margins required to exceed a given percentage of 'availability' in the short sector. This will vary according to the statistical curve that the signal variation is most closely approximated by. Table 5.2 shows a summary of three relevant statistical distributions – the lognormal, Rayleigh and Ricean for given probability values exceeded – on the basis of measured standard deviation.

As an illustration, assume that in order to work at an acceptable level of performance, an analogue system requires an operating signal level of −90 dBm at the receiving antenna. This would mean that if the signal level predicted at a specific point is −90 dBm, then 50 % of the signals in the short sector would be below this level. For planning purposes, this is too

**Table 5.2.** Fading probability values; note that for the Ricean case, *k* has been converted to logarithmic units.

| Probability exceeded (%) | Log normal (dB) | Rayleigh (dB) | Ricean $k =$ 10dB (dB) | Ricean $k =$ 0dB (dB) | Ricean $k =$ −10dB (dB) | Ricean $k =$ −20dB (dB) |
|---|---|---|---|---|---|---|
| 50 | 0 | 0 | 0 | 0 | 0 | 0 |
| 90 | 10 | 8.2 | 8 | 7.5 | 2.8 | 0.9 |
| 99 | 20 | 18.4 | 18.4 | 17.5 | 6 | 1.5 |
| 99.9 | 30 | 28.4 | 28.5 | 27.6 | 9.4 | 2 |

low a service level for most applications. If we assume that the required availability on the edge of coverage is 90 % and the Rayleigh distribution is most appropriate, then a correction of 8.2 dB should be added as a margin. This means that in the planning tool we should consider the edge of coverage not to be at −90 dBm, but rather at

$$\begin{aligned}\text{Signal level required} &= \text{minimum operating signal} + \text{fade margin}\\ &= -90 + 8.2 = -81.8\,\text{dBm}\end{aligned}$$

This should then be used to define the edge of coverage in the propagation prediction. Note that this has no effect on the prediction used itself; it simply modifies the threshold value for acceptable service. Any increase in the threshold value will normally result in a decrease in the service area and a decrease in the radius to the edge of acceptable coverage around the base station.

In summary, the process of determining the actual value to be used in a link budget is as follows:

- Identify the performance metric of the task; e.g. 95 % locations availability.
- Determine the relevant statistical distribution to best represent reality.
- Calculate or read from tables the standard deviation coefficient to achieve the probability that the wanted value has exceeded (for example $2\sigma$).
- Determine the value of standard deviation to be used in the calculation. This will be based on measurement data.
- Multiply the standard deviation by its coefficient to achieve the wanted margin in dB. For example if the standard deviation is 5 dB and the coefficient is $2\sigma$ then the answer is $5 \times 2 = 10$ dB.
- Add this to the required field strength or signal power at the antenna (note that technically, using field strength means that the receiver antenna characteristics do not need to be considered at this point).

## 5.4 Receiver Antenna Environment – Body Loss and Other Factors

Fast fading will affect the field strength available in the vicinity of a receiving antenna, but there may be other characteristics that will affect the energy actually incident at the antenna itself. For mobile elements, losses due to the human body or a vehicle may have an additional effect. In terms of pure attenuation, Figure 5.10 shows some typical losses for antennas at head height and waist height on a 'typical' person. Such figures are normally based on an '8-position average'. This is obtained by measuring loss from a sample of people at angles of $360/8 = 45$ degree increments and then producing an average figure both for the directions and for each person. This is necessary because in general, the orientation of the mobile antenna with respect to the base station with which the mobile is communicating (or other mobile system) will be random.

Of course, such figures are merely 'typical' rather than absolute, since the human body is a highly variable factor, and there will be users who cause far higher or lower losses. In some circumstances where the typical network subscriber will be different from the 'typical' case shown in Figure 5.10, it will be necessary to derive suitable figures for incorporation into the

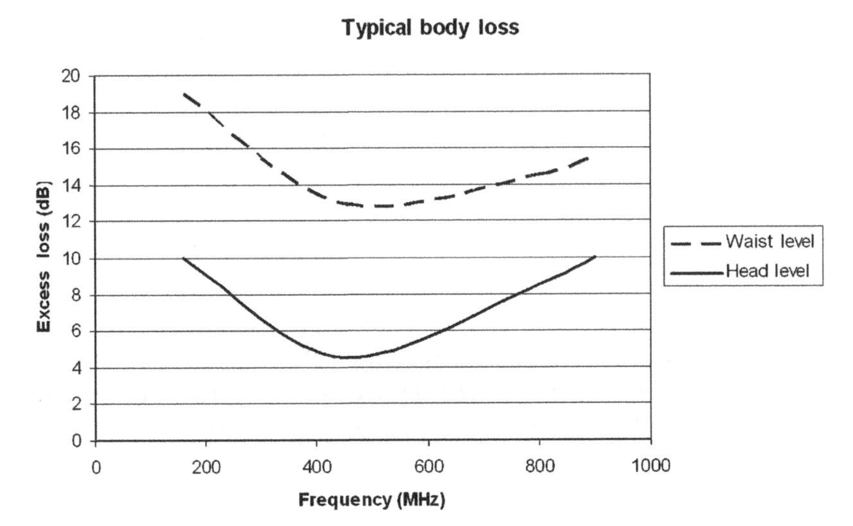


**Figure 5.10.** Typical body loss at waist and head height.

link budget; for example fully kitted out soldiers will have different characteristics from business people in suits.

Body loss is another factor that must be incorporated into the link budget, unless the 8-position average figures are already incorporated into the antenna loss figure (some systems, such as pagers, will have performance specified when next to a body).

## 5.5 Elements of a Radio Link

### *5.5.1 Generic Link Diagram*

The link diagram for a generic radio system is shown in Figure 5.11a. Not all elements will be present in all radio systems, but it is broadly representative; tuning units in particular may not be necessary unless the radio has to be capable of tuning over a wide frequency range. If this is required, the tuning unit helps to match the antenna impedance to the rest of the system. Figure 5.11b shows how the signal level may vary from the transmitter to the receiver, although not to scale. In general, specific sub-components downstream of the transmitting radio output will be connected via connectors and feeder cables. Both connectors and feeders may cause a loss in signal strength each time they are used. Radio systems that are able to operate over a wide frequency range (such as military radios and some modern multi-band radios) may have a tuning unit that is used to match the antenna impedance and thus maximise signal radiated from the antenna. If a tuner is present, it may also feature an embedded amplifier. Many systems will not have a tuning unit, but may have an amplifier that boosts the signal level before it enters the antenna.

Energy entering the antenna at RF is then propagated into the environment, and some of this energy is picked up at the receiving antenna. At the receiver, there may be one or more antennas to pick up the signal. A hand-portable system will have a single antenna but a vehicle or aircraft may have several to combat fast fading. This is known as a diversity antenna system. If diversity antennas are present, then the individual signals are picked up by

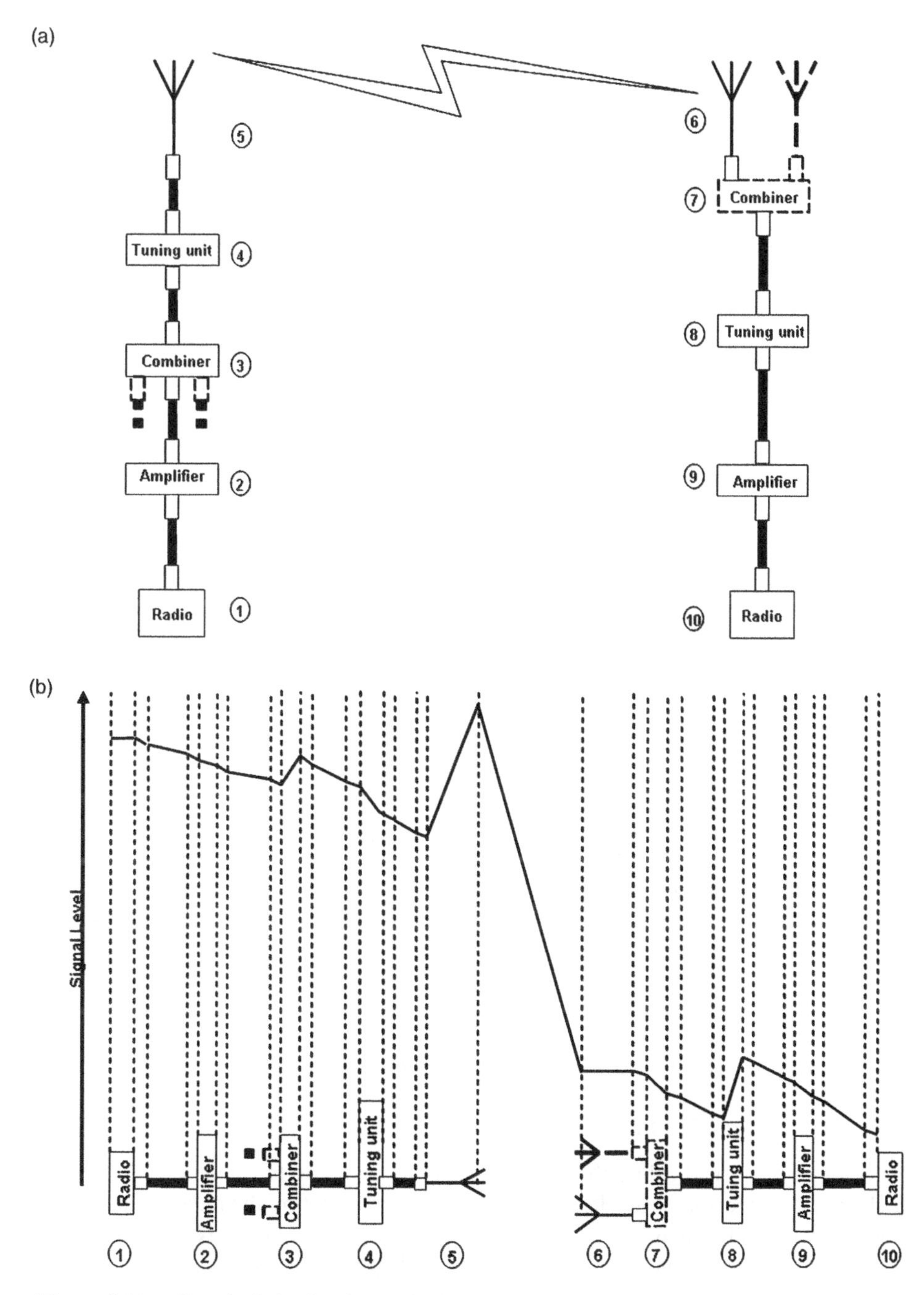


**Figure 5.11.** Generic link, showing typical elements that may be present in a Radio system.

a combiner or switch, which generates a single signal to be conducted through the rest of the system. This may be amplified (although it will also amplify any noise or interfering signals as we shall see in Chapter 13) and may again pass through a tuning unit if one is present before finally arriving at the receiving radio.

Figure 5.11b shows the same components of a radio link, but this time spread out in a linear fashion with a representation of the signal strength as it travels from the transmit radio output through to the input of the receiving radio. The signal will generally drop slightly as it travels through connectors and feeder cables, particularly if they are long. It will also be enhanced by amplifiers and possibly antenna gain (as shown in this example at the transmitter end). There may also be a gain effect if there are multiple receiving antennas (diversity gain), compared to the single antenna fading condition. Other elements in the link may cause losses. Eventually the signal reaches the receiving radio. If the signal that arrives is higher than the minimum operating level required, then the original message should be retrievable at the receiver. If it is not high enough, then the original message or part of it may be lost.

The process of determining a link budget is composed of two main elements; determining the elements within the link that cause gain or loss between the radios and the antennas, and also determining the minimum acceptable signal level that must be available at the receiving antenna as it moves through the service area, both at the fast fading and slow fading scales.

We now look at some of the elements shown in Figure 5.11 in a little more depth before going on to develop some example link budgets and show how they may be represented in planning tools.

### 5.5.2 Nominal Power

The nominal power will be the power output from the radio that best represents the performance of the system. Power is normally quoted in units of Watts (W), dB relative to a Watt (dBW) or dB relative to a milli-Watt (dBm). Of these units, the decibel versions are best suited for link budget calculations, because all other units will be quoted in dB; for direct comparison with received signal strength, dBm is the best unit. The simple conversion formulae are:

$$Power(dBW) = 10\log_{10}(Power(Watts)) \quad [22]$$

$$Power(dBm) = Power(dBW) + 30 \quad [23]$$

$$Power(dBW) = Power(dBm) - 30 \quad [24]$$

### 5.5.3 Feeder and Connector Losses

Feeders are the electrical cables that connect the radio output to the antenna, via connectors at each end. If connectors or cables attenuate the power output by the radio, then the power presented to the antenna will be less than otherwise and consequently transmitted radio energy will be reduced. These losses should not be large in most systems and, if there is a significant loss, it would be normal for an amplifier to be fitted to counteract this loss.

### 5.5.4 Tuning Units, Amplifiers and Combiners

For radios that cover a wide proportion of the radio spectrum (of the order of an octave or more), the antennas used will not be matched over the entire range. Uncorrected, this would cause higher loss for those portions of the band that are more mismatched. To overcome this, the radio may be fitted with a tuning unit. If so, the gain of the tuning unit, or the overall gain or loss of the tuning unit, mated to the antenna must be taken into account (being careful not to double count).

For some systems, an amplifier may also be fitted to boost signal strength close to the antenna so that a higher power is fed into the antenna unit. If an amplifier is fitted, then the affect of this must also be included in the link budget.

Combiners are used when signals from different systems are fed into the same antenna for transmission. This may be done for reasons of efficiency or of space on a given antenna. Given the further development of so-called MIMO (multiple-in multiple-out) systems and the general reluctance in many areas to allow more masts to be built, this will be ever more present for future network designs. Combiners may involve a loss to each of the systems feeding into the system.

Each of these systems may be present in either the transmitting or receiving end of the link.

### *5.5.5 Base Station Antennas*

Antennas are the elements that transform conducted energy into radiated energy. The conversion of energy from electrical power into radiated power occurs as a result of the movement of charged particles along the antenna at radio frequencies. There are a wide variety of antennas for different applications. The VHF/UHF performance of a particular antenna is normally expressed in terms of a reference antenna, which for mobile systems is normally a dipole. The gain of a dipole antenna can also be referred to a hypothetical antenna that radiates equally in all directions, known as an isotropic radiator. Table 5.3 shows the relative gain of some fundamental antennas, referenced to an isotropic antenna.

The isotropic antenna radiates in all directions equally, but physically realisable antennas such as those shown in Table 5.3 do not. This is illustrated in Figure 5.12. The whip antenna shown is a type of monopole antenna. The loop antenna is another common type.

Antennas may suffer losses due to inefficiencies, mismatches with the feeder system and being used to transmit signals away from the optimum transmission frequencies (in which case it is important to represent the antenna gain or loss at the actual frequency used and not just the centre frequency supported). For link budgets, and to allow calculation in propagation models, an antenna used in a link budget calculation is referenced against a common antenna type.

The antennas in Table 5.3 are a small subset of those used in mobile radio networks. In particular, many antennas have directional polar responses, so that energy can be directed in a given sector. This is used to manage the capacity of networks (discussed in Chapter 10) and

**Table 5.3.** Gain of common antennas reference an isotropic antenna.

| Antenna type | $G_i$ (dBi) |
|---|---|
| Isotropic | 0 |
| Hertzian dipole | 1.75 |
| $\lambda/2$ dipole | 2.15 |

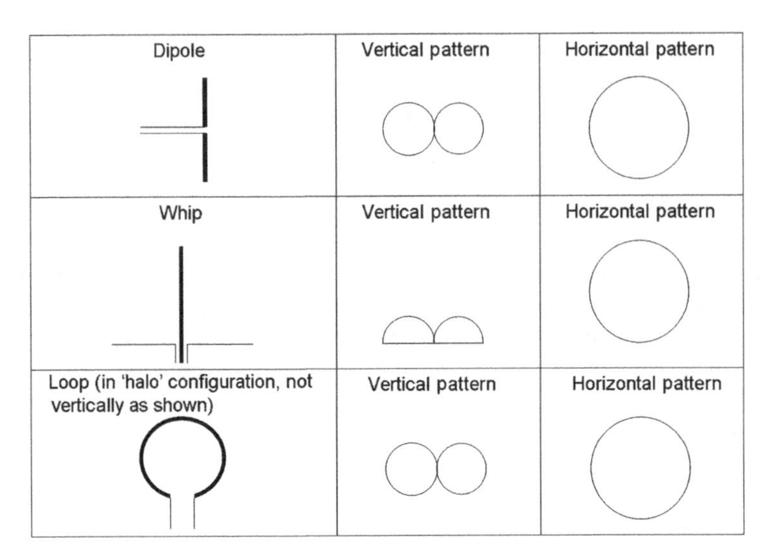


**Figure 5.12.** Typical antennas and their polar responses.

to modify the coverage of individual sites. If this is the case, then any link budget must include the antenna gain or loss in the direction of the receiver.

An illustration of a directional antenna is shown in Figure 5.13. This only shows the front part of the directional response; there will be radiation from the back of the antenna as well, but far less than that on the main lobe. When determining link budgets, it is essential to determine the gain in the direction of the receiver.

When used at the receive end of the link, base stations may also have more than one receiving antennas to exploit a mechanism called space diversity. In this case, two or more antennas are used to receive the signal from a mobile station. The antennas will be sited

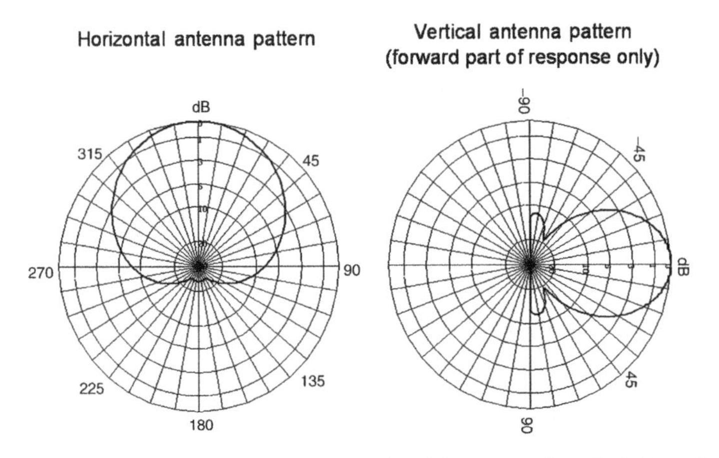


**Figure 5.13.** Sectored antenna response in horizontal and front part of vertical plane. Reproduced by permission of P & D Missud.

**Table 5.4.** Typical VHF mobile antenna losses in 150–174 MHz range.

| Antenna type | Typical losses (dBi) |
|---|---|
| Handheld (vertical) Telescopic $\lambda/4$ | 1 |
| Handheld (vertical) Helical | 4–5 |
| Handheld (tilted) Telescopic $\lambda/4$ | 5–6 |
| Handheld (tilted) Helical | 8–9 |
| Belt Telescopic $\lambda/4$ | 20–30 |
| Belt Helical | 10–20 |

fairly close together but will be far enough away to de-correlate fading between the mobile and each antenna. In this way, the effects of fades can be reduced. In link budget terms, this manifests itself as a gain in the mobile-base station link.

### *5.5.6 Mobile Antennas*

The mobile antenna must fulfil a wide range of criteria for reasons of practicality. It must be small, light, unobtrusive and inexpensive. This means in general that the antenna used for the mobile end of the link will be less than optimal from an engineering point of view. Tables 5.4 and 5.5 show some typical portable antennas and their loss compared to an isotropic antenna, for bands in the VHF and UHF ranges; the figures are fairly general and depend on frequency.

This loss will be experienced when the mobile is both receiving and transmitting, and so must be considered for both the uplink and the downlink.

### *5.5.7 Receiver Sensitivity*

The performance of a radio when it is not subject to external interference is governed by its sensitivity. This is the required input level to achieve a given degree of performance. This description is important; it identifies that there is no such thing as a single sensitivity, but rather that there will be different values for different degrees of performance. In some cases it will be necessary to determine sensitivity from graphs of bit error rate (BER), but in most practical cases, the engineer will be using either commercial equipment with published technical characteristics, or figures defined in a standard such as that for TETRA, TETRAPOL, GSM, UMTS and so on.

**Table 5.5.** Typical UHF mobile antenna losses in the 450–470 MHz range.

| Antenna type | Typical losses (dBi) |
|---|---|
| Handheld (vertical) Telescopic $\lambda/4$ | 2–4 |
| Handheld (vertical) Helix | 8–10 |
| Handheld (tilted) Telescopic $\lambda/4$ | 12–15 |
| Handheld (tilted) Helix | 17–20 |
| Belt Telescopic $\lambda/4$ | 25–35 |
| Belt Helix | 8–10 |

**Table 5.6.** Corrections from 1 MHz for thermal noise.

| Bandwidth (kHz) | dB Correction | kTB (dBm) |
|---|---|---|
| 6.25 | −22 | −136 |
| 12.5 | −19 | −133 |
| 25 | −16 | −130 |
| 100 | −10 | −124 |
| 200 | −7 | −121 |
| 1000 | 0 | −114 |

For digital equipment, the sensitivity used must be that for the conditions prevalent at the antenna, particularly whether it is moving or not. Digital receivers typically have resilience against fading built in for the situation when the fading is occurring rapidly, and thus there will be a static sensitivity and a dynamic sensitivity assuming a given vehicle speed. This is only strictly true for the speed identified, and if the digital system is to be used outside of its specified speed range, then it may well be necessary to perform measurements to determine its performance at the wanted speed. It is important to determine whether the sensitivity values are for unfaded conditions (in which fading must be accounted for additionally in the link budget) or for the faded condition (in which case this has already been accounted for and should not be added again).

The fundamental limit on receiver sensitivity is set by the thermal noise limit. This is the energy caused by random movements of charge and current due to the movement of electrons in the radio receiver. Thermal noise is equivalent to kTB, where $k$ is Boltzmann's constant ($1.38 \times 10^{-23}$), $T$ is the receiver temperature in degrees Kelvin (often taken to be 290 K) and $B$ is the receiver bandwidth. This is often approximated to −114 dBm/MHz as an easier figure to work with. For example corrections from this value for different bandwidths are shown in Table 5.6; see Chapter 13 for a discussion on receiver front-end noise.

These figures are not achieved in practice, so the concept of a 'noise figure' is normally added to this value to determine the minimum achievable value in practice. The noise figure is modelled by adding noise of a given number of dB to the thermal noise value to equal the actual value at the input to the receiver. So for a noise figure of 10 dB, the minimum achievable value for a 25 kHz bandwidth would be $-130 + 10 = -120\,\text{dBm}$.

For any non-noise input to be registered in the receiver, it must be higher than thermal noise plus the noise figure. In addition, any modulation scheme will require that the signal level is a given number of dB above noise (or noise plus interference, as we shall see in the next section). This varies according to modulation scheme and desired level of performance, often quoted in terms of bit error rate (BER). If this value is, say, 12 dB above noise (or noise and distortion), then the minimum receiver sensitivity for a 25 kHz system with a 10 dB noise figure would be:

$$\text{Minimum sensitivity (dBm)} = -114 - 16 + 10 + 12 = -108\,\text{dBm}$$

In practice, for most commercial radios, the minimum sensitivity figures will be contained in the radio data sheet and thus can be read off directly rather than calculated, although it will also be necessary to consider the level of noise present at the receiving system to identify whether the system is likely to be de-sensitised compared to its design value by that noise. Noise is discussed in further depth in Chapter 13.

### *5.5.8 Sensitivity and Noise*

We determined minimum sensitivity in the absence of noise in the previous section. In many cases, the system will not operate at the minimum sensitivity but will instead be de-sensitised by noise. We will be looking at noise and interference in Chapter 13, but in the mean time it is important to recognise that in this case, the receiver will require a signal strength that is above the combined noise plus interference. For mobile elements, this will probably change as the subscriber moves around the service area, but for fixed installations, the noise and interference present can be measured or predicted as long as the interferers and their characteristics are known. When ground clutter attenuation is used to represent ground usage in the propagation model, then a portion of this can be attributed to typical noise values found in those environments.

This works in the following manner; if an equivalent figure of, say, −70 dBm (referenced to a dipole antenna) is predicted as the median value for a particular location and the clutter loss is 15 dB, then the value stored by the planning tool will be $-70 - 15 = -85$ dBm. This value can then be compared to the minimum equivalent signal value required, which will be the receiver sensitivity plus the fade margin (and possible shadowing margin, for point-to-area models) to determine whether the signal level is high enough to provide an acceptable service. So if we use the value of −108 dBm, with a 10 dB fade margin (and no shadowing loss), to give a minimum acceptable value of −98 dBm, the predicted value is −85 dBm, giving a margin of $(-85)-(-98) = +13$ dB, so the network does offer an acceptable service at this location since the margin is positive.

The wanted value will depend on the service; thus a digital system may require a minimum value of $E_b/N_o$ whereas an analogue system will require a given signal-to-noise ratio. It is also important to note that the sensitivity is always quoted against a required performance target; it makes no sense on its own. Thus 'a signal-to-noise ratio of 12 dB to achieve a raw BER of $10^{-3}$' makes sense, whereas '$E_b/N_o$ of 12 dB' makes no sense unless it is further qualified. There must always be the concept of a certain value input that will lead to a certain value output, where the input value may be a power level or a signal-to-noise ratio and the output may be SINAD, output power, BER and so on.

## 5.6 Building a Link Budget

### *5.6.1 Introduction*

Link budgets can be constructed in different ways to specify link characteristics or to calculate an unknown value to be used then in the design process. In general, the process involves the same basic considerations; determine the point in the link process for which the answer is required, and then calculate from either the transmitter or receiver end to compute the required value. For the following examples we will use a simplified form of Figure 5.11 to illustrate the principles, shown in Figure 5.14. If the path includes more elements from Figure 5.11, then it is only necessary to account for the additional losses and gains of each part.

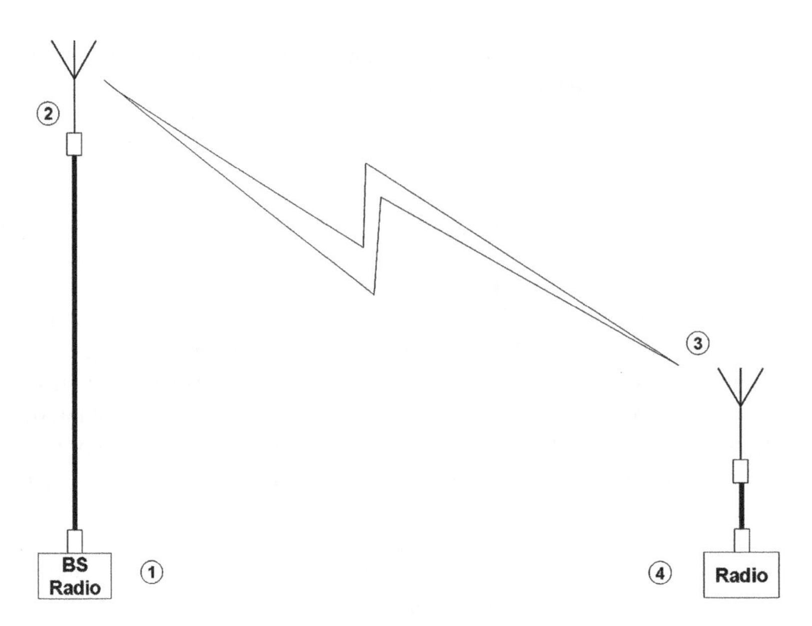


**Figure 5.14.** Simple path illustration.

### *5.6.2 Link Loss Calculation to Determine Level at Receiver*

To determine the power at the transmitting antenna, the nominal power output by the transmitting radio is modified by the losses in gains due to the equipment between the radio and antenna, and then the gain or loss of the transmitting antenna relative to a standard antenna (typically a dipole) is included. If we then know the path loss for a given path, we can include the effects to determine the level of the received level at the receiver. This is illustrated by example in Table 5.7.

**Table 5.7.** Calculation to determine level at receiving radio.

| Example link budget | | | |
|---|---|---|---|
| Transmitter elements | Value | Units | Calculation |
| (A) Nominal power at radio output | 41.5 | dBm | |
| (B) Total Tx feeder loss | 3 | dB | |
| (C) Total Tx connector loss | 1 | dB | |
| (D) Transmit antenna gain (reference a dipole) | 2.5 | dBd | |
| (E) Effective radiated power | 40 | dBd | A − B − C + D |
| (F) Path loss | 135 | dBd | |
| (G) Level at receiver antenna | −95 | dBd | E − F |
| (H) Receiver antenna gain | −3 | dBd | |
| (I) Total Rx conector loss | 1 | dB | |
| (J) Total Rx feeder loss | 0.5 | dB | |
| Level at receiver input | −99.5 | dBm | G + H − I − J |

The losses for feeders and connectors in this illustration are shown as positive, so they must be subtracted from the available input levels, since the power available at their output must be lower than on the input side. The receiver antenna gain is shown as a negative value, but this negative value must be added to the input power otherwise a double negative would be a positive, and thus the antenna loss would incorrectly be calculated as a gain. This is typical of path budgets; it is always important to determine which values should be added and which should be subtracted. In practice, it is often useful to draw out a diagram of the link in the form of Figure 5.14 for simple links, or like Figure 5.11 for more complex links.

There is another potential pitfall as well; does the path loss refer to a median value or is it something else? Does the calculated path loss include the required fade margin for the wanted availability, or has this not been added (in which case it must be)? Thus, another key aspect of link analysis is to determine the exact origin of each term to ensure that all terms are accounted for, and none are double-counted.

If all is well, then the figure for received signal level in dBm can be compared with some desired threshold value to determine whether the link will work or not (as a binary decision), the margin above or below the required threshold in dB, the probability of successful operation (PSO) or the expected BER or SINAD as required.

Of course, in many cases we will not know the path loss but will instead wish to know the maximum allowable loss that can be tolerated in the system before the performance falls below that considered acceptable.

### *5.6.3 Link Budget to Determine Maximum Allowable Loss*

If we need to determine the maximum path loss that can be tolerated in a link to achieve a given minimum acceptable performance, then we can calculate the effective radiated power from the transmitter, calculate the level required at the receiving antenna and take the difference between the two, as illustrated in Table 5.8.

In this case, the calculation at the transmitter end is the same, but we calculate back from the receiver radio to determine the equivalent signal required at the antenna. Note that in this

**Table 5.8.** Calculation to determine maximum allowable path loss.

| Example link budget | | | |
|---|---|---|---|
| Transmitter elements | Value | Units | Calculation |
| (A) Nominal Power at radio output | 41.5 | dBm | |
| (B) Total Tx feeder loss | 3 | dB | |
| (C) Total Tx connector loss | 1 | dB | |
| (D) Transmit antenna gain (reference a dipole) | 2.5 | dBd | |
| (E) Effective equivalent radiated power | 40 | dBm | A – B – C + D |
| Receiver elements | | | |
| (F) Receiver sensitivity | –104 | dBm | |
| (G) Total Rx connector loss | 1 | dB | |
| (H) Total Rx feeder loss | 0.5 | dB | |
| (I) Receiver antenna gain | –3 | dBd | |
| (J) Minimum required signal level | –99.5 | dBm | F + G + H – I |
| Maximum tolerable path loss | 139.5 | dBd | E – J |

**Table 5.9.** Loss required at street level to meet required performance; the tolerable path loss must be defined in terms of the wanted availability or another system performance metric (such as dynamic sensitivity for digital systems).

| Factor | Value | Units |
|---|---|---|
| Maximum tolerable path loss at antenna | 139.5 | dBd |
| Body loss | 6.5 | dB |
| Building penetration loss | 20 | dB |
| Fade margin for required availability | 10 | dB |
| Street level median loss value from prediction | 103 | dBd |

case, we need to add the connector and feeder losses to the required sensitivity in order to determine the signal required at the output of the antenna. The antenna has a −3 dB gain, which is a 3 dB loss, so we need 3 dB more at the input to the antenna. Thus in this case, the antenna gain is subtracted (resulting in a net addition) from the figure at the antenna output. As can be seen, the calculation of link budget is not in itself difficult, it is just necessary to ensure that terms are combined in the right way.

It is important to be careful about the path loss calculated. This must be the maximum path loss when including the effects of fading for the required availability. Also, if the system under consideration needs to take into account such as body loss, building penetration or both, then these need to be accounted for before the median field strength calculated by a propagation model can be compared to it. This is illustrated in Table 5.9, which shows the correction to apply to obtain the equivalent maximum tolerable path loss as calculated by a prediction model that calculates a median loss for 50 % of locations at street level, assuming the loss figures used.

### 5.6.4 *Link Budget to Determine MMOFS*

For practical use in a planning tool, we may be interested in the equivalent field strength that is required at the antenna taking into account all effects between the antenna and the receiver input. This is often referred to as the median minimum operating level (MMOL) for conducted power level or MMOFS (median minimum operating field strength). To this value, we will need to add the additional margin required to give the wanted degree of performance. Many tools will report the results of path profile or coverage predictions in terms of field strength, and thus it will be necessary to compute the equivalent value needed in the correct form. The required value is illustrated in Figure 5.15. The required median field strength required is obtained by adding the losses and gains between the receive antenna (including its own loss or gain value in the direction of the transmitter) and the receiver and converting to the equivalent field strength.

When the wanted value in dBm has been calculated, it needs to be converted into an equivalent field strength. For an impedance of 50 Ω (which is the typical value for most mobile radio systems), this can be determined by the following equation:

$$\text{Field strength (dB}\mu\text{V/m)} = \text{Power (dBm)} + 20\log f + 77.2 \qquad [25]$$

where $f$ is the frequency in MHz

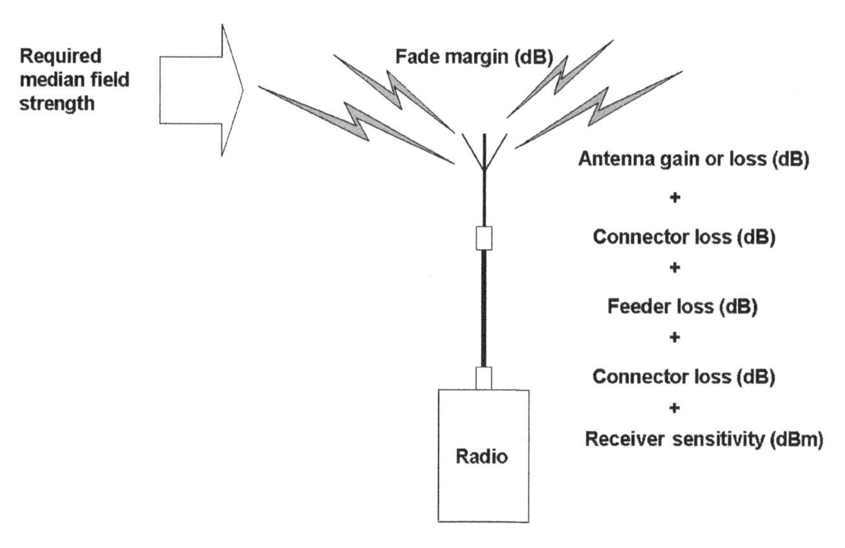


**Figure 5.15.** Required field strength at the receiving antenna.

An example is shown in Table 5.10, for a frequency of 390 MHz. Note that the mobile antenna has loss, not gain (we have shown it in a different way in this example because this is the way it can be expressed in some link budgets, and we want to reinforce the need to check the meaning of each term).

This is effectively one part of the link budget, broken down into a form that allows us to configure one part of the planning tool. This is typical, and it is often the case that different parts of the link budgets are stored in different parts of the planning tool. For example all of the aspects relevant to the transmitter may be stored in the characteristics of each individual base station (in case there are differences), whereas the mobile element is not explicitly modelled (since coverage plots effectively assume that the mobile moves throughout the entire service area) and is therefore modelled by setting a global characteristic such as the field strength value, plus some representation of antenna height above local ground.

**Table 5.10.** Equivalent field strength required at antenna.

| Example | Units | Calculation |
|---|---|---|
| (A) Unfaded radio receiver sensitivity (dBm) | −111 | |
| (B) Connector loss (dB) | 0.5 | |
| (C) Feeder loss (dB) | 2 | |
| (D) Connector loss (dB) | 0.5 | |
| (E) Mobile antenna ***loss*** (dBd) | 6 | |
| (F) Equivalent power required @ antenna (dBm) | −102 | A + B + C + D + E |
| (G) Fade margin required (dB) | 10 | |
| Value to be converted to field strength (dBm) | −92 | F + G |
| Equivalent field strength @ 390 MHz (dBμV/m) | 37.0 | Equation [25] |

### 5.6.5 Other Factors in Link Budgets

Depending on the technology in use and the structure of the link, there may also be other elements in a link budget. This may include such elements as follows:

- Additional connector and feeder losses.
- Amplifier gains on transmit side or on receiver side (in which case receiver noise figure may need to be adjusted).
- Combiner losses.
- Interference and noise at the receiver, which will de-sensitise the receiver threshold (see Chapter 13).
- Processing gain for CDMA systems such as UMTR UTRA and CDMA 2000.

These features will all manifest themselves as either gains or losses, and if in doubt it is usually a good idea to draw the link out on paper and work through it to ensure that elements are being accounted for correctly and not double-counted. Additional information for some technologies such as TETRA, GSM and UTRA is also available in the specifications, which usually contain sample link budgets.

It is also worth bearing in mind that a link budget may need to be constructed for each *type* of subscriber to account for differences in antennas used, feeder and connector losses and the sensitivity of the radios, and that it will be necessary to build a link budget for both the uplink and downlink direction. Most modern technologies are designed to provide balanced links (so that the losses in each direction are the same), but this may not be representative of reality. This is due to the different environments in which the base station and mobile are likely to be found, in which the noise floor may be different. Thus even though the theory of path reciprocity (path loss is the same between any two antennas, irrespective of which end is regarded as the transmitter) holds true, in practice it may be necessary to adjust the figures to reflect the true situation. Also, it will be necessary to determine link budgets for each type of service to be offered if required receiver sensitivity changes between these services.

*The fundamental factor to bear in mind is that if the link budgets are calculated incorrectly, then every other activity that occurs in the design process is in error.*

## 5.7 Expressing the Link Budget in a Planning Tool

In some simple planning tools, there may be a dialogue screen in which the link budget is entered in a single area, but for tools that allow handling of multiple sites with different characteristics, it is often the case that elements of the link budget have to be entered in different places in the tool. Thus, for example:

- Elements of the link budget relevant to the base station are entered in the dialogue box for each station on an individual basis.
- Environmental aspects, such as building penetration loss, may be entered in a dialog screen applicable to the propagation model and used globally for simulations.

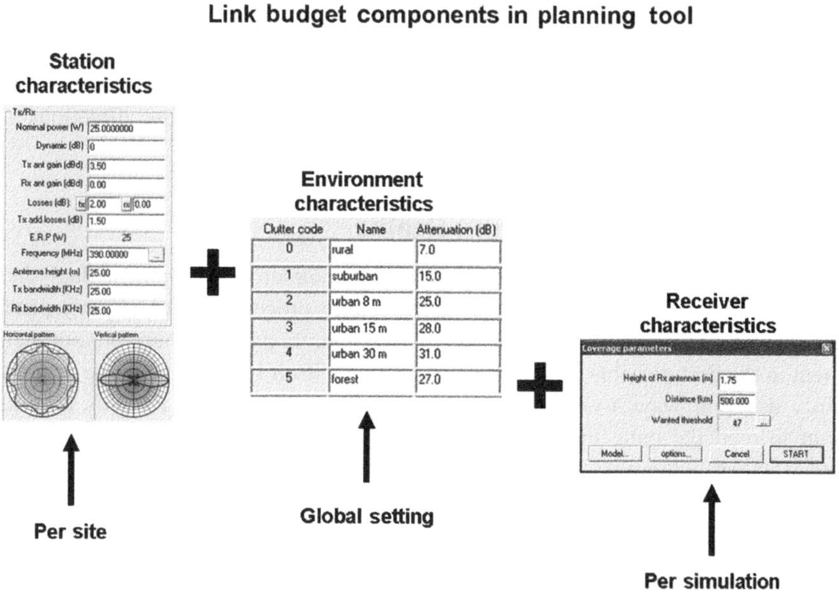

**Figure 5.16.** Expressing link budget features in a planning tool. Reproduced by permission of P & D Missud.

- It is likely that receiver characteristics will be entered as a calculated figure such as a wanted field strength or equivalent power into the antenna for a given simulation. Therefore the receiver part is not wholly expressed in the tool (only the antenna height). In other cases, there may be a specific receiver dialogue box, particularly when receivers are being represented by subscribers in a specific location rather than as virtual subscribers that exist in potentia anywhere within the planning tool project. The receiver characteristics are typically changed when calculations are performed for each type of subscriber.

Data entry in a specific tool is illustrated in Figure 5.16.

Entering the data in the form shown allows different station power, feeder and connector losses, gains and antennas to be modelled in the same simulation, while keeping the environmental and receiver characteristics the same for the whole simulation so that the results are consistent.

## 5.8 Balanced and Unbalanced Links

Often, radio equipment is designed in such a way that when the link budget for the uplink and downlink are put together, they have the same maximum allowable path loss in both directions. In this case, the links are said to be balanced, and a radio link from a base station to a mobile element should have the same probability of being established as the link from the mobile element back to the base station. Sometimes, however, links are not balanced in both directions, often due to power limitations on the mobile side. In this case, one link will be stronger than another, and thus it may be possible to detect messages without being able to successfully send messages, or vice versa. In order to establish full two-way communications, it will be necessary to plan the network coverage based on the weaker of the two links. Calculating the link budget in both the uplink and downlink directions allows determination of which is weaker.

## 5.9 Equipment Data Sheets and Reality

One note of caution must be sounded when using commercial equipment data sheets to determine technical parameters to use in the planning process. Each company producing radio equipment is in competition with other manufacturers and will therefore put their own equipment in the best possible light. This may manifest itself by expressing the performance in slightly different ways to enhance the figures as best they can. For this reason, it is important to read the data, units and conditions of test carefully. It is also important to consider whether equipment performance may drop during its life and mean that old equipment may not work where new equipment would.

Another key aspect is that antenna polar diagrams are valid when the antenna is measured in an anechoic chamber, away from any nearby metallic structures. When the antenna for a base station is mounted on a mast, it is likely that the pattern may vary from the ideal. Sometimes, measurements will be provided. For example Figure 5.17 shows a comparison of an antenna pattern for a single omni-directional antenna from a data sheet with the typical pattern of three omni-directional antennas mounted around a tower. The ripple in the performance in this example is about 2 dB. Naturally, to be accurate, equipment parameters must represent the actual rather than nominal figures. Often this is difficult to achieve, and thus it is normal to err on the side of caution when choosing values to be used.

Also, as a final note, remember that it is important to identify the exact meaning of the terms used in order to ensure that all factors are included. For example it is important to

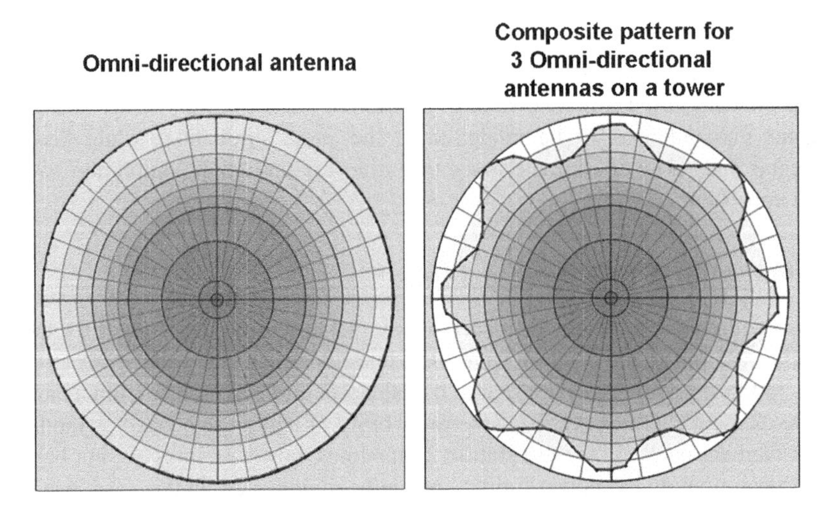


**Figure 5.17.** Real omni-directional antenna patterns. Reproduced by permission of P & D Missud.

determine whether receiver sensitivity is expressed as a faded or unfaded value, and thus whether a fade margin needs to be added or not.

## References and Further Reading

[1] ITU-R PN.1057: Probability Distributions Relevant to Radiowave Propagation Modelling
[2] ITU-R P.341-5: Transmission loss
[3] Antennas and Propagation for Wireless Communications Systems, Simon R. Saunders, John Wiley & Sons Ltd, Chichester, 1999, ISBN 0-471-98609-7
[4] Introduction to Mobile Communications Engineering, Jose M. Hernando, F. Perez-Fontan, Artech House Publishers, London, ISBN 0-89006-391-5
[5] Mobile Communications Engineering, William C. Y. Lee, McGraw-Hill Book Company, New York, ISBN 0-07-03039-7
[6] Mobile Communications Design Fundamentals, William C. Y. Lee, Howard H. Sams & Co., Indianapolis, ISBN 0-672-22305-8
[7] Land Mobile Radio System Engineering, Garry C. Hess, Artech House Publishers, Norwood, MA, ISBN 0-89006-680-9

# PART TWO

# 6

# The Radio Network Design Environment

## 6.1 Introduction

In Part One, we looked at some fundamental material essential to radio network design and also at useful background information to put the implementation and use of radio networks into a global context. In Part Two, we will delve into the issues of radio network design from a practical perspective. This will introduce advanced modelling techniques and how they should be used. We will also look at how the radio network design process fits into a larger project to implement the network and operate it effectively. It is necessary to have this wider view of the overall activity in order to set up appropriate interfaces between the relevant components of the project, and also to ensure that where appropriate, the expectations and concerns of those who may influence the project are managed and not ignored (to the later peril of the entire project).

It is a common failing of engineers that they tend to focus exclusively on the technical aspects of an activity and relegate other facets to a lower level of priority and interest. In some cases, this can include a reluctance to spend much time on 'soft' issues such as project management, the processes to achieve particular solutions and the means of communicating information to others in the project that need to know. In fact, these issues are vitally important and can determine whether the project will succeed or fail. It should always be at the back of the engineer's mind – if not at the front – that the project only exists because of some business driver that justifies its existence. By 'business driver', we do not limit ourselves to commercial reasons – business refers to the activity of the organisation that needs the network – so noncommercial drivers are included. For example business drivers include the following:

- A new UMTS network is required so that the company can compete in the 3G market.
- The company is extending its presence in a new area and needs a new network to support its operational staff.

- The legacy network is nearing the end of its effective life. As an unencrypted analogue system, it is also vulnerable to interception by unauthorised receivers. A new digital system is required to provide increased security.
- The government has committed to meeting the ICAO intention of implementing CPDLC (Controller Pilot Data Link), and therefore, new network infrastructure is required to serve this.
- A military exercise is to be held and communications will be required to support the various command requirements during the activity and for safety reasons.

In each case, the business need is outside the influence of the network design team; they must use their knowledge and skills to support this wider need as far as possible and to come up with methods that do not unnecessarily constrain the business aims. The aims of each part of the business will be important in achieving overall success, and there are many individual parts in any sizeable projects as illustrated in Figure 6.1.

Whether the design team are working onsite or far away in remote offices, it is important never to lose sight of the business need – it is the easiest way for the network design to diverge from that required and the project become imperilled.

It is also important to recognise that the overall business activity will involve a wider variety of people than those that the designers will tend to come in contact with, and those managing the overall project should be prepared to engage these groups in order to ensure that the needs are met, that fears are calmed and that objections are overcome. The

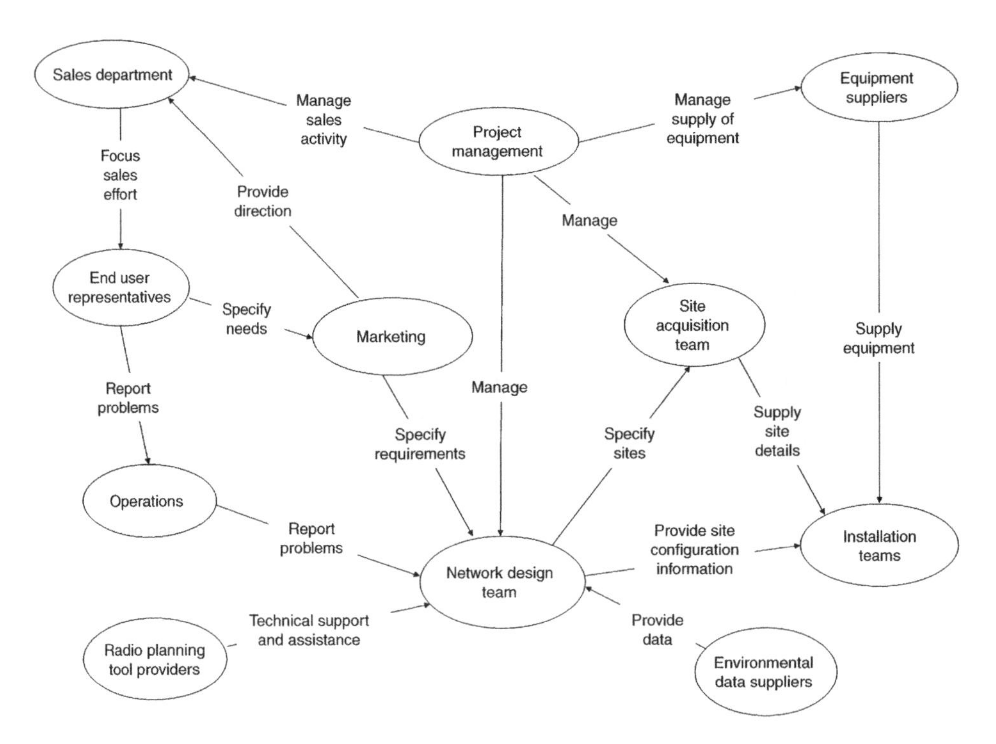


**Figure 6.1.** Design process in the commercial environment.

designers may have a role in doing this, but this should be led by the overall management team in general, bringing in expertise as necessary. It can sometimes be surprising how many people can be affected by the implementation of a network, and this is described in greater detail in Section 6.3. Figure 6.2 illustrates the stakeholders that may be present in

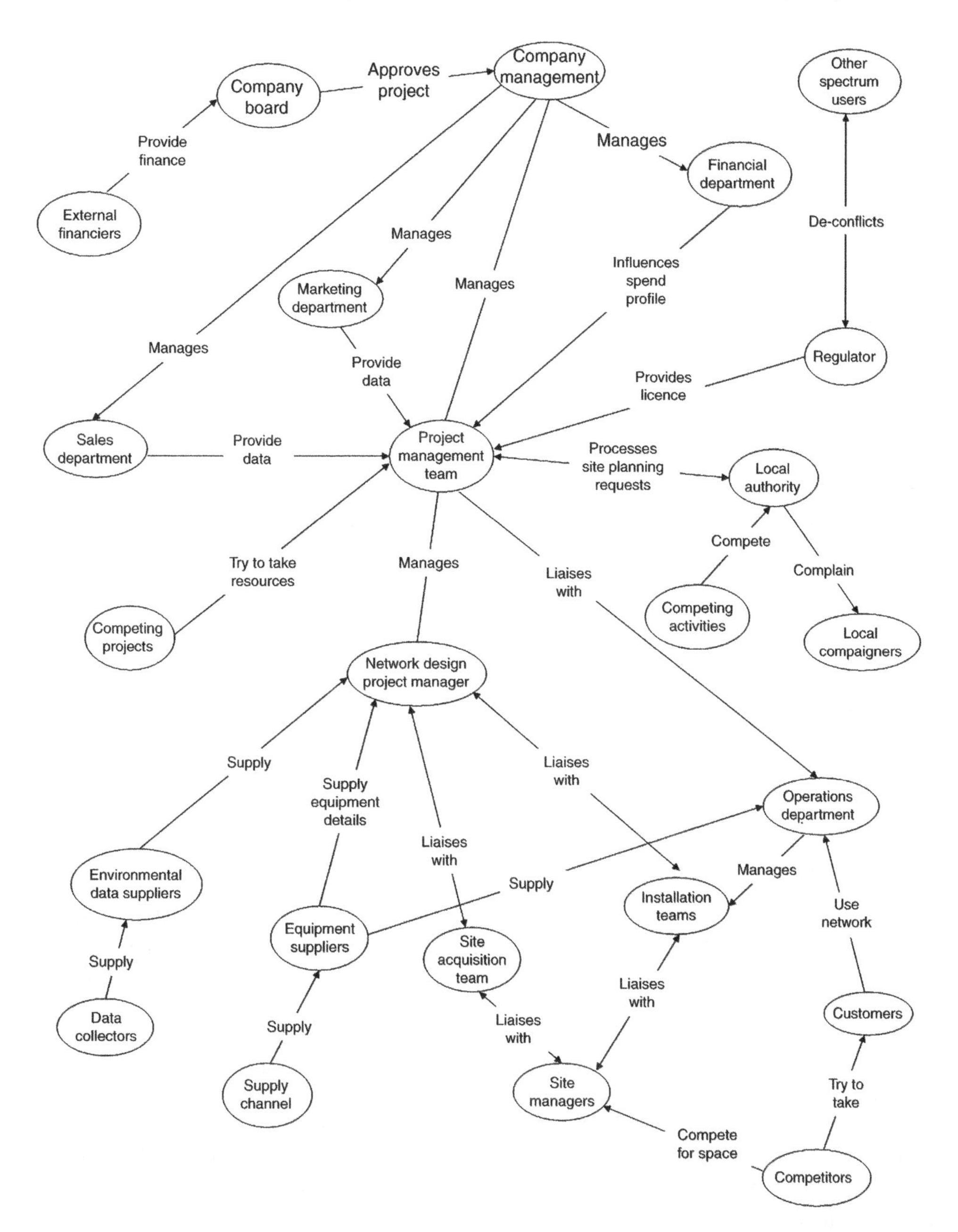


**Figure 6.2.** Stakeholders affected by or affecting a radio network project.

a particular project. In practice, the stakeholders involved will be different for every project.

It may seem that we are taking things too far to include people like local campaigners and the supply chain of the equipment suppliers into this diagram, but they can have a major impact on whether the project will be completed on time, implemented in a cost-effective manner or even completed successfully at all. In fact, such diagrams can be extended almost infinitely to include the financial condition of the country (perhaps, government funding might be pulled or lack of consumer interest takes the demand away for the network) and many other aspects. Although it is not credible that designers and engineers involved in a project can influence all stakeholders, it should be recognised that for success, the delimiting feature between those 'inside' and 'outside' should be the difference between those who contribute to the project's success and those who oppose it; the enemies are these people, not the marketing team, whom we have never got on with!

If we accept the contention that the design activity supports the wider business need, then that necessarily also means that the design project itself must be structured to ensure these business needs are met. Particularly, candidate designs must be tested against as close a model of the business needs as possible; this should be strongly represented not only in the design of test cases but also in the project life cycle adopted, the design rules developed to reach those candidate designs and the specifications used to determine whether the needs are indeed met or not. We look at all these issues in Section 6.4, to the end of the chapter.

Later in Part Two, we will return to the practical aspects of setting up a project, selecting the right project resources and then how we go about actually performing the project and seeing the design through to completion. But before we consider the other stakeholders and the activities that go towards a successful project design, it is worthwhile to examine what we expect from the design team itself. This will include a project leader, possibly leaders for individual sub-projects, technical experts and the designers themselves. Ideally, the project managers should also be skilled network designers; the ability to multitask in this environment is usually highly desirable and effective. We will start this section in detail by looking at our definition of radio network designers and engineers and broadly what we would expect them to know and to be able to do.

## 6.2 Network Design Professionals

To paraphrase Orwell, not all network designers are equal; some are more equal than others. It is worth considering the types of skills and knowledge we would expect of different levels of network engineer. To illustrate this, we will adopt the terms of junior network design engineer, senior network design engineer and network designer (or consultant) to differentiate between those engineers working in a project. This is broadly applicable to the levels adopted in many organisations, although there may be additional differentiation and qualifications. The definitions we propose are set far higher than accepted by some organisations, but setting them too low leads to low productivity and too many mistakes. Unfortunately, this is too often encountered in the industry and we argue that the descriptions below are achievable and desirable, particularly, in an organisation that ranks network design as one of its core activities.

### 6.2.1 *Junior Network Design Engineer*

A junior design engineer may have started directly after graduating with a first or further degree or may have been working in another engineering discipline beforehand. They should be familiar with good engineering practice and have sufficient knowledge in mathematics, physics and study principles that will allow them to be able to learn their trade without excessive supervision. By the time they are full-fledged junior design engineers, they should be able to meet the following criteria:

- Understand the radio networks they are expected to work on, in terms of the technology, design constraints and considerations and, importantly, what is expected of the network once it has been designed. It is not enough just to understand the technical aspects but an understanding of what makes a network successful is essential.
- Be computer literate and capable of using the design tools used by the company either under supervision or following working instructions. This will include understanding which parts of the program configuration should not be changed since they will adversely affect the quality of the output.
- Be able to interpret the results produced by the planning tool and describe the salient points to another engineer or to a manager. This involves the ability not only to describe the results themselves but also their implications.
- Have sufficient grasp of the overall design process to be able to alert senior engineers of any important unexpected issues that may arise.
- Be able to learn new tasks as required by the dictates of new projects.

At this stage, the junior engineer should not be expected to be customer-facing or to make important decisions. They will, however, be able to take on relatively simple design tasks such as producing coverage plots without significant supervision. A person hoping to make a career as a network designer would probably be able to reach this level within about six months of starting, depending on the complexity of tools in use and the complexity of the projects to be undertaken. This timescale will also vary depending on the breadth of tasks to be performed, the number of radio network technologies and the variety of radio design projects. It may be necessary to 'qualify' on each type of project before being able to progress.

### 6.2.2 *Senior Network Design Engineer*

A network design engineer should ideally start as a junior network design engineer. In some areas of engineering, it is possible for experienced engineers without domain knowledge to be able to quickly and successfully lead other engineers in projects; however, radio network design is quite a niche activity and often the devil is in the detail. For this reason, it is not normally appropriate for engineers to come in at this level without having learned the basics. This has been done in some engineering companies in the past, sometimes with poor results. This is not to say that it never works, but that if it does work, it is more by luck than judgement and in any case the project risks are both too high and also unquantifiable. For that reason, we advise against it.

Above and beyond the capabilities of the junior network-planning engineer, the senior network design engineer should be able to meet the following criteria:

- Have substantial domain knowledge about radio networks, their applications and design criteria, and be able to develop and enhance this knowledge as technology and market demands evolve.
- Be able to operate radio design tools unsupervised and to be able to use them flexibly and if required, in innovative fashion. This includes designing work processes to meet the project requirements using the tools at hand.
- Be able to generate specific tools where there is no existing system. This does not imply the ability to do advanced software programming, but it would include being able to generate complex spreadsheets and ideally should at least include the ability to prototype simple tools in Basic, Python or another high-level language. This ability is an enormous force-multiplier for the engineer's skills.
- Be able to design the methods to perform specific projects. This also includes the ability to document these processes concisely such that junior network engineers can work from them.
- To be able to spot project risks and to determine methods to militate against them.
- Work on all but the most complex of projects unsupervised, having been properly briefed on the project requirements.
- Lead teams of junior network design engineers in the network design activity.
- Recommend link budgets and be able to confidently calculate them (see Chapter 5 – this is not a simple task).
- Understand in depth different propagation models, their application to specific applications and be able to usefully contribute in discussions on the same.
- Be able to engage with the customer, explain engineering principles and being able to effectively communicate the radio technology limitations that will determine what is practically achievable and what is not. In doing this, a senior network engineer should be an influential member of the overall project team.

From this description, it can be deduced that in most companies, senior network design engineers will be the highest level of competence required. For consultancy companies and for those for which radio network design is the principle activity, it will be necessary to have more experienced and capable engineers to act as consultants or as we term them radio network designers.

### *6.2.3 Network Designers*

A professional radio network designer capable of leading an entire project and bringing it to a successful conclusion must extend the capabilities of the senior network design engineer substantially. Such a person needs to embody a wide range of ability, skills and knowledge including, but not limited to, those in the following list:

- A wide understanding of the radio network environment, the stakeholders involved and their main issues and concerns.

- The ability to interact with other project stakeholders in a consultative capacity.
- The ability to plan and manage major complex projects.
- The ability to effectively engage and maintain the enthusiasm of project staff throughout the entire project.
- A good understanding of state-of-the-art design methods and the circumstances in which they can be applied.
- An ability to develop novel design methods whenever required. This may be by adapting existing methods used in other applications or it may mean generating entirely new methods.
- A solid understanding of all the technology issues involved.
- A continually refreshed understanding of the latest technologies, design methods, design environment and the way the 'market' is going. This is achieved by continual analysis of emerging techniques published in papers and new books, attending conferences and working bodies and networking with other design professionals.
- The capability to able to write ground-breaking papers, speak at conferences and participate in the working and study groups of internationally acknowledged radio-related organisations (e.g. ITU, ETSI, COST).

It should be clear from this list that a detailed understanding of radio technology itself is fairly far down this list. This is not to say that it is unimportant but rather that it forms only one aspect of radio network designer's mental toolkit. Unfortunately, one of the most common failings in engineers wishing to move into this field is a lack of willingness to move away from the books and into the more pro-active aspects. The problem with this is that unlike, say, designing an electronic board or an algorithm, successful radio network design is a multidisciplinary activity that necessarily involves a lot of interaction with other people. Often, inexperienced engineers regard such activities as less attractive than what they consider 'proper engineering', requiring less skill and therefore to be looked down on. In fact, the reverse is generally true and there is no way of being a successful radio network design engineer without being capable of performing all these activities. Without them, the best that one can achieve is to be a specialist member of a larger team with limited prospects of personal development.

While the purely technical aspects of designing radio networks are often constrained by the technology being used, the other aspects will vary from project to project according to the circumstances. The key, therefore, is not to learn a single set of methods by rote but rather to develop a particular practical approach that can be adapted to a wide range of technologies and situations. This is the focus of the rest of the book, and for the rest of this section, we will be focussing on the issues that are the concern of the radio network designer rather than the more limited concerns of the radio network engineer. Before we move on to this, we will, however, briefly consider two competing strategies for organising a company's network design capability.

### *6.2.4 Network Design Capability Strategies*

If a company wishes to establish or refine an organic radio network design capability, then consideration must be given to the approach to be taken. In the industry, currently there

Table 6.1. Comparative cost of two project approaches.

| Comparative Project Costs | |
|---|---|
| **Approach A – low-tech / personnel heavy** | |
| Costs: | Cost (see note 1) |
| 10 × planning tool licences | £100,000 |
| Planning tool data | £100,000 |
| Project Manager | £80,000 |
| Planning engineers x 10 | £500,000 |
| **Total** | **£780,000** |
| **Approach B – high-tech / personnel light** | |
| Costs: | Cost (see note 1) |
| 2 × planning tool licences | £100,000 |
| Planning tool data | £150,000 |
| 2 × senior network designers | £200,000 |
| **Total** | **£450,000** |

Note 1: The figure for costs is the total required, e.g. for staff it includes salaries, employer contributions to taxes, overheads etc.

appear to be two distinct approaches with the split often occurring between those involved in mobile phone network design and everyone else. It appears that in those following the mobile phone designer approach use larger numbers of less capable engineers, who operate relatively rudimentary planning tools; in our previous discussion, these engineers would broadly correspond to our junior radio network design engineers (at best). The tools involved provide the ability to perform coverage predictions and simple interference analysis, with some automatic or semi-automatic frequency planning capabilities, but little else. Although these tools develop to meet new requirements, there appears to be little enthusiasm to adopt the most state-of-the-art methods.

The alternative approach is to exploit the very best planning tools, combined with smaller numbers of more capable engineers. This may involve higher capital outlay on the planning tools and the data required to support them (but often does not), but since there will be fewer staff employed, this will be recouped by reduction in employee costs. Table 6.1 illustrates this with a very simple example. The figures included are indicative, but do show a reasonable comparison.

Clearly, the high-tech approach using a more sophisticated approach is less expensive, so why is this sometimes not adopted? The reasons may be that there is a perception of higher risk associated with having fewer staff and a higher dependency on software tools that may be incompletely understood by management. Also, a degree of isolationism and traditionalism may lead management not to seek alternative approaches to configure the radio network design capability. Thirdly, there is a dearth of highly qualified engineers at the senior radio network design engineer level, which may mean that it is simply not possible to recruit suitable staff to allow this approach to be taken. Whatever be the approach adopted, it is worth periodically reviewing the latest techniques and tools to see whether departmental reorganisation or acquisition of the latest tools to replace aging ones. The capital outlay involved in this should be outweighed by improved productivity and enhanced capability so long as that capability meets the business needs.

Next, we will look at the various people involved in a project in more depth, each of whom has a different set of requirements and concerns. The design engineers form one group of people with an interest in the network, but there are many others. The term stakeholder has come to use to refer to those who have an influence on a particular activity or entity, and this is the term we shall use. These stakeholders will belong to a specific group that has its own view of the network. The term stakeholder is justified because it firmly establishes that each group has some definite interest in the project, no matter what form that interest takes, and that interest has to be addressed if everyone is to be happy with the outcome. As we shall see, this does not necessarily imply that their desires are to be met or that all stakeholders are intent to make the network project a success.

## 6.3 Network Stakeholders

### *6.3.1 The Concept of Stakeholders*

Every entity or activity is associated with its group of stakeholders, each of whom have some influence on it. The term stakeholder should not be confused with shareholder; shareholders are stakeholders, but staff, departments, customers, suppliers and many other groups are also involved. We use the term stakeholders for any entity that has an influence based on their decisions. Thus, there are stakeholders who are supportive to the project; the engineers, project managers and staff working on the project. There may in some cases be stakeholders that are actively working against the project; individuals or groups who do not want the network to succeed due to internal politics, competitors (who are stakeholders of other organisations, but who influence the project and must, therefore, be considered), campaigners who do not want radio masts in their back yard and so on, and there will also be other groups whose allegiance will be less clear cut; customers, industry regulators and potential investors, for example. So it can be seen that stakeholders in a business are not restricted to those who work for or invest in a business. In the same way, the stakeholders in a project to build and run a new radio network can be many and varied and can have considerable influence on the network design. Although in many cases the design engineer may not meet some of the stakeholders, it is important to understand that they exist and that, to some extent or another, their views need to be considered (although not always agreed with and the designer may in some circumstances be determined to thwart their activities). An extreme example of this would be to consider the effect of the enemy on a network during a military operation; the enemy may seek to destroy, disrupt or exploit the network. The enemy can hardly be considered a part of the design team, but they have a huge influence on the design, since the designer must consider how to prevent them achieving their aims. They are still stakeholders since they have designs on the success of the network – they want to destroy it! In the civil environment, a growing resistance to the installation of base stations (BS) may cause the same effect in a slightly less dramatic way. In fact many stakeholders will have the power to stop a network project in its track at various stages during the process. This includes management, financiers and other projects, which will be competing for finite resources within the business. The issue for the project manager is to ensure the maximum buy-in from as many stakeholders as possible and to undermine the influence of those who oppose the project (preferably using socially acceptable methods!).

We will now illustrate this further by examining the stakeholders likely to be involved with a range of different project types.

## *6.3.2 Stakeholders in Typical Projects*

### 6.3.2.1 The Internal Project

In many cases, the business need resides within the organisation that will also design and implement the network. This will be true for organisations that have their own organic design capability, normally because there is sufficient demand to justify having a permanent department, security considerations require that only staff members have knowledge of the systems used or that the domain is so niche that external consultants would not be able to gain sufficient understanding of the requirements to support effective and efficient radio network design. In this case, the principle stakeholders might include the following groups of people depending on the organisation's role and type (some stakeholders listed here will not be relevant to some organisation):

- The managers of the organisation, such as:
  - The Managing Director or equivalent.
  - Financial Director or other individual responsible for budget.
  - Engineering Director.
  - Operations Director.
  - Design department head.
- Higher levels of authority or command:
  - The Board.
  - Parent company.
  - Higher level of command.
  - National regulatory bodies.
  - International regulatory bodies.
- External financial stakeholders:
  - Venture capitalists (providing funding to the company).
  - The company's bank.
- The project team:
  - Project leader / network designer (depending on role split).
  - Project engineers.
  - Technical experts.
- Other departments involved in the project:
  - Marketing department (who may have built the business case).
  - Sales department (who will have to sell the service).
  - Operations department (who may have to physically install the department).
- Commercial stakeholders:
  - Selected equipment vendors.
  - Site acquisition experts (in some projects).
  - Radio site managers and landlords.
  - Telecom providers (for backhaul).

- Other external stakeholders:
  - Competitors.
  - Regulators.
  - Pressure groups.

This list is by no means exhaustive, and it would have to be adjusted to a specific project to be accurate and complete, but it does show that the range of stakeholders involved in even a fairly small project may be quite extensive. Although many of them will probably have minor influence in the overall project, some of them may have the power to intentionally or otherwise affect the project. Some examples might include the following:

- If management are not convinced by the business plan, they may cancel the whole project.
- The Financial Director may successfully argue that the benefits on offer will not justify the expense.
- Venture capitalists may pull the plug if the project is not progressing satisfactorily.
- The marketing department may have grossly over-estimated demand dooming the project from the start.
- Vendors may fail to deliver as promised delaying project rollout.
- Competitors may successfully convince the market not to use the service offered or may get there first.
- Pressure groups may prevent BS being deployed on environmental or health grounds.

It should be possible to identify many other influences that these stakeholders may exert on the network and consequently its design.

#### 6.3.2.2 Projects for External Customers

For projects where the design team are in a different organisation than the customer (or other organisation that is actually paying for the network to be implemented), then the list shown above for the internal project is augmented by a second, similar list but looked at from the aspect of the customer organisation. It may also include other notable stakeholders such as the customer's own consultants brought in to judge the designer's work. In some circumstances where other competing consultants have lost this particular project but retain a relationship with the customer, it is also possible that these competitors will work diligently to undermine the confidence of the customer in the designers. This is somewhat underhand, but it does happen. The design team must be aware of this possibility and be able to negate their efforts.

Another factor in this type of project is that although stakeholders such as vendors are present as they are in the internal project, the relationship that the designer will have with them may be substantially different. Since the vendor may be working directly to the customer, the designer may experience less support in terms of the provision of vital technical data, for example. In large projects, the dynamics of the relationships between stakeholders may become very involved and complex. The network designer will not be involved in all these relationships, but it is important to maintain an understanding of the key relationships and to influence them as required. This is far from easy task, and even though it is not a technical activity, it is still a vital one and one that the network designer must be prepared to undertake and capable of doing so.

#### 6.3.2.3 Projects in the Noncommercial Civilian Sector

The discussions in this section have focussed on the commercial project, but in fact there is relatively little difference for noncommercial projects in general, except that the names will change. Many of the aims of the noncommercial project are identical to those of the commercial projects, such as the desire to meet the stated requirements within the committed budget and within the planned timescale. We can replace some of the stakeholders in the list of the commercial project by merely changing the names:

- The managers of the organisation, such as:
  - The Director, Director General, Chief constable, or other relevant title.
  - Budget holder(s).
  - Engineering Director.
  - Operations Director.
  - Technical department head.
- Higher levels of authority or command:
  - Regional Director.
  - National Director.
  - Home office (for equivalent level body).
  - National regulatory bodies.
  - International regulatory bodies.
- External financial stakeholders:
  - National budget holders.
  - Departmental comptroller.
  - The treasury.
  - The Audit Commission or equivalent.
- The project team:
  - Project leader/network designer (depending on role split).
  - Project engineers.
  - Technical experts.
- Other departments involved in the project:
  - Operational department (on whose behalf the network may be implemented).
  - Operations department (who may have to physically install the department).
- Commercial stakeholders:
  - Selected equipment vendors.
  - Site acquisition experts (in some projects).
  - Radio site managers and landlords.
  - Telecom providers (for backhaul).
- Other external stakeholders:
  - Adversaries (e.g. criminals, terrorists).
  - Regulators.
  - Pressure groups.

Some of these are similar to the stakeholders in the commercial project but with different names. Others will be unique to the noncommercial sector, but the principle of dynamic relationships between discrete stakeholders is still valid. Note that there will always be some commercial interest in this type of project – equipment vendors or external consultants, for example.

#### 6.3.2.4 Projects in the Military Sector

Even military projects will have different stakeholders potentially pulling in different directions. The actual stakeholders in a particular project will vary substantially based on the type of project and the circumstances.

Larger projects such as fixed infrastructure projects carried out in normal peacetime conditions will normally follow the noncommercial structure described in the previous section. Military forces, being large geographically dispersed organisations with a large mobile element, do use such networks either employing specialised military equipment or using civilian technologies such as TETRA. These projects are managed by dedicated project teams (such as the IPT [Integrated Project Team] in UK) using project structures similar to other governmental practices.

Projects carried out during actual military operations will have other dynamics and objectives. There will be competition for scarce resources normal in military operations and external stakeholders such as enemy communications electronic warfare will be of higher importance. On the plus side, however, there will be no sales and marketing departments to worry about.

#### 6.3.2.5 A Stakeholder Generalisation

The purpose of identifying stakeholders in a project and involving them in the design process is to better ensure that the network that eventually emerges from the design and rollout is one that meets the target user needs. It is not to meet the needs of every stakeholder, as has already been discussed; some of them may not want the network to succeed. It is possible to summarise some aspects of stakeholders, which are as follows:

- Stakeholders have influence over the network and its design.
- That influence may be positive, neutral or negative, and it can change during the project.
- Stakeholders are differentiated by their view of how the project affects them.
- The designer will have to work with various stakeholders; ignoring their existence is a risky approach.

We will be looking at the influence of stakeholders in later sections, and seeing how the networks design process can benefit or suffer from their involvement. We now turn to how some of the stakeholders can be brought together in an approach that benefits the design process.

## 6.4 A 'Business-Centric' Approach to Design

In the previous section, we discussed the concept of stakeholders and identified their importance in the design process. In this section, we will look at how this approach can help us to keep the design focussed on the key user needs. This is based on the idea that way that stakeholders interact forms the business environment within which the designer works. By 'business' we are referring to any shared endeavour, rather than simply to a commercial organisation; the same principles apply to emergency services, civil aviation and military environments, and they all are engaged in some business or another. Stakeholders within the business work together to achieve the business aims, and this is also true at different levels within the business, so for example a particular department or project team will also consist of stakeholders with their own responsibilities and concerns. In some businesses, inter-departmental rivalry or prejudice may make it difficult for them to work together, but this can be very counterproductive and it can lead to poor end results. It is usually easier to meet the business aims when departments work together, particularly if the interfaces can be correctly defined and set up.

The principle aim of the business-centric approach is to ensure that all parts of the business work together to meet the business aims in the best possible way. To illustrate the principle, let's look at an alternative approach. For a particular opportunity, assume that the process occurs as follows, with each department working alone, without reference to the other departments in the company:

- The business development team identify an opportunity to design a radio network for a small operator and despatch a sales team to follow the opportunity up.
- The sales team work on the opportunity and convince the customer to buy.
- A project team is created from the engineering department and briefed on what the company has sold to the customer.
- The project team carry out the design and supply it to the customer.

This is a simple scenario, involving only two departments within a company. What can possibly go wrong? Well, here is a short list of possible problems:

- *Inappropriate solution offered.* Due to lack of knowledge of the customer's domain, the sales team misunderstand what the customer is trying to do and propose a solution that will not work. The customer, being nontechnical, is relying on the supposed technical expertise of the supplying company and does not realise this. The sale is carried through by the selling expertise of the sales staff alone. This has set up a disastrous situation for both customer and the company, which will probably become apparent only when the engineers are belatedly brought to the project.
- *Scale of project underestimated.* The sales team propose a technically feasible solution that could meet the customer's need, but due to lack of knowledge of the design complexity, it underestimates the effort required to design the network and therefore under-quote the job. This may not become apparent until the engineers become involved, by which time the company has signed the contract with the customer. The company then faces the prospect of absorbing the extra cost, or of letting the customer down, either of which is highly undesirable.
- *Insufficient handover.* The sales team do a sterling job, proposing a very good, properly priced solution based on a solid understanding of the customer's requirements. They then hand the project over to the engineering department who are briefed on the project quickly before the sales team disappear to follow another opportunity. The engineering project team design the network and take it to the customer. The customer reviews the design and quickly sees that it does not meet their needs and is not in line with their expectations. They reject the work and will not pay for it. This is clearly undesirable; it

has resulted in a very unhappy customer, who may have other plans now in jeopardy because the design has not materialised on time. The company has lost revenue and if news of this project gets out into the market, its reputation will suffer.

- *Lack of communication with the customer.* The job is properly handed over from the sales department to the engineering department, with detailed briefing and with transfer of all documents and notes of all discussions carried out between the sales staff and the customer. The engineering project team disappears into their office and emerge sometime later with a design that meets the requirements as stated in the material received from the sales team. They take the design to the customer, who has heard very little from the company in the meantime. The customer complains that their business has evolved since the original order was signed and the design, although it meets the initial criteria, is no longer suitable. The customer is obliged to pay, but is not happy. The network design is never used, and therefore the company cannot use it as a reference for future work. Again, it may come back to haunt the company if the story gets round the market.

These may seem the sort of fundamental errors that no one would ever make, but in practice they happen very often and when working as consultants, we often come across these sorts of problems – if we are brought into a project late, we turn up expecting this kind of problem. Thus it is that the most crucial aspects of projects can be human in nature rather than technical, and of the various failings, the inability or unwillingness to work of stakeholders to work with other stakeholders is often the root cause. With a bit of thought and planning, and a genuine undertaking to commit to working together, these problems can be easily overcome. If we look at the original problems illustrated above, we get easy ways of preventing them happening:

- *Original problem.* Inappropriate solution offered. *Solution*: the sales staff must bring in some expertise from another source to provide the essential domain knowledge. This may be present in some other part of the company, perhaps on the marketing side, or it may be necessary to bring in external expertise. Bringing in external consultants will cost – but nowhere near as much as original situation. In essence, it means bringing in other stakeholders who have the necessary expertise.
- *Original problem.* Scale of project underestimated. *Solution*: this is a clear case where the expertise of the engineering department will have an immediate and positive effect. If they are brought in review the proposal before it is submitted to the customer, then the problem should be spotted before it becomes an issue. An even better solution would be if the engineering department contributed to the proposal from the start, which means this particular problem should never arise in the first place.
- *Original problem.* Insufficient handover. *Solution*: this is a compounded error, where the project team do not understand the original requirements and worse, fail to correct the original mistake by failure to communicate with the customer sufficiently. It merely starts with the ineffective handover. In practice, this kind of problem is more likely to cause friction between the sales department and the engineering department than anything else. It will in any case have a negative impact on the efficiency of the project.
- *Original problem*: Lack of communication with the customer. *Solution*: this is a more difficult one, in that what the customer has asked for is what is delivered, but this is not what the customer now wants. In contractual terms, all of the requirements have been met and the company is entitled to be paid for the work it has completed.

However, an unhappy customer is still an unfortunate outcome. Better communication with the customer during the project would perhaps have allowed negotiation to come up with a more suitable solution. This would have been a better outcome for all concerned.

These types of problems are fairly common, although often they are commonly less severe than the examples illustrated above, and they can still affect the efficiency and effectiveness of a project. In principle, they are all due to insufficient or ineffective communications, which would on the surface appear to be an easy problem to resolve. In practice, however, it can be far more difficult to fix due to entrenched opinions, resentment and mistrust. These issues are likely to be beyond the scope of the design engineer's job, and are an issue for the company's management. This underlines the fact that effective network design necessarily relies upon management buy-in to the principles of the business-centric approach in order to work. If that commitment is made, then major strides forward can be achieved. We will expand on this theme in the later sections of the book, but first we will move on to look at the typical phases of a radio network design project.

The principal method of improving a design project is ensuring good, timely communications between all internal stakeholders in the company. The flow of information from those who know best to those who need the information to make the best choices is extremely important. The type of information flow required is illustrated by example in Figure 6.3. In this diagram, time flows downwards, and the elements across the top are discrete business elements. Information flowing between elements is shown in the form of horizontal lines. In this example, we have assumed that the marketing department have the best knowledge about potential customer behaviour. Therefore when the business development department are seeking to propose the project, they seek both their input to gain an understanding of potential subscriber behaviour and also that of the design team, who can advise on technical feasibility and the type of resources that will be required. Bringing in these external elements to what is effectively an internal business development activity helps to ensure that the business development team have neither invented a customer need nor invented the technology to serve that need (believe it or not, this has happened). This flow of information continues throughout the activity between the elements until the project is complete. Figure 6.3 is a simplification of the requirements, but does demonstrate the principles.

None of the elements discussed in this section are technology-based engineering, but they are as essential to the success of the project as selecting the right technology, getting link budgets and other technical parameters correct and performing suitable frequency assignment. As such, in the sense of 'completing the engineering activity to bring into service a new radio network', they are indispensable parts of the design activity.

## 6.5 Design Elements

Before we consider how we structure the project, it is worth reflecting on what it is exactly that the designer can influence during the design. This does not include every aspect of the radio network. In general, the designer faces external constraints caused by the business needs, the characteristics and the limitations of the technology to be used, and regulatory constraints such as the available spectrum, service area constraints and interference minimisation rules.

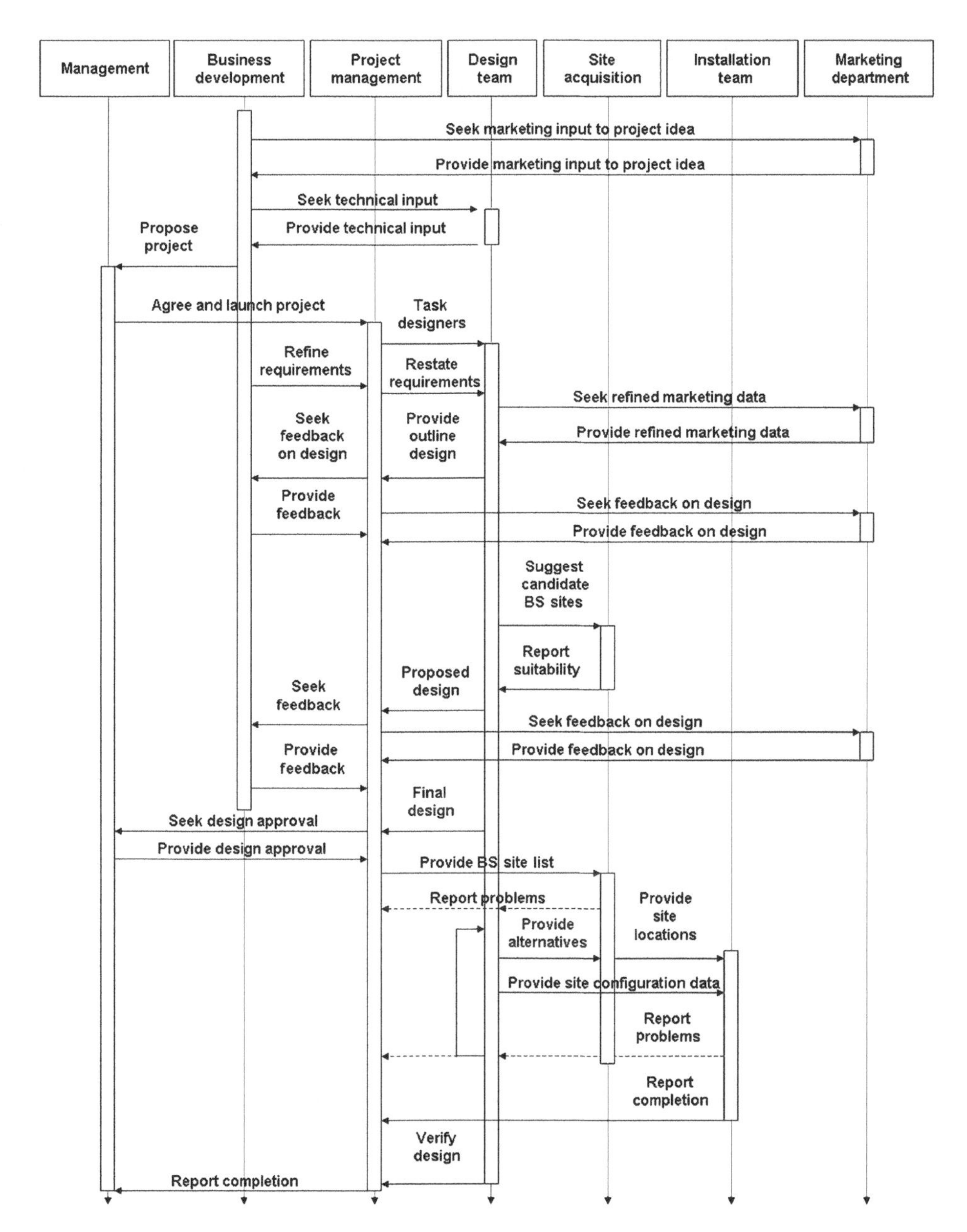


**Figure 6.3.** Information flow through the business; time flows vertically downwards, information flows across departments and activities are shown as vertical boxes.

For example apart from exceptional cases, new equipment types will not normally be designed for a particular network. All equipment will normally be either bought directly off-the-shelf or modified in some small ways. There is therefore little point in spending effort wishing that equipment could be more powerful, be more sensitive to smaller signals or work in frequency bands not supported, unless it is likely that such performance is likely to be forthcoming. Because of this, the radio network designer spends most effort on producing an optimal design of physical rather than electronic aspects of network design. These typically include the following:

- Selection of a suitable technology to use, usually from a list of technologies known or expected to be able to provide the functionality required, where this is an option.
- Selection of equipment from particular vendors.
- Selection of antenna types, based on performance and ability to be configured to meet the network coverage and traffic needs.
- Configuration of the variable aspects of radio equipment, for example power to be used at specific BS.
- Physical positioning of radio infrastructural in terms of geographic location, antenna orientation and height above local ground.
- Frequency planning.
- Optimisation to eliminate interference or in some cases to move it to a noncritical location.
- Optimisation of launch delay or other technology-dependent factors used to optimise network performance.

These factors are designed to configure the in-air aspects of the network – in other words, the parts between the two or more antennas in any link. The radiated power, antenna characteristics and type of the transmitted signal are important, the receiving antenna and the ability of the receiving system to receive a signal are important and the path in between is important. An additional design constraint is to ensure that sufficient backhaul capacity exists to support the mobile traffic, but apart from this, the mobile design engineer has little interest in the other line-based aspects of the network, since they do not affect what he or she does. Unfortunately, this also means that the vast majority of books on radio engineering have little to interest the radio network designer, since they focus on the internal workings of the technology – things the radio network designer cannot change! This is what we intend to address in this book. We will be investigating all of the aspects listed above throughout the rest of this book, but first we need to take a step back from the technical aspects and look at how radio network design fits into the business process.

In doing this, we will be focussed on the nontechnical aspects of a project that can determine whether a network will succeed or fail. This includes the people involved in a project, the structure and focus of the activity; the design criteria used to ensure the project will result in the desired outcome and the various stages in a typical network design project. This will be further extended in later chapters, as we look at each part in greater depth.

## 6.6 Project Phases and Project Life Cycle

First, we can look at the typical phases of a project, while bearing in mind that the actual phases may be different depending on lots of factors, some of which will be discussed in later sections of the book.

A typical project for a radio network design may fall into the following main phases:

- Building the business case.
- Outline planning.

- Initial coverage design.
- Detailed coverage planning.
- Capacity planning.
- Frequency assignment.
- Interference analysis and optimisation.
- Rollout (which the designer may be involved in, but will essentially be a physical activity).
- Verification by survey.
- Bringing into service.

These phases may be carried out sequentially, may be iterative in nature or more likely a mixture of the two. The selection of a suitable project development life cycle will be important, and in available literature there are few reference covering radio network design project life cycle paradigms, but practical projects can be built around the methods derived originally for software development. In fact, the network design process has many parallels and thus following the applicable tenets of the software design life cycle can be beneficial. The wealth of literature on this subject offers far more useful reference material for the project development life cycle designer. To illustrate this, we can look at three useful models for the life cycle, the Classic, Prototyping and Spiral.

### *6.6.1 The Classic Life Cycle*

The classic life cycle can be adapted to the project phases we have identified earlier in this section. It is effectively a sequential approach, although there is often some iteration between steps. However, the main aim is the sequential completion of steps; for example it would be poor practice to bring the network into service and then go back and iteratively re-visit the 'building the business case' step.

This approach is probably the oldest one used in business (hence the name), and it has the benefits of being straightforward and ensures that each step is completed before moving onto the next stage. For the phases such as building the business case and outline planning, this is highly desirable, but it can be less useful for some of the later steps in which a degree of iteration and cross-reference between activities may improve the efficiency of the design process and may reduce project risk. The classic life cycle also has a number of other drawbacks. Firstly, it is somewhat inflexible, and if one part of the project is held up, the whole project is held up; this will increase the risk that the project is not completed on time. Secondly, it is often the case that the network requirements are not fully understood at the beginning of the project, and so if a linear life cycle is followed, there is no explicit option to refine the design as the requirements become more evident. Thirdly, this approach leads to intermediate deliverables being unavailable until that entire step has been completed. For example the customer (in the case of designing the network for an external customer) or management (in the case of an internal project) may wish to review the design, but with this approach they will have to wait until the design is complete. If the customer is unhappy with the design at this stage, significant amounts of re-working may have to be done.

The classic life cycle is therefore useful for certain parts of network design, but is not ideal for other areas where iteration may be desirable or where the requirements are not fully understood at the beginning of the project. A typical classic life cycle is shown in Figure 6.4.

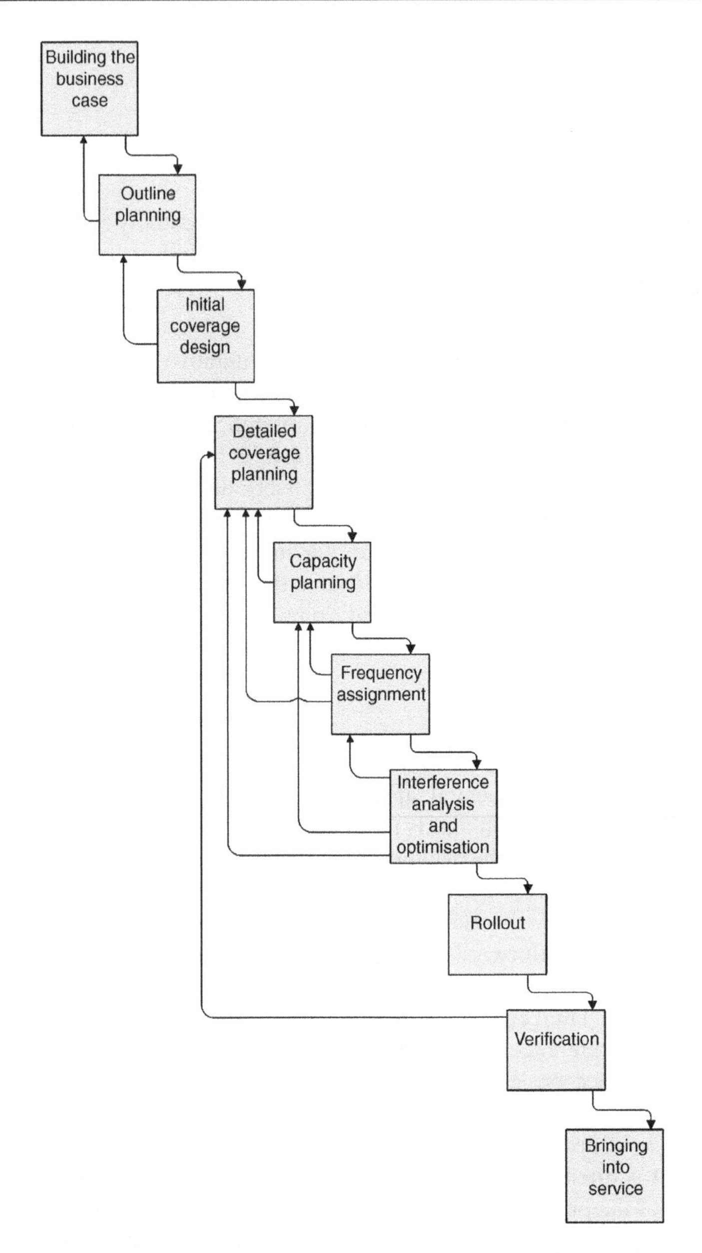


**Figure 6.4.** Classic life cycle applied to network design. The arrows shown going upwards indicate that the output of one process is checked against the output of a previous one. It does not show iteration between process.

In radio network planning terms, it is useful for replicating earlier designs into new areas, where the activities and risks are fully understood and the project design project is essentially a 'turn-the-handle' activity. It does not inherently require major contribution by the customer.

### 6.6.2 *The Prototyping Life Cycle*

What happens if the performance of a network cannot fully be explored until the design is completed, and the service that can be offered to customers cannot therefore be defined until that stage? If the project is led by a business need based on satisfying customers, how can the project aims be refined in sufficient detail to allow the design to be completed? In this case, both the service provided and the design to meet the requirement must be a trade-off between requirements, developed as an iterative approach during the project.

Where the full network requirements cannot be fully expressed at the beginning of the project, a prototyping method can allow the design to proceed on the known information, while continually refining it. Using the project phases we have identified, it will be possible to build an outline business plan based on the incomplete requirements of the network. An outline design can be produced against this, and rough performance metrics obtained. The outline design and associated metrics can then be used as a template for further discussions with the customer to refine the requirements further. In principle, they act as a method of testing the true requirements; the customer can identify those things that are genuinely needed, while dismissing those that are not. They also promote further and more detailed discussion on the actual requirements so that they can be further refined for the next iteration. This can then be repeated a number of times until the full design requirements are fully expressed. The same process can also be used at other points in the process such as proving network performance and reducing technical risk by building pilot networks over a small part of the service area or a separate test area. The lessons learned from this can then be applied to the full design and rollout. This is illustrated in Figure 6.5.

The prototyping process is cyclic, so the phases of the design are repeated many times with a more refined version of the design. This cannot be applied to every aspect of network design – for example to the actual rollout of the network – it would be ludicrous to rollout the network, then go back and do it again to a slightly different design. However, for refining the design, it is often a useful approach.

This process of prototyping allows the design to move forward even though not all aspects are understood and as such, it does not suffer the delays that would be caused if a classic life cycle is used instead. It has a role in reducing technical risk by providing a mechanism to test the riskier elements before applying them to the whole network. In a consultative-type project it also allows for increased but manageable dialogue with the customer during the project and, as we will see later, this is normally desirable, since it is a mechanism for managing the customer's expectations and for identifying potential problems at an earlier stage.

One disadvantage of the prototyping life cycle is the potential extra resources required to produce prototypes of the design or of the network itself. Consideration must be given as to whether there is sufficient justification for developing extra deliverables such as interim designs or pilot networks. These inevitably add extra expense to the project, and if they are not required, then this extra expense is for no benefit. The prototyping project is generally less efficient for low-risk, well-understood projects. It is also sometimes difficult to do in practice, particularly in the management of discussions with the customer; unless very carefully managed, the increased level of dialogue can lead to 'mission creep' – where the customer uses the dialogue to change (and generally enlarge) the project scope. If this has not been agreed in advance or properly covered by an official change request, together with

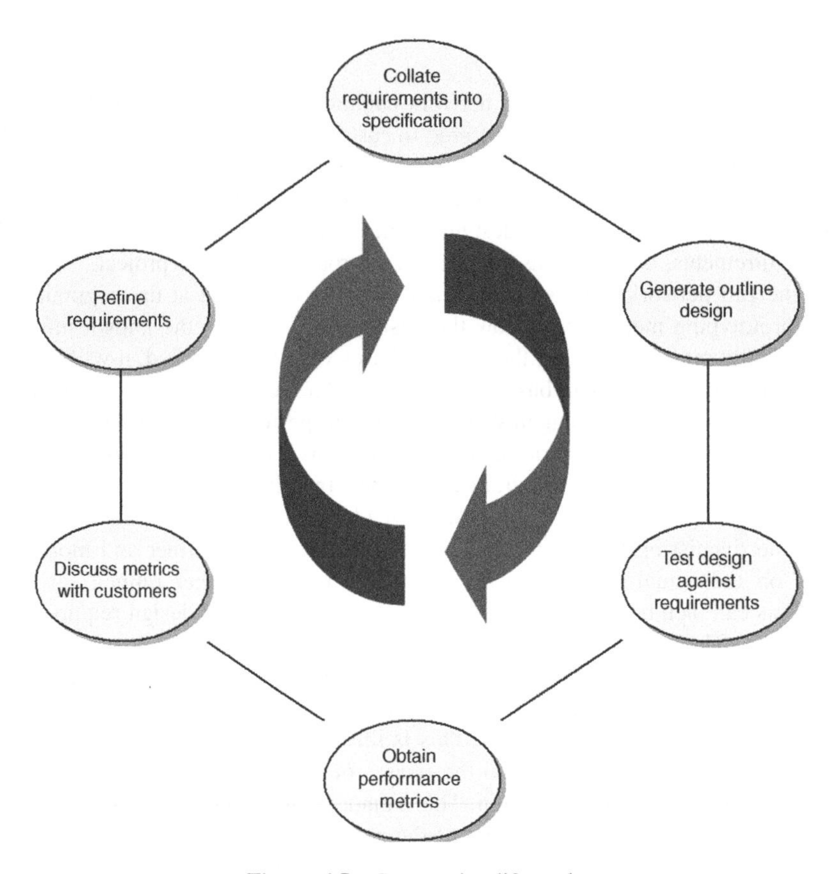


**Figure 6.5.** Prototyping life cycle.

the funds to pay for the changes. The management of customer expectation is not a trivial issue, and in general this should only be carried out at project manager/network designer level. It should not under any circumstances be carried out by project team members who are not authorised to commit the company to any changes agreed.

In terms of radio network planning, the prototyping life cycle has greatest benefit when working on new, potentially risky designs. Ideally, it should only be attempted in projects where the designer and customer have a healthy relationship and where the project is being carried out in a jointly consultative manner rather than where the relationship is cool or customers are stand-offish. It can only be applied in circumstances where the customer is prepared to commit to the principle and to apply sufficient resource from their organisation to contribute to it.

### *6.6.3 The Spiral Life Cycle*

The spiral model is a compromise between the classic life cycle and the prototyping life cycle, and it also includes management of risk, which is not inherently present in the other two methods. There are four phases of the spiral life cycle; planning (and improving the

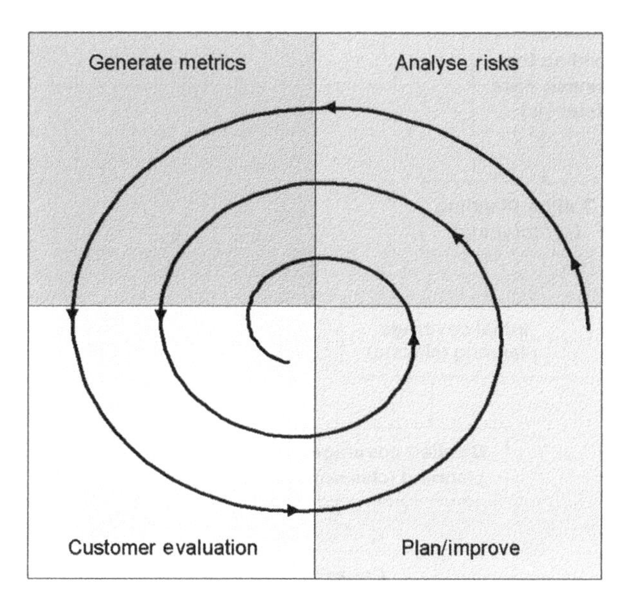


**Figure 6.6.** Spiral life cycle.

plan), risk analysis, generating metrics to determine the quality and characteristics of the design, and customer evaluation. In terms of radio network design, the planning phase corresponds to the determination of objectives (network requirements). This is followed by risk analysis, to identify technical or other issues that may prevent these objectives being made, before generating information to present to the customer. At this stage, the project is subject to customer or management review to ensure that it really does meet the customer's needs. This process is then repeated a number of times until the project is complete, spiralling in towards the actual solution, as shown in Figure 6.6, which shows the spiral going in an anti-clockwise direction.

In effect, the spiral method uses the prototyping approach to manage risk throughout the project. This can be particularly useful for risky projects, although it does require skilful risk analysis and the ability to properly identify and scope the risks themselves.

The spiral approach is similar in concept to the prototyping model, but because it has greater emphasis on risk, it is slightly different. It can be applied to large, risky projects but it requires significant management attention and the customer also needs not only to buy-in to the approach but also to provide sufficient resources to ensure that it can be completed correctly. This is likely to be higher than required for the other two methods.

### *6.6.4 Combining Project Life Cycles*

We have seen that each of the project life cycles has advantages and disadvantages, and that they can be appropriate to some aspects of the project but not others. In practice, a combination approach which uses the most appropriate project life cycle for each phase or sub-component of the project can be highly effective. Thus, where it is necessary to complete a phase before continuing, we use the classic life cycle. For example we really

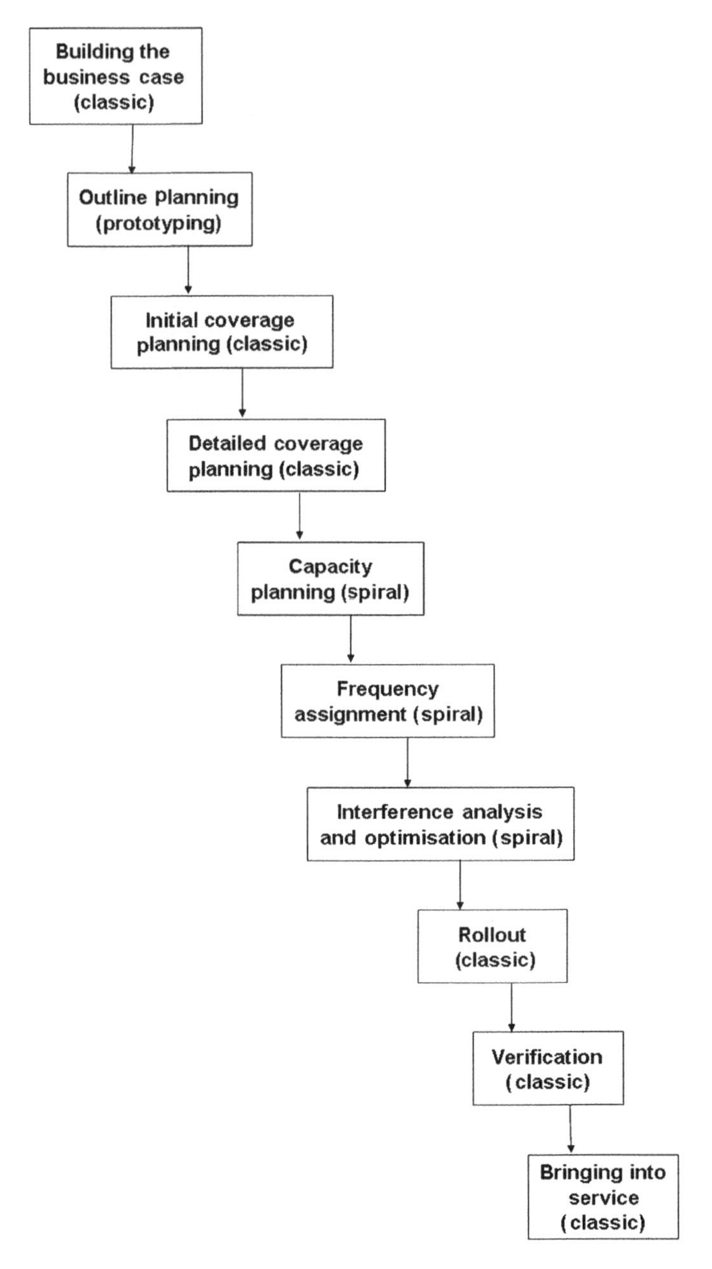


**Figure 6.7.** Application of life cycles to project phases.

need to generate an accurate business case before deciding whether to proceed with the rest of the project, so the interface between this phase and the next should really follow the classic model. However, within the activity to build the business case, it may be appropriate to adopt the prototyping method to refine the business case until everyone is happy with it.

During the detailed design phase, the spiral method may well be the safest to use, so that risks can be identified quickly and managed. This is illustrated in Figure 6.7.

The decision of which method to use for a project or part of it that should be undertaken by the project manager is a very important decision. Next, we need to look at how we can determine when we have met the objectives of a phase of a project, or of the project itself. First, we will briefly identify those things that the network designer can and cannot influence in the design.

## 6.7 Design Specifications

### *6.7.1 Our Approach to Projects*

The approach we normally use, and present here, is to spend significant effort at the beginning of a project in order to technically de-risk it, identify exactly how we intend to approach it and agree what we are doing precisely with the customer. This method has evolved over many years and appears to have significant benefits over traditional methods employed by others and even by ourselves many years ago. The intention is that by expending extra effort at the beginning ('front-loading' the project), we can then perform the actual design activity in a more ordered fashion. It can also mean that, if necessary, work instructions can be prepared to allow more junior staff to safely work on the projects while consultants manage a number of projects. Figure 6.8 illustrates what we intend to happen compared to traditional methods, and what we believe happens in the majority of cases where there is a direct comparison. The graph shows both risk and effort on the same graph and there is no scale – it is meant to be illustrative rather than quantitative.

In our method, we put significant effort in at the beginning, and thus project risk is managed from the start. By going into potential project risks and resolving them as far as possible at this stage, the risk reduces from the beginning of the project (as long as we have identified all risks). We can subsequently manage effort to bring the project to a successful completion. A traditional method may not involve such an initial amount of effort, and may focus on other aspects then de-risking the project. In this case, project risks do not decrease, they continue to rise, but this may not be identified till later in the project until suddenly it is

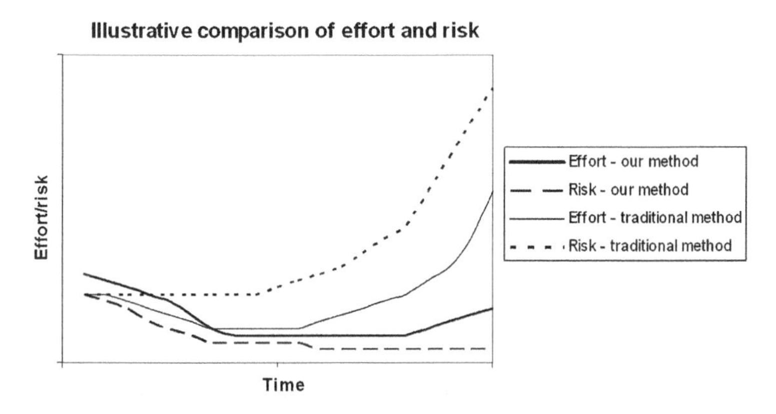


**Figure 6.8.** Our approach to projects.

realised and more staff and effort are put into the project at a late phase. However, by this stage it is too late. Overworked staff and new staff brought into the project only serve to make things worse, and thus both effort and risk increase till the end of the project, whereupon it fails.

We are not suggesting that our approach will prevent this sort of thing ever happening, but we hope to minimise the risk of it occurring and so far, we have managed to avoid getting into those problems, while observing others make the same mistakes. Because of this, we believe our approach is 'battle-proven' and are confident to present it here. Much of our approach is based on the production and maintenance of a number of key documents that have relevance throughout the project, and this is what we go on to now.

### *6.7.2 Specification and Documentation within the Project*

Thus far, we have considered the project phases, life cycle and the design factors that are actually under the control of the designer. But we also need to consider how we can determine whether we are ready to move from one phase to another, and whether our design meets the exact requirements. This is a difficult thing to achieve in practice, and considerable thought must be put into properly defining the decision criteria to determine this. These criteria must also be agreed by all relevant stakeholders well before the decision time; otherwise it is likely that there will be disagreement about what was actually promised, or what the actual requirements were. This is relatively common in projects where insufficient thought has gone into this aspect of the project, and it leads disputes that lead to customer dissatisfaction or even legal action. More commonly, it results in 'mission creep', where the designer is forced to perform extra tasks not included in the original contract, since it is not possible to prove that they were not included in the offer. The only way to achieve this is to document everything in sufficient detail to capture all of the requirements in an unambiguous manner. This is true for both commercial and design documents used throughout the project. We can identify a number of important documents during the project life cycle:

- The business case.
- The invitation to tender (ITT) or other customer-produced statement of requirements.
- The contract (for external projects, also sometimes for internal projects).
- Method statement (MS).
- User requirements specification (URS).
- Functional specification (FS).
- Detailed design document (DDD).
- Test specification (TS).
- Acceptance certificate.

We can use these documents, and the specifications within them, to determine whether the requirements are met or not. This is illustrated in Figure 6.9, which shows two projects; one an internal project for internal customers (left hand fork), and the second for a competed external project (right hand fork). This is an illustration of specific projects; others will vary in structure and content.

In defining these documents, we are adopting a similar approach to that used for software engineering projects. This is intentional, and it is because we can draw some aspects of the

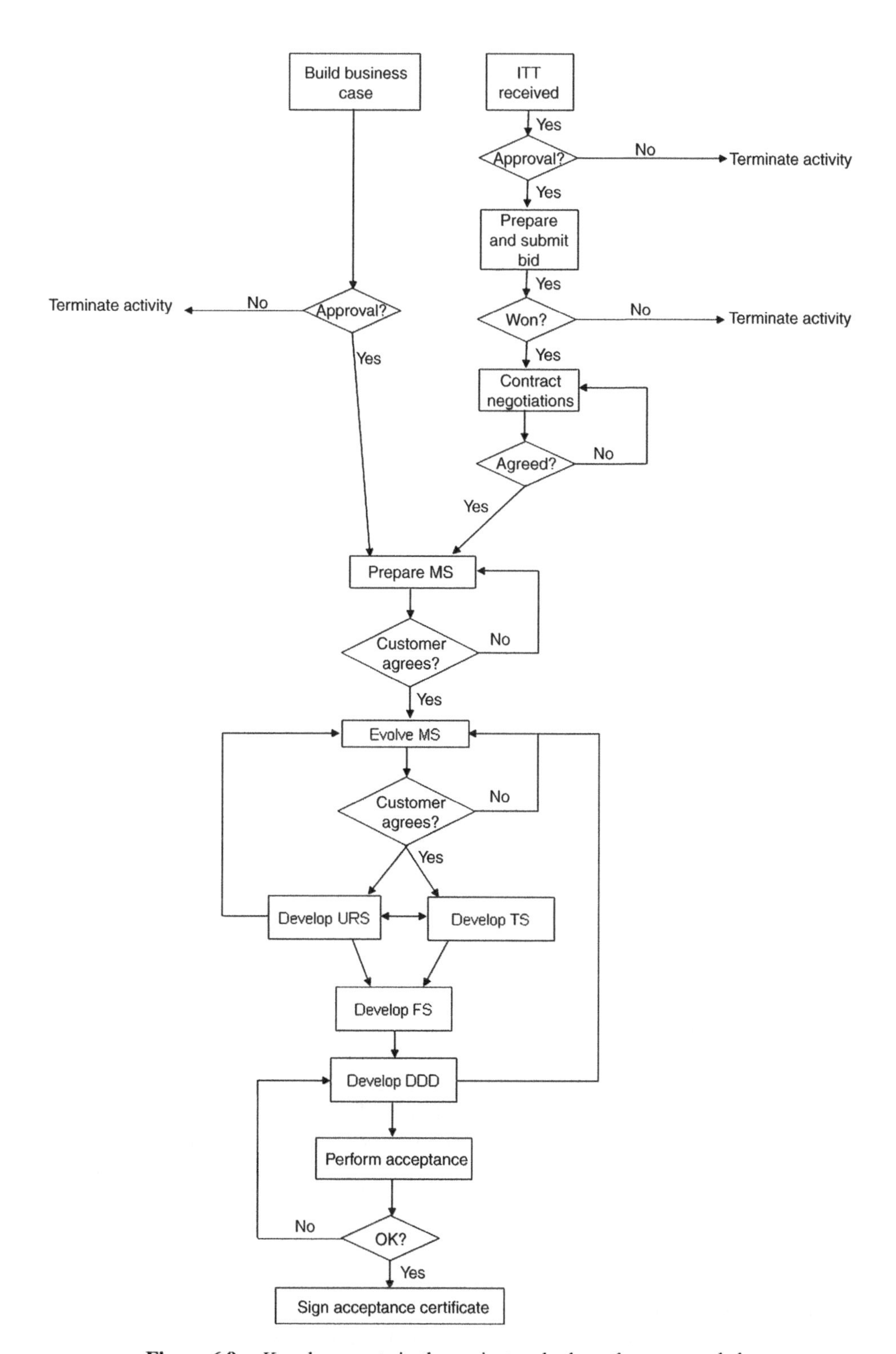


**Figure 6.9.** Key documents in the project and where they are needed.

methods used in software design projects and re-use them in the network design project. For example in a successful software engineering project the software delivered at the end of the project meets the demands of the user community. This does not happen by accident, but is because of the following:

- The URS has been validated against the actual user requirement. It is highly detailed, precise and unambiguous, and it captures the features required by the user in the form of technical metrics that can be measured against in a practical manner. It is also self-consistent and does not contain any conflicting requirements.
- The FS has been thoroughly validated against the URS. It captures all of the features of the URS in detail, and correctly converts nontechnical metrics into entirely measurable metrics. In many cases, the elements of the FS may well be incorporated into a more developed URS and a separate FS will not be necessary.
- The design has been thoroughly validated against the FS and case-tested against the requirements in the URS. It has been captured in the DDD and this may be reviewed by the customer before the design is formally adopted. Since this is the primary document defining the design, it essentially *is* the design.
- The resulting network has been verified against the URS in sufficient detail to convince the customer that it meets the need. This testing has been carried out against a TS generated against the URS, ideally generated early in the project and agreed by the customer. The TS contains a series of tests designed in such a way as to ensure that the results are unambiguous; the system either clearly passes the test, or it clearly failed. Ideally, there is no middle ground; otherwise there will be scope for dispute.
- At the end of the project, the customer signs the acceptance certificate. This is proof that the customer has accepted that part of the project, so that invoicing can take place.

The process of validation and verification is perhaps the most crucial aspect of the software project. Validation can be described by the question, 'Are we building the right system?' Verification can be described by the question, 'Are we building the system right?' Between them, and expressed in terms of measurable metrics, they are undoubtedly the most important question in software design. We would argue that these questions are equally valid in a network design process, and by extension, the properties of the successful software design project are equally desirable in a successful network design project. This is at the core of the project model we propose, and it has been used in the successful design of a great many projects in recent years, covering a raft of radio technologies, including both fixed and mobile.

We in the following section discuss these documents. They may not all be used for a specific project, and indeed in a consultative project when working in a structured fashion with the customer, we would tend to use the MS as an organically growing document throughout the project that obviates the need for the URS, FS and TS. However, since this is not always the case, we also describe the URS, FS and TS.

### *6.7.3 The Business Case*

The business case it likely to be developed by a management team, and it may well be complete well before the designer ever becomes involved in the project. On other occasions, the designer may well be involved in the production of metrics to be used in the report, or to comment on the technical viability and technical risks. Whatever the case, this may form a key document for the design later in the project. Even if this is the case, it is often important to probe the requirements further rather than to rely on the contents of the business case on its own; it is unlikely to be sufficiently detailed to allow full design against it. As we will see later, it is essential to probe the requirements in as full detail as possible so that the network can meet this design.

The designer is unlikely to have influence over the contents or style of the business case document, so its use in the project as a reference will vary between projects; it is entirely possible that the designer will never see it.

### *6.7.4 The Customer Statement of Requirements*

This may be part of the invitation to tender or request for quotation (RFQ) from a customer who wishes to put the design or implementation out to commercial tender. For internal projects, it may instead be an internal statement of requirements. Again, in either case it is unlikely that the designer will have any influence over its style, content, precision, accuracy or consistency; although sometimes designers can be asked to contribute. Sometimes, detailed requirements are drawn up by those who have little understanding of what is physically achievable or desirable and this can cause confusion or contention later in the project, so it is vital when the tender is not clear or consistent that these issues are probed in more depth at this stage – certainly no company should accept an order it knows it cannot meet the requirements of, or when the requirements are not clear. Unfortunately, in the rush and enthusiasm to win the business, this does happen fairly regularly. Typical reasons include the over-enthusiasm of sales staff, who may gloss over important details or too readily agree to conditions expressed by the customer that have a major impact on the work to follow. Also, there may well be a reluctance to probe the customer in case they should take offence, or there may be a feeling that asking questions may give the impression that the company does not understand that requirements in the way that they should. In general, both these qualms should be overcome and in most cases we have come across, the customer is pleased to have their requirements discussed in detail. On the rare occasions that customers are reluctant to provide further information, this may be a sign that the customer themselves are not giving the project the best chance of success, and that rather than become embroiled with a project likely to fail, it is better to step away.

The usefulness of the ITT document will depend on how well it is constructed, but if it is not sufficiently detailed, the requirements expression should be captured in another way. This is the material that must then be used as the basis of the commercial agreement between the customer and the designer's company.

Again note that although we have expressed the use of this type of document in a commercial context, the process is equally valid for noncommercial projects, where the ITT may instead be the user-expressed requirements, as put together by a management or command project team. Again, the designer may have no or little influence over the scope or contents of this document.

### *6.7.5 The Contract*

For commercial organisations, the commitment to the project by both customers and designers is likely to be expressed in terms of the contract. This may be sufficiently detailed to capture all of the technical requirements of the network, but this is rare. It is more normal for the contract to refer to external documents or to a process whereby the requirements are agreed once the project is fully underway. It will describe the top-level requirements and the methods whereby success can be established, but again, this will normally need to be enhanced. The contract will, however, normally determine project timescales and commercial conditions. The difficulty for complex engineering projects is that often it will be necessary for the designer to consult with the customer in order to capture the requirements properly, and this will be part of the project, not started until after the project has commenced – and therefore the contract has been signed. In this case, it may be appropriate to build the ability to generate the full contractual terms during the project. This would most often be expressed by the response to the customer's requirements, where the details of the offer are described. It is important to ensure that this aspect is included in the response because it will be difficult to negotiate later.

### *6.7.6 Method Statement*

At the beginning of the project, there will be insufficient information to describe the whole network design, or in some cases even general details. In this case, an MS can be used as a commercial document to agree between the customer and designer exactly how the latter intends to carry out the project. As the name suggests, the MS should detail the exact methods to be used during the project. To avoid later disputes, it should be as detailed as possible at the start of the project, and it should include mechanisms to allow changes to be agreed if this becomes necessary later in the project as a result of better information coming to light. Examples of the sorts of things that an MS should include are as follows:

- A detailed description of the service area that the network must serve.
- A detailed description of how this network service area is delineated and modelled, such as the exact environmental data to be used.
- Propagation models to be used in network simulation, including all settings to be used.
- The link budgets to be used.
- The channels to be used in a frequency assignment activity.
- How interference is to be modelled, quoting references to make the method unambiguous.
- Test methods to be used.
- The methods to be used for changes, when they should be used and the procedures that should be followed.

The MS may also be used as an organic design document that incorporates those aspects that would otherwise be captured in the traditional URS, FS and TS documents. The authors use the MS as the primary document in any project, and it has precedence over every other document other than the contract (or it may supplant parts of the contract, subject to agreement with the customer). Where we have external URS, FS and TS documents, these are always referenced from the MS and are effectively sub-components of it.

We will look further at MS in the Chapter 8.

### *6.7.7 User Requirements Specification*

Since all the documents we have mentioned in this chapter are important, the URS is arguably the most crucial, and it can be the hardest to create properly. The purpose of the

URS is to ensure that the network design criteria are expressed in the form of metrics that can be used to create the FS or if a separate FS is not to be created, then it should be sufficiently refined. To illustrate the principle, consider the following customer aspiration:

*The network designers must prove that the design provides in-building, mobile and hand-held voice communications over the whole of the UK mainland.*

At first sight, this statement appears to be straightforward, easy to interpret and eminently reasonable. But is it? In fact, this is the type of statement that easily rolls off the customer's tongue, but is a massive pitfall for all concerned. We can state with confidence that no practical network will ever be able to achieve this seemingly simple request, and also that it is not what the customer actually needs (or if they do, they must expect to pay vast sums of money and overcome probably unsolvable objections). So what exactly is wrong with this statement? Without going into every detail, we can identify a number of issues that must be discussed further before a realistic set of statements can be made. For example:

- The statement calls for coverage over the whole United Kingdom. This is a vast area (although a lot smaller than many other countries). There are areas with very low population density or indeed no one at all. So who needs the coverage? In fact the ratio between the densely populated areas and those lowly populated is very low, so if we provide ubiquitous coverage for all these less-populated areas, the chances are that there will be no or little demand in these areas – is this really what the customer wants? Also, do they mean all of the islands in the United Kingdom, even the uninhabited ones, just mainland UK, or somewhere in between?
- What is meant by in-building coverage? Buildings vary from small structures to massive skyscrapers and to underground car parks – do we need coverage there? This will make an enormous difference to the network design and cost – it may well make it unworkable. Underground bunkers are buildings, but are they intended to be included in this statement?
- What is meant by voice communications? The term is vague in terms of the probability of call success – all engineering must be designed around percentages and margins – any term that states – or more likely implies – 100% availability, 100% of the time has no place in a URS. This is a situation where it may be necessary to educate the customer with a dose of reality (sensitively done, of course!).

There are many other issues, and issues that arise from those issues that would need to be discussed and determined in depth. It is the role of the network designer/consultant to perform this task and by dialogue and analysis, come up with an achievable, realistic set of requirements. It is probably the most important role of the consultant, and it will be discussed later in the book in some depth. Although it may seem unnecessarily pedantic to go into such detail about every aspect, and it may be difficult for the customer to understand the need; the degree of analysis at this stage is directly proportional to the probability of success of the project and so it is a very worthy activity. For the moment, the crucial pointer to take from this is that the creation of the URS is far from simple and involves asking questions that in some projects have never been asked – unfortunately. Again, we will be discussing the URS in later sections of the book.

### 6.7.8 *Functional Specification*

The functional specification is a document drawn directly from the software engineering paradigm. In software engineering, it is a specification that precisely describes what the system must be in terms of software design, describing exact data input into the system, how this data should be processed and what outputs are required. In the radio network design context, we need to specify a number of similar aspects, which we can either include in the functional specification or incorporate into the MS or the URS. In either case, the type of characteristics that can be captured would include such things as:

- Link budgets to be used for each type of link in the network. This will typically include BS to mobile and the return path, BS to handheld (outdoor) and the return path, BS to handheld (indoor), and link budgets for backhaul if required.
- Propagation models to be used, described in sufficient detail to be unambiguous. Since this will typically be configured in a radio-planning tool, it should be expressed in terms of suitable to describe the actual settings, values and checkboxes to be used. It should also be described in terms of the ITU recommendation or other reference to allow external comparison if required. It will allow later auditing if necessary to ensure that these instructions have been followed.
- Values for excess loss due to ground obstructions, or other method of accounting for radio clutter/increased noise floor due to environmental factors. Again, this should be detailed in terms of the settings to be used in the planning tool and references to the academic sources used to determine which model to use. The same is true for any other values for noise or any other corrections to the core propagation model.
- Any atmospheric correction factors to be used in planning, for example any variations to the effective Earth radius and, if appropriate to the frequency band, any corrections for precipitation, excess absorption or ducting effects.
- A detailed description of how the physical environment is to be modelled, including the terrain and clutter databases, representations of mobile user community and usage of the network. This has to be fully determined at this stage; a statement such as, 'a 50-m resolution database will be used' is insufficient. This says nothing about where the data comes from, how accurate it is, how old it is, etc. There is therefore too much latitude in this description; ideally it should identify the specific dataset used so that another person can obtain the same data at a later date if necessary. For example the 'the UK Ordnance Survey Panorama database, 2002 edition'.
- An exact description of the service area. Again this should be sufficiently detailed as to allow another person to obtain the same data again at a later date.
- A specific list of radio sites to be considered in the design (where existing sites are to be used). This may not be fully determined at the start of the project. If so, then instead a set of rules should be established to identify how candidate sites are to be arrived at should be included.
- A set of defined rules to be used in site selection from the candidate list. This again should be as detailed as possible (see Chapter 8).
- Detailed descriptions of the rules to be used for interference analysis, frequency assignment and determining the number of channels to use to cater for traffic demand.

As can be seen this list can get very long, and the document will be large for large projects. It can also be seen that rather than expressing functional requirements, this document is more closely akin to a description of configurations, data and design rules against which the network will be designed. It is entirely appropriate to incorporate this information within the MS, or if not, then the MS will refer to the FS as the primary reference document for these aspects. In any case, it is important that only one document has primacy on any single factor, otherwise confusion may arise if there is a discrepancy, and document configuration becomes far more difficult.

### 6.7.9 *Detailed Design Document*

Just as the DDD in a software project precisely identifies how the functional specification is met by the design adopted, the DDD in the radio network captures all the design proposed in sufficient detail to allow the network to be actually built. The degree of detail included will depend on the requirements of the project. Sometimes, the document will contain only information relevant to implementation that can be given to the installation team. This will include such aspects as follows:

- A list of each site to be used for fixed antennas.
- For each antenna on each site:
  - The height above ground.
  - Orientation in azimuth and tilt.
  - The power to be emitted by the antenna.
  - The frequencies for transmission and reception.
- The frequency plan for the entire network.
- Neighbour lists and other technology-specific characteristics.

On other occasions, the document will also include other materials to help to prove that the network will meet the requirements. This will usually include the following:

- Coverage plots.
- Best server plots.
- Interference plots.
- Coverage statistics.

The metrics in the DDD can then be used as the baseline for measurement of network performance, as detailed in the TS, and they should also be used as the basis for acceptance testing. Naturally, the metrics in the DDD must match the initial requirements as embodied in the MS or FS (which would be referenced from the MS), and these in turn must be an appropriate representation of the URS and hence back to the performance agreed at the beginning of the project.

### 6.7.10 *Test Specification*

The TS can also be incorporated into the MS or kept as a separate document. Historically, one of the most common mistakes in setting up projects has been leaving the design of tests to prove the network performance until late into the project. This is generally too late. The main reasons for this are as follows:

- Without having specific tests in mind, the network cannot be critically engineered to meet the required objectives. Naturally, the tests must represent the desired performance in sufficient detail that meeting the test criteria necessarily meets the performance criteria; this means that designing to the tests is the same as designing to the entire FS.

- The tests must be used as the main acceptance criteria. Because of this, they must be agreed between the involved parties as early as possible. Doing this as early as possible, to prevent some potential issues arising, ones that are difficult to resolve once they have become an issue. This includes the following:
  - It will prevent misunderstandings being propagated through the network design. The customer may not be able to fully grasp the implications of some of the design factors in the URS and FS, but the tests will be closer to the customer's understanding of how the network must perform. This means that the TS has a role in managing the customer's expectations and ensuring there is no expectation disconnect between designers and customers.
  - Designing the TS early on in the project means that the tests will be agreed against the original requirements. During projects, it is typical for the customer to try to modify the requirements as the design progresses. While this is fine so long as there is a mechanism for change and that the modifications are not carried out to the detriment to either party (typically the engineers, who are effectively asked to carry out extra work for free), it is highly undesirable if it leads to contention. This will particularly come to the fore in designing and agreeing tests for acceptance if they have not already been agreed. The customer will want to set the tests against their latest (and possibly continually evolving) set of requirements. The designer will want to set the tests against the originally agreed requirements. This can clearly lead to contention and is so best avoided where possible by agreeing the TS as early as feasible, and for including the tests to be carried out in any change that may be agreed between the two parties.

The TS must be sufficiently detailed to properly represent all of the network requirements, but it cannot be so onerous that the test process takes an inordinate length of time or is too difficult to achieve readily. Statistical methods and sampling must be judiciously used; putting an effective TS together is by no means a simple task, and should not be underestimated. We will look at testing in Chapter 15.

### *6.7.11 Acceptance Certificate*

The acceptance certificate may take the form of a single sheet of paper with a place for the customer to sign the network off. Conversely, it may well be a form of the TS, with places for designers and customer representatives to sign off each individual test. Whatever the case, it is essential that the acceptance test is related to – and only to – the TS, and not to other, less well defined criteria. This again means that the form of acceptance must be agreed at the beginning of the project, together with the understanding that it forms the sole criteria for acceptance and that nothing else should be taken into account.

## 6.8 Design Deliverables

The project design activity will result in a number of discrete deliverable documents. These will be created at different parts of the project, and some – particularly the MS – may well evolve during the process. Although for specific projects, customers may ask for different deliverables; we would highlight the following deliverables to be appropriate during the project. For the sake of completeness, we have added some of the internal deliverables that may be required before the business is won, just to show the entire set of deliverables, a design and sales department may be required to produce the following during an entire substantial project (this does not include commercial documentation not related to the design).

- On receipt of ITT or RFQ:
  - Initial feasibility study.
  - Documentation of bid/no-bid decision (assumed bid in this description).
  - Information in support of bid document.
  - The bid document.
- On receipt of order:
  - Initial MS, to be agreed before progressing.
- Early in the project:
  - URS.
  - FS.
  - TS.
  - MS, incorporating new documents.
- On completion of design:
  - DDD.
  - Acceptance certificate (incorporating completed TS).

## References and Further Reading

[1] Software Engineering: A Practioner's Approach, Roger S. Pressman, McGraw-Hill, ISBN 0-07-707936-1.
[2] Taming the Wild Project, Mobile Radio Technology Dec 2003, Intertelc Publishing Coden: MRTEFP ISSN: 0745- 17626 SICI:0745-7626(299312).

# 7

# Selection of Engineering Tools and Data

## 7.1 Introduction

In the last section, we looked at the nontechnical aspects of the radio network design project. Now we turn our attention to the technical aspects, and we start by looking at the radio network design tools and associated data necessary to plan a modern mobile radio network.

In the modern radio network engineering world, traditional methods using paper and pencil are no longer appropriate; they are too slow, risky and expensive compared to modern methods. Occasionally, the argument is made that the capital costs of modern radio planning software tools and associated data are too expensive and that it is cheaper to use traditional methods, but in all except the very smallest of projects this is difficult to justify. The extra labour involved in using traditional methods quickly offset the cost of the tool, even without considering the extra effort required for re-working and for correcting mistakes that are more likely to arise when using older methods. Even if the purchase of a tool is not justified for a project, there are other methods of obtaining their benefits; by renting the system for the project, hiring external consultants who will use a sophisticated tool on the customer's behalf, or using a bureau service that allows customers to direct the activities of a skilled operator working with a modern planning tool.

We will look at the capabilities, benefits and limitations of modern planning tools. The radio network designer must be able to identify the functionality required for different phases of the project, and be able to select the right tool or tools to fit the overall project requirements. For organisations that expect to perform a variety of projects, the required feature set is likely to be broad and should be fully considered before selecting a tool to buy or rent. The selection of the right tool is a complex issue, and one must consider not only the functionality offered but also a wide variety of other factors. These includes the support offered to customers by the manufacturer, how well-proven the system is, the data the tool can use and which radio technologies the system is capable of modelling.

Besides radio planning tools, we will also look at the environmental data the tools need in order to work effectively. This is a large subject in itself, and one that has often been poorly understood. We will look at Digital Terrain Maps (DTMs), Digital Elevation Models (DEMs), ground clutter maps, ground conductivity maps, building maps and the map images used by the tool operator to relate the results of simulations to important social features such as towns, roads, railways and so on. We will also look at methods of modelling mobile elements within the network. This whole area has hidden levels of complexity, and the radio network designer must understand the issues in order to be able to make the right decisions and offer the right consultancy support to customers.

To put this whole subject into context, we need to take a brief look at the history, current status and future trends within the realm of radio network planning tools. After the introduction to radio network planning tools, we will consider the key features under specific headings. This is intended to be a reference for readers when considering which radio planning tools to use for a specific project or to obtain for more general use.

## 7.2 Engineering Tools for Network Design

### *7.2.1 History of Planning Tools*

The earliest computerised radio-planning tools were path profile prediction tools, which became available in the 1980s. These allowed an operator to apply a propagation model to some representation of the terrain between two known locations, either entered manually using information extracted from paper map contours (a paper tape was laid along the path and marks made for each contour line – this data was then entered using a digitisation tablet to produce a path profile) or by an internal digital terrain model previously generated and stored on the computer. These allowed operators to determine the viability of proposed fixed links, and they were often used in the military environment for positioning the nodes and relays of trunk radio systems such as Ptarmigan (in the UK). Generally, the tool would report the predicted signal strength at the receiver terminal, or another metric that was understood by the operator, such as link margin or Probability of Successful Operation (PSO). This allowed planners to determine whether links were likely to work, and also allowed basic parameters such as receiver height above ground, frequency of operation and antenna type to be changed in order to determine whether the link could be improved. In general, their primary benefit was to allow unworkable links to be dismissed without actually going to the site to do a physical survey; for the proposed links, it was still appropriate to do some kind of survey on site. Some simple frequency assignment tools also started to appear at this time, as an aid to manual assignment or with some simple assignment methodology.

The second generation of radio planning tools added the ability to perform multiple profile simulations on an area, thus generating what has come to be known as a coverage prediction. This is typically displayed as a coloured region around a base station. This allowed network planners to consider the mobile scenario, in which the second terminal was not fixed but instead roamed throughout the service area. Very simple tools would only show the coverage from a single site at a time, whereas more sophisticated systems would show composite coverage from multiple sites, best server and interference displays. They typically allowed prints to be created for comparison of the coverage to the wanted service area and to the coverage of one potential site against another, usually by manual comparison of plots. These started to appear in the very early 1990s. Over the decade, tools continued to improve and provide a wider range of functionality.

### 7.2.2 Current Planning Tools

Current state-of-the-art planning tools provide a wide range of functionality. Over the last few years the trends have been along the following lines:

- Additional functionality for specific technologies as they arise, such as modelling of dynamic links and network usage simulation for 3G technologies, and the ability to model interference issues between dissimilar technologies operating in the same band.
- Ability to handle larger and more complex network design activities with reduced operator intervention. Current tools are often able to handle networks with thousands of base stations, and perform interference analysis and frequency assignment across the whole network.
- Increased interoperability with other systems via direct access to databases and other data sources. This helps to promote the concept of business-centric operations, where all departments contribute to, and receive information from a single-managed data structure.
- Improved propagation models with extended frequency ranges and additional correction factors, capable of handling the demands of new technologies and improving the prediction of older ones.
- Automatic design methods for network dimensioning, site optimisation, frequency assignment and other network characteristics. These can then be operated by less skilled staff, and they can perform tasks not practical for human operators or by less complex software systems.

These improvements have vastly increased the capability of the radio planning engineer, but in general have come at the cost of higher training requirements for operators, since in order to get the best from such tools, it is essential to have a thorough grasp of their capabilities, together with a solid domain knowledge and the ability to use the tool imaginatively.

### 7.2.3 Future Trends in Planning Tools

At present, a lot of design work is going into improving the capability of radio planning tools to meet future demands. Apart from the continual enhancements necessary to cope with new radio technologies, there is also a desire to provide the advanced functionality of state-of-the-art planning tools to a wider audience and with significantly reduced training requirement. This can be achieved by automating complex tasks and by providing a user interface that depends less on the technical aspects of network design and more on a statement of the end user requirement in operational terms. For example, consider the following situation and then we will examine how this can be solved by a current tool, and then by the next generation.

*The overall requirement is to provide mobile coverage over a particular service area, which is defined as the urban sprawl of a town. The current task is to identify candidate base station locations for further analysis.*

This can be achieved by a current standard tool by performing the following tasks:

- Select appropriate environmental data manually from storage, which may be on CD-ROM or from a central data storage system. This will take time and involves the risk that the operator will select inappropriate data, or fail to find the required data.
- Create a project within the selected radio-planning tool (a 'project' is a workspace to perform the required operations). This involves the correct identification of the service area and configuring the project to include all of the wanted area. There is a risk that the operator may select the wrong area, choose too small an area (which may prevent the task being completed normally), or choose a large area that will add substantially to the processing time.
- Define the service area using a polygon. This will require identifying the correct menus or controls to allow this task to be completed, possibly from a large number of available menus and controls.
- Select appropriate propagation model and associated settings. This is an area fraught with risk for the inexperienced operator, unless the tool has very strong mechanisms for detecting inappropriate settings. If these settings are not configured correctly, then the output of the whole task will be wrong. Any actions taken on the basis of the output will therefore also be flawed.
- Configure technical characteristics of radio equipment to be used, including the calculation of the link budget. Again, any errors here will invalidate the results.
- Use automatic site finding functionality to select candidate sites. This requires locating and configuring the available site finding functionality. The complexity of this task and margin for error will depend on the characteristics of the particular tool, but may be significant.
- Manually check and refine the automatic design. The effectiveness will depend entirely on the skill and experience of a particular operator.

Clearly, the operator of such a tool has to be capable of identifying the steps involved and their sequence, and also has to avoid errors that may make the results invalid. A more advanced system may perform in the following way:

- The operator selects the area to be studied by drawing a polygon on the wanted area. The system automatically looks up its environmental data catalogue, selects the most appropriate data and loads it. This system will offer a simple initial interface (perhaps no more than a display of the world), and all the data selection tasks will be performed automatically. The risk of the operator selecting the wrong data or not being able to find it is much reduced.
- The operator selects the task to be done, from a small list of potential tasks each suited to the needs of the business. The number of available tasks will be smaller than those for the nonprocess driven tool described previously and thus the risk of incorrect selection is significantly reduced. On selection, the system.
- Configures the propagation model and associated settings. This removes the risk that the operator selects an inappropriate configuration.
- Configures the technical characteristics of the radio and calculates the link budget automatically. Again, this removes the risk of manual error.

- Performs the automatic site finding function. The system automatically configures the settings based on the needs of the specific task.
- Refines the design until the design meets the performance criteria. This is performed automatically according to preset rules. The dependence on operator skill and experience is removed.
- Configures the results display and stores the required design. The results will now be configured for further use without manual intervention.

In this case, the majority of the technical aspects have been removed from the operator's task list. Having been programmed previously by a domain expert, the system performs the tasks automatically. The risks involved in the task are much reduced, and likewise the training requirements for the operator are reduced. This is increasingly what is wanted by modern network operators and network planners, and the current trend in business is for organisations to reduce the number of their technical staff while retaining expertise in their own core business; understanding their operational requirements in terms of the business aims. The obvious conclusion is that future demand will be providing radio-planning expertise to nonexperts via systems that incorporate some of the knowledge of domain experts (effectively, what are known as expert systems). With such a system, an operator can express the requirements in terms of the business activity and get an answer without requiring that the operator have any deep technical knowledge. So, future systems will automatically perform tasks and will require operator input only when necessary and in nontechnical terms. Of course, still there will be a need to of experts to construct these systems and set them up for use by nontechnical staff, but there will be fewer of them and more nontechnical staff.

## 7.3 Benefits of Using Design Tools

Any radio-planning tool is only worthwhile if it provides business with tangible benefits. The principal benefits are reduction in cost, reduction in time to complete the design task, ability to perform tasks not otherwise possible and management of risk. If these are not achieved then having a radio-planning tool might be nice, but it is not productive or cost-effective for the business. In many cases, however, planning tools do provide distinct benefits to a business.

Typical benefits include:

- The time to complete each task in the design process is much reduced compared to older techniques and require fewer personnel. This improves design efficiency and reduces overall project costs.
- Minimisation of the need for physical survey of potential fixed sites. In particular, poor sites can be quickly tested and rejected without the need to send personnel to do physical surveys. This saves money and time by reducing the number of survey staff required and the time they must spend surveying. This can improve time-to-market or, in military terms, time to deploy in addition to reducing costs.
- Rapid identification of the best sites to use, and of alternative sites if the best ones are not available. This allows the site selection process the maximum flexibility, by identifying a number of configurations that work. This can then be fed into the site acquisition activity, and then the site acquisition team can look at other aspects such as site cost and

desirability in order to determine the configuration that best meets the overall business need. The increased flexibility in the task improves efficiency and by offering alternatives in sites, also offers cost reductions.

- Enhanced capability to perform tasks. Some tasks cannot be easily achieved by manual means, such as complex frequency assignment and highly iterative design, such as that needed for CDMA systems. In this case it is necessary to have a computerised method of solving the problem, and the options are to design one in-house, get a software development company to design a bespoke system or buy a Commercial-Off-The-Shelf (COTS) system. In most cases, it will be cheaper to buy a COTS system, and if an appropriate, well-established system is selected, it will have proved effective before and thus risk is minimised. If no appropriate tool exists for a specific task, then it will be necessary to configure an existing tool or possibly even create a new one.
- Risk management is a key aspect of any project. In general, it is highly desirable that risks are understood as fully as possible and as early as possible in the project to prevent unpleasant (and potentially expensive) surprises later on. Typical ways in which planning tools can help specify risk include:
  - Proving whether the selected technology is likely to achieve the business need in terms of being able to provide working links to the mobile subscriber. There have been examples where entire businesses have been set up on the basis of the expectation that a new technology could provide innovative service to customers, and these have failed when it became apparent that the technology was not capable of delivering. This could have been prevented by modelling of the proposed system by radio planning tools.
  - Providing the ability to make early, but still relatively accurate, estimates of the number of fixed sites that need to be deployed to meet the business needs. From this, rough costs can be estimated early in the project and thus the business model can be improved.
  - Identifying whether allotted or assigned spectrum will provide sufficient capacity to meet the expected demand. This can be achieved even early on by performing frequency assignment on a nominal network that shares common features with the expected network design.
  - Early identification of areas for which service provision will be difficult. This may include features that lie on adverse terrain such as deep, twisting valleys and coastal regions that lie below convex hills, for example. Since the design challenges for this type of situation are more pronounced than for other parts of the network, early identification allows the allocation of sufficient resources and time to resolve the issue within the project plan and budget.
  - Identification of areas of potential interference and also the scale of interference likely to be encountered in the network design. Again, this allows mitigation methods to be established before the problem actually arises in the full-scale radio network design.
- The ability to perform scenario modelling allows the performance of the network and its evolution to be analysed in depth before critical design decisions are made. This allows a design to be created that not only meets the current perceived need but also builds in resilience to expected changes, such as increased customer take up at a later stage or a change to the service offered, and also unexpected changes, such as partial network failure. The requirement for this type of analysis will depend on the type of network and

its use; it may be highly important in networks designed for emergency services use, or it might be too much of a luxury for commercial networks.

The particular benefits provided by the planning tool will vary by project complexity and may also vary through its life; once a tool has been purchased and paid for by one project, its use on other projects are at a lower cost. In any case, the system must provide appropriate functionality for the task at hand. In the next few sections, we look at the important features of network design tools.

## 7.4 Radio Network Design Tool Fundamentals

Planning tools are fundamentally based on the prediction of radiowaves. They work by bringing together three distinct elements, as shown in Figure 7.1. These are the modelling of the environment in which the radio network is to operate, the implementation of propagation models such as those described in chapter four and the modelling of radio network elements for both the transmit and receive paths. All these elements are necessary to allow accurate prediction of radio propagation.
It is also important to recognise that planning tools effectively have a hierarchical nature; some functions depend on the presence of other functionality and data in order to function themselves. This is illustrated in Figure 7.2, which shows an example hierarchical structure. The shaded boxes indicate data sources and the white boxes indicate functions. Continuous links indicate reliance from one function to another, while dashed lines indicate requirement for data.

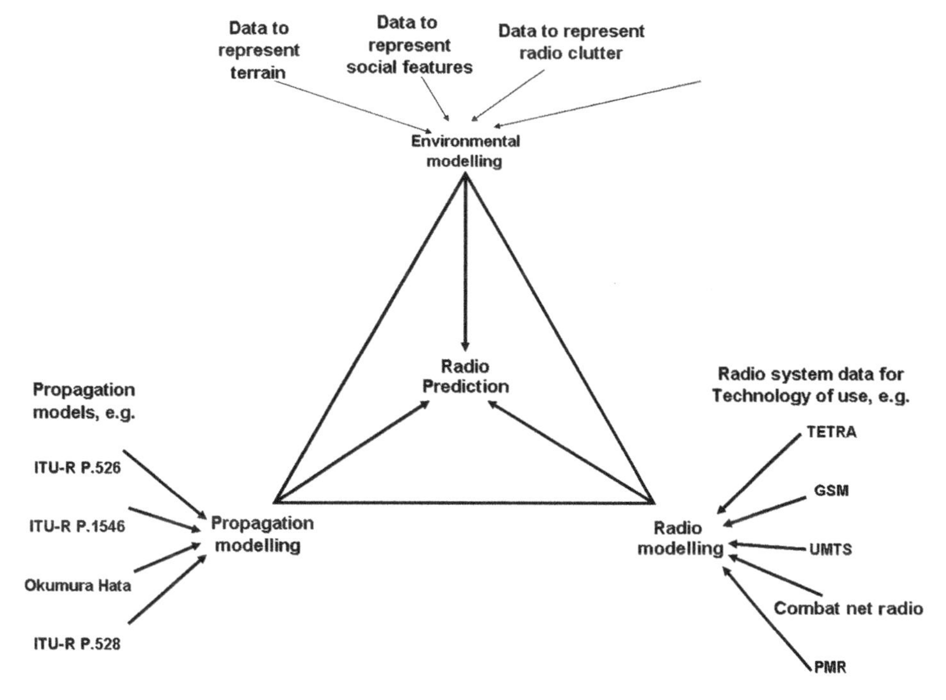


**Figure 7.1.** Essential elements of a radio planning tool.

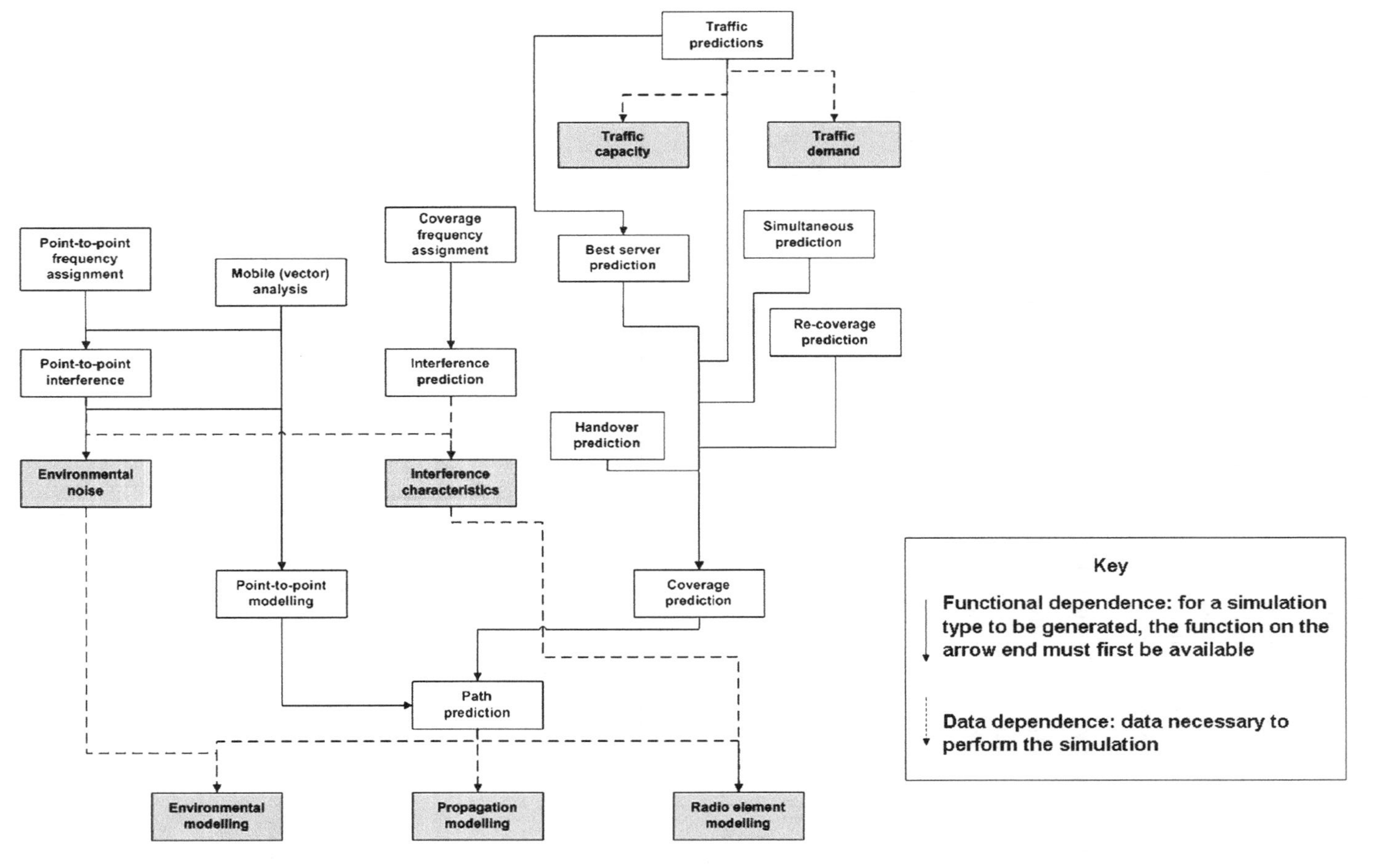


**Figure 7.2.** Hierarchical nature of planning tools. The concept of this diagram is that for a simulation such as traffic predictions to be present, the best server function must be supported. For best server to be present, coverage predictions must be present, and coverage predictions depend on the ability to perform path prediction.

The fundamental function is the ability to perform path predictions. Coverage predictions are based on the ability to perform multiple path predictions, and handover, best server; simultaneous and re-coverage predictions can be calculated from coverage predictions. Traffic predictions require (in this example – it might be different in some cases) best server predictions, plus data for traffic demand and capacity for each base station. Interference predictions for either point-to-point or coverage analysis require interference characteristics data. The exact requirements will depend on the interference prediction being performed. Frequency assignment will require interference predictions in order to be able to function, and so on. As we will see later in the chapter, this has an effect on the data required to perform specific functions, which are also described later.

## 7.5 Geographic Information System (GIS) Functionality

The term Geographic Information System (GIS) has come to encapsulate software systems capable of handling data that represent maps and displays them to an operator. Early systems displayed digitised versions of paper maps, but modern systems are capable of showing map images, terrain databases, social feature maps, such as population density, postcode sectors, towns, roads and even the outlines of individual buildings. Advanced systems offer the ability to perform many geographically based functions, such as identifying how many people live in a certain area or finding the quickest direction from one location to another, for example. Many examples of GIS systems are available on the Internet and embedded in many applications.

Most modern radio planning tools provide some type of GIS functionality, and some provide advanced features. The degree of GIS functionality required for a specific project will vary, and in some systems without a Graphical User Interface (GUI), there may be no GIS functionality at all. For most radio planning tasks, such as the positioning of radio sites, determination of the predicted coverage and population covered, a system will generally need a GIS system. For this purpose, we can identify some core functionality that will be required in almost all situations, and some other features that are highly desirable. We can sub-divide these features into the following categories:

- Import and handling of available environmental data.
- Geographic projection and re-projection of environmental data to match the projection of potential site lists or other important data for the project.
- Ability to view different data layers (such as map images, terrain and clutter models), and to manipulate those views in 2D and 3D.
- Mensuration (the ability to measure distances and bearings).
- Statistical and deterministic reporting of results.
- User-defined data manipulation.

We will look at each of these in more detail next.

### *7.5.1. Import and Handling of Available Environmental Data*

Despite some effort throughout the industry, there are as yet no universally accepted standards used by all GIS systems. This is unfortunate since if it is not possible to import environmental data for the area of the project due to wrong format, then the data is unusable.

If it is not possible to determine exactly what format the data to be used in the project will be, then it is essential to select a system capable of directly importing as wide a variety of data formats as possible. Sometimes, the manufacturer of the radio planning tool will also provide data sourced from third-party companies, but converted into the format of the radio planning tool. On other occasions the tool itself or a companion product, will feature an environmental data import facility. Finally, environmental data suppliers may supply the data in a variety of formats. If none of these conditions are met, then there is no alternative but to develop a bespoke data import facility. This is likely to be expensive and may be difficult to achieve in practice; some formats are subject to licensing and others are poorly documented.

When selecting a radio-planning tool, consideration needs to be given as to what type of data will be required and how the system will be able to import it. In some cases, for example where the environmental data will always be supplied by the radio tool manufacturer or by organisations that offer that format as standard, the issue is of little importance; in others, it will make a huge difference since it is entirely possible that it will take more time and resources to get the data correctly into the tool than is required for the network design task itself. Consideration should be given to:

What type of data will be required? This may include some or all of the following (if some of the terms are unclear, further information can be found towards the end of this chapter):

- Digital terrain or elevation model.
- Radio ground clutter map.
- Map images.
- Satellite or aircraft photography.
- Site schematics.
- Rainfall maps (required for backhaul planning above about 8 GHz).
- Ducting probability maps.
- Maps of roads, train lines, airports and other important social features.
- Population models.
- Postcode/zip code maps.
- Geo-coded business directories.
- Prediction result overlays (signal received, best server coverage, interference etc).

It may also be beneficial to consider the following points:

- The system must be capable of handling some values concurrently, such as the terrain altitude value and clutter code value. This is required for the propagation prediction process. Thus, the ability of the system to cope with multiple data concurrently for each point is also important.
- What resolution and precision of data will be required (see Section 7.6)?
- Who are the likely suppliers for this data? What data can they provide and in what formats will they provide the data in?
- What will be the import process? How complex and time-consuming will it be and what level of skill is required for the import process? Can it be left to an office junior or does it require a highly skilled operator?

Asking these questions before selecting a tool is likely to prevent problems afterwards.

### *7.5.2 Projection and Re-projection of Environmental Data*

Spatially organised data will normally be referenced via a geographic coordinate system that provides unique values for unique locations. This is of course true of geographic data, and most people will be familiar with some form of geographic coordinate system. Examples include:

- Latitude and longitude in degrees, minutes and seconds, for example 50 degrees, 35 min 12 s North, 3 degrees, 45 min and 5 s West. This is the most common format used for aeronautical and maritime applications.
- Alphanumeric national format such as UK National Grid Reference (NGR), for example TQ 123 456.
- Numeric national format such as the UK NGR numeric format, such as 512 300, 145 600.
- Military Grid Reference System (MGRS), such as 29SNC018630, for example.
- Universal Transverse Mercator (UTM), such as RHMF4658.

It is important to recognise that there are a wide variety of formats, and sometimes conversion between these formats will be required. Data conversion between geographic reference systems is typically complex, relying as it does on the description of the geographic reference system and the underlying datum (an example of which is World Geodetic System 1984 [often abbreviated to WGS-84] the default used by the GPS system). Ideally, the selected tool should include the mathematics necessary to automatically convert between specified projection systems and should also support sufficient projection systems to be able to describe any location that may arise. As a minimum, we would suggest that any planning tool to be used should be capable of supporting the following projection systems, and to be able to automatically convert between any of them:

- The national grid system of the country in which the project is based. Since it is usually fairly straightforward to convert between the alphanumeric and numeric formats, it is not strictly necessary that both are provided.
- UTM (North and South).
- Degrees-minutes-seconds with WGS-84 datum.
- Degrees-decimal degrees with WGS-84 datum.
- For military systems, MGRS should be supported.

In our experience, these are the projections most commonly encountered.

### *7.5.3 Data Views*

The radio-planning tool must also provide the operator with effective views of environmental data, as plan (2D view), a synthetic 3D view, or both. The system should allow the operator to explore the project area effectively, allowing important 'salients' (features of note) that will affect radio propagation and social features to be viewed in sufficient

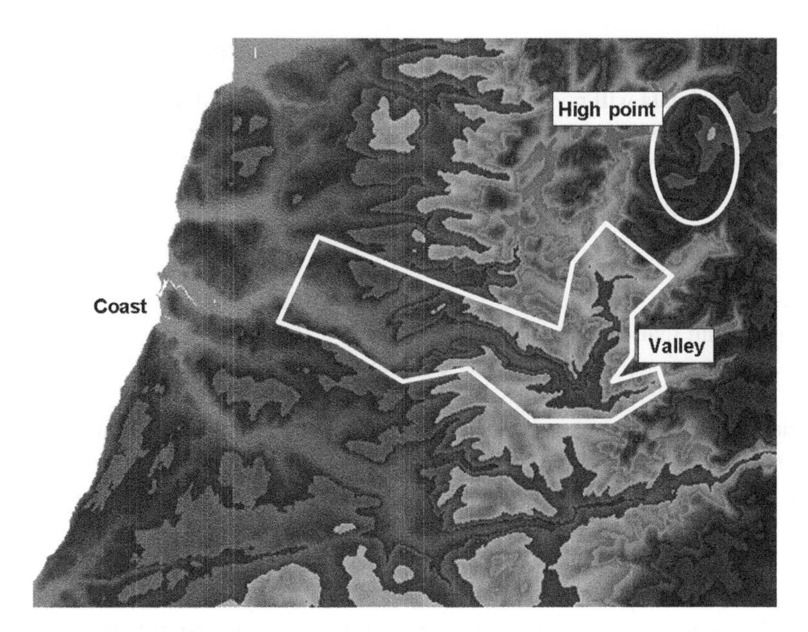


**Figure 7.3.** Example of 2D DEM view, in this case the DEM has been processed from terrain data captured during the Shuttle Radar Topographic Mission (SRTM) carried out by NASA using the space shuttle. Reproduced by permission of P & D Missud.

detail to allow the operator to make important decisions such as antenna positioning. Types of detail that the system should allow the operator to distinguish should include:

- Terrain details, such as highpoints, valleys, dips and coastal regions. Figure 7.3 illustrates some of these points. The different shades show different altitude bands and the flat light area to the left is the sea.
- Radio clutter features such as open areas, agriculture, bodies of water, forestry (ranging from light, mixed woodland up to dense jungle). An example is shown in Figure 7.4. A conversion code relates the shade to the meaning of the clutter, which is normally used in the radio prediction algorithm.
- Social features such as towns, villages, roads, railways, airports, ports and so on. The exact features required will depend on the needs of the project. Figure 7.5 shows an illustration of a typical map image, showing villages, hills and altitude contours, rivers, forestry and lakes. The squares are used for scale, and are a kilometre on a side. The ability to see these features allows the operator to relate coverage predictions to the towns, roads and other areas that require coverage.
- An example of a 3D display is shown in Figure 7.6, which shows a pseudo-3D view of the clutter map overlaid on the DTM.
- Such images help the operator understand the operational environment in greater depth than simple 2D images can.

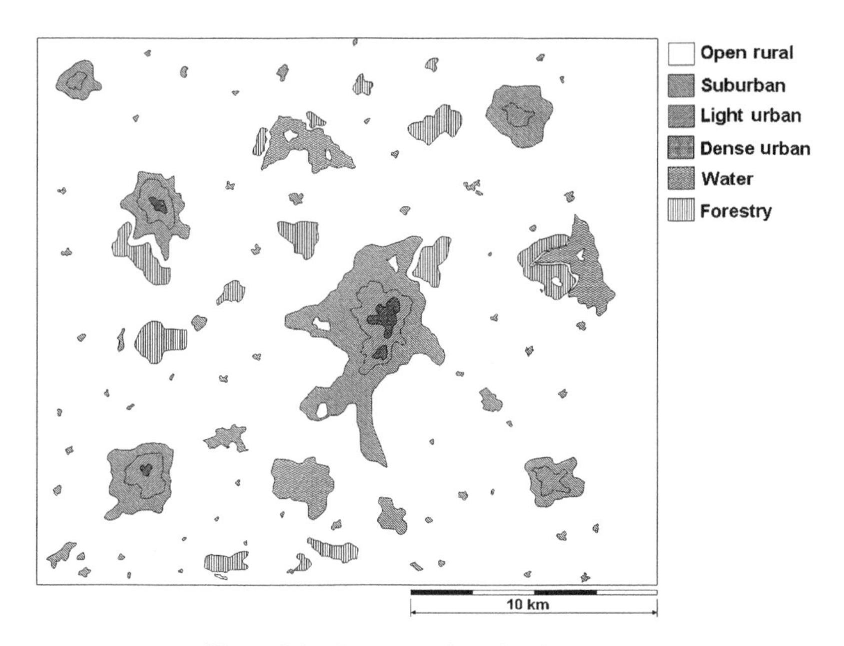


**Figure 7.4.** Representation of a clutter map.

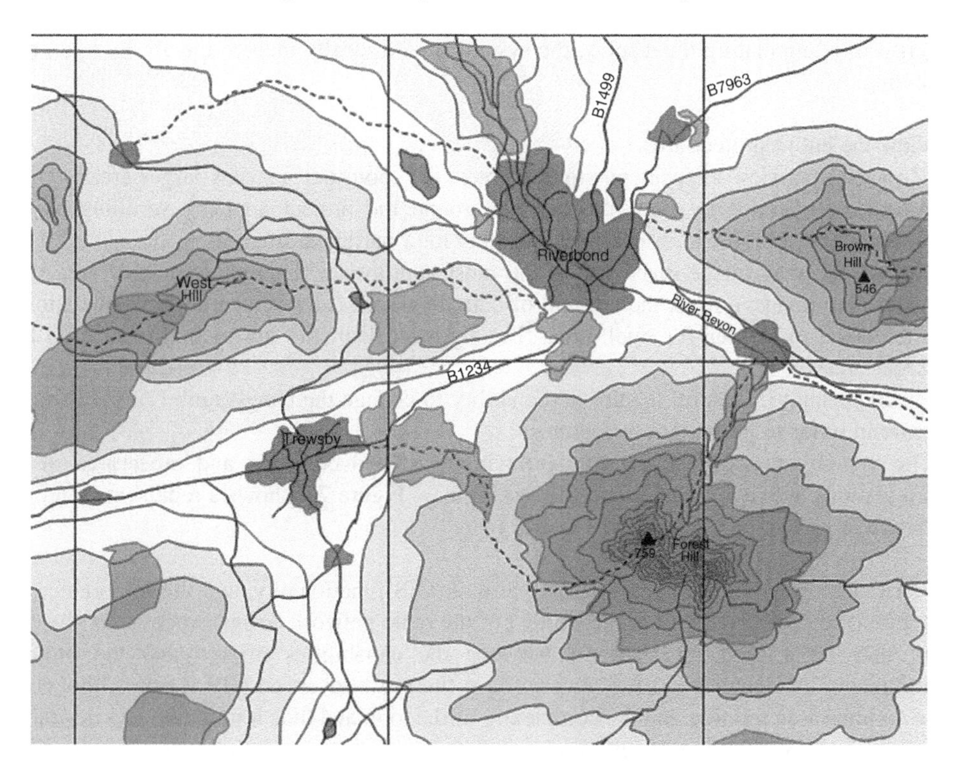


**Figure 7.5.** Illustration of a map image.

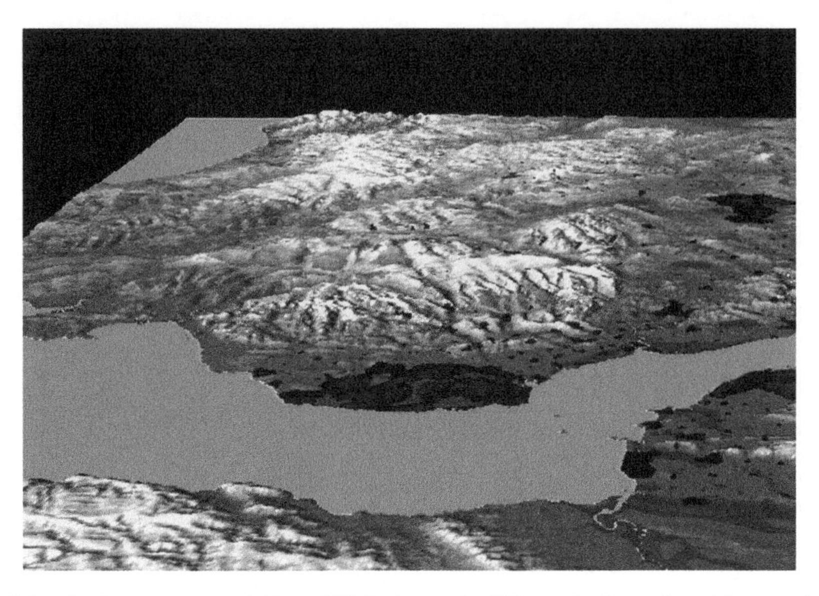

**Figure 7.6.** A clutter map overlaid on DTM, shown in '3D' mode. Reproduced by permission of P & D Missud.

In terms of manipulating the display, the system should ideally include the ability to do the following:

- View the entire project area.
- 'Zoom in' the view to focus on smaller areas, or 'zoom out' to view larger areas.
- Allow the operator to move the display around the project area, by scrolling or by allowing the operator to re-centre the display on a particular area.
- Change the map image displayed to the most appropriate for a given display area (for raster map images) or set the amount of data displayed (for vector images). This allows great detail to be seen for small areas, but a greater degree of abstraction to be used for larger areas.
- For 3D displays, a useful feature is the ability to change the exaggeration of the vertical scale in order to probe terrain features.
- The system must be capable of overlaying predicted coverage and other area-based simulations over map images and the terrain map. Figure 7.7 shows a representation of a coverage prediction overlaid on a social map.

Modern radio planning tools vary widely in their GIS functionality, and the importance of this should be considered within the context of the radio network design process. As always, there may be a need for trade-off between the capabilities offered and the project requirements. Although it is often nice to have the most advanced GIS features, this does have an impact on training and the complexity of the tool, and thus it might be less desirable in practice. GIS handling is often a major part of a project, and therefore the GIS functionality offered by candidate systems have to be carefully considered as part of the process of selecting a tool.

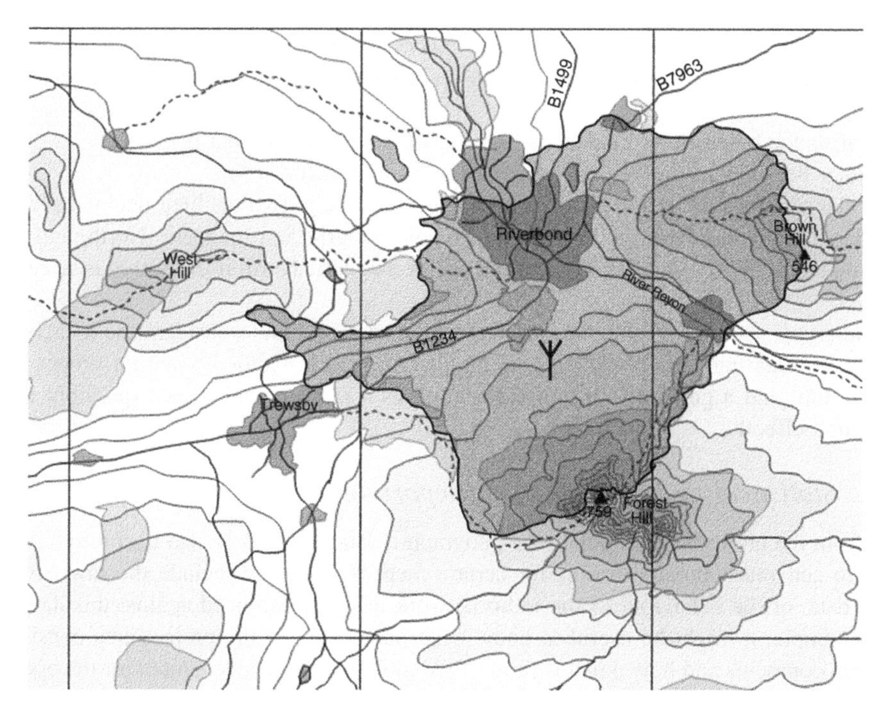


**Figure 7.7.** Coverage overlaid on a map image.

### *7.5.4 Mensuration*

Mensuration is a term that refers to measurement, and in the case of GIS systems, it means the ability to measure distances and angles between two points. Standard GIS tools work on the absolute distances between two points, and may be very accurate in terms of optical angles and physical distances. However, radiowaves do not travel along the same paths as visible light (see Chapter 4), but instead along a curve dependent on atmospheric refractivity. This means that the path length experienced by a radio transmission is not the same as the physical distance as could be measured on the ground. The distance actually travelled with also vary with conditions, and for narrow antenna beams (such as those used in microwave backhaul), it might mean that on some occasions the main beam of the transmitting antenna may not be pointing directly at the receiving dish. In general, for mobile systems this tends to be less critical.

Apart from the refractivity issue, it is also important to ensure that the distance reported between two antennas is the slant range including the difference in antenna height.

When looking at measurements provided by a radio-planning tool, typical functionality that should be available includes the following:

- All distance measurements for long-range paths are based on the great circle range, not a Cartesian measurement. On the Earth, the great circle path is the shortest distance between two points. It often appears as a curve on a map for long paths, but this is a function of the map projection rather than reality.

- All distance measurements for radio paths account for k-factor.
- The distance measurements should be to a sufficient precision to allow decisions to be made.
- The distance between two antennas is the slant range, not the flat distance.
- Angular bearings in the vertical plane account for k-factor.
- Angular bearings are reported to sufficient precision (e.g. two or three decimal places).
- Horizontal angular bearings should be available for grid and magnetic North, and there should be the ability to automatically correct for magnetic North if the option is selected.

In summary, it should be established whether the proposed tool considers the k-factor for distance and bearing; this is likely to be the discrimination factor between a proper radio planning tool and a general GIS tool being used as a radio-planning tool (possibly not to much good effect).

### *7.5.5 Statistical and Deterministic Reporting*

Apart from the ability to manipulate the geographic data views, it is also useful to have the ability to generate tabulated results for certain metrics. This can include the raw environmental data, or the behaviour of the radio network design referenced against this data.

For example, it might be useful to know what percentage of the project area consists of urban environments and how much is rural. If the tool provides a mechanism for the operator to define a region using a polygon, it might be useful to know how many people live inside the polygon, and the postcode regions included. Also, if the predicted coverage can be filtered, then reports detailing the total service area, the population served and the percentage of each different clutter category served are all very useful metrics for assessing how effective a potential base station site will be.

Reporting this type of information is what standard GIS tools are all about, and sometimes it can also be appropriate to have a separate, dedicated GIS system (which may also have other functionality such as command-and-control, vehicle fleet management, network management etc embedded within it), which can accept coverage data and other results produced by the dedicated network planning tool. This allows the best of both worlds; accurate radio planning results combined with the power of advanced GIS reporting. If this is required, the ability for the two systems to communicate may be vital.

### *7.5.6 User-defined Data Manipulation*

The GIS features of the tool must also include the ability to perform tasks in the project area as well as viewing and manipulating the environmental data. This includes the ability to:

- Place radio sites by cursor and by entering geographic coordinates in a number of projection systems.
- Move existing radio sites.
- Delete existing radio sites.
- Place radio links by specifying two radio sites and identifying that a radio link exists between them.
- Delete radio links.
- Select deployed sites; view and modify technical parameters.

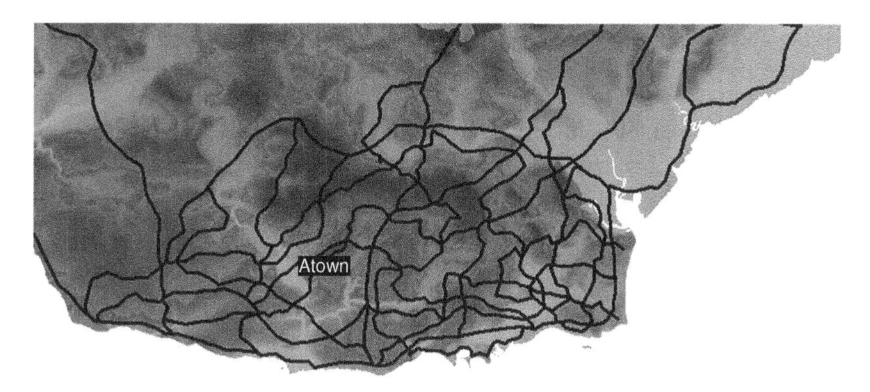


**Figure 7.8.** Vectors on DEM; these are used to show the routes of roads. Reproduced by permission of P & D Missud.

- Select and manage multiple sites, using selection by polygon or list or other criteria such as channel, network ID, site owners and so on.
- Add user-defined vector data for display over the fixed data. Figure 7.8 shows a simple example where roads that are important to the project are overlaid onto a DTM. Although map images may be available for this, it is useful to be able to show only this feature at times, and it may also be possible to restrict results to these vectors for some operations.
- Use graphical tools such as polygons to modify fixed data, such as changing terrain height, radio clutter characteristics and specifying service areas (we will see more of this in the next chapter).

In small systems, the manipulation of sites and other features will be relatively straightforward; for larger systems and larger network design projects, it will be far more complex.

## 7.6 Propagation Modelling

We discussed the various types of propagation models in Chapter 4. It goes without saying that a radio planning tool to be used for a radio network design project must have models that are suitable for the technologies and tasks required, for example based on the models already discussed or, if not, then on equally well-referenced and proven models. This is a fundamental feature and one that differentiates a GIS tool being used for radio planning and a dedicated radio-planning tool. Often, general GIS tools used as radio network planning tools will feature propagation models that are limited in scope and do not really encapsulate the full requirements of appropriate models. Typical shortcomings include:

- Not having appropriate models for each task to be performed. Different models are likely to be needed for planning, interference modelling and coordination, even for a single service. Planning tools for many technologies will need a range of models.

- Not having all of the settings or sub-components of propagation models incorporated (see for example the options in ITU-R P.452 to understand how they may not always be incorporated fully).
- Not understanding model boundary conditions (ITU-R P.526 is a good example of a model with complex boundary conditions).
- Not having appropriate treatment of radio clutter such as urbanisation, forestry etc. This again is a complex issue, particularly when considering issues such as in-building coverage and different levels of urban categories.
- Most important of all, a software implementation of a propagation model has to be validated against the original model specification, and that model has to be extensively verified against physical measurements. Any model that has been put together without this attention to detail and effort may be more troublesome than it is worth.

In order to determine whether a system truly has appropriate models for radio network design, it may be appropriate to ask the following types of questions:

- Is each model that may be used fully referenced and tested? It is possible that a new model has been developed and therefore it may not be referenced. This is fine, so long as extensive testing has been carried out – sufficiently extensive so that the model is fully proven for the application it will be used for.
- Does the system feature the breadth of models that may be required for every task to be carried out during the project? It is highly unlikely that one model will be suitable for all tasks. Particularly, as explained in Chapter 4, radio-planning models are not the same as interference and coordination models.
- What facilities exist to allow the comparison of models with measured results? Even proven models may benefit from adjustment for the specific application by taking account of measurements taken in conditions that approximate to the operational configuration of the network. A model tuning facility is therefore also an important feature.
- Will there be a requirement for different types of models for different projects? Consultants in particular will require tools that feature a wide variety of models so the widest possible types of tasks can be addressed. This will include models for different radio technologies, different frequency bands, and different tasks such as interference prediction, frequency planning, coordination and others.
- What capabilities are there for adding new propagation models if required? Is it necessary to wait until the manufacturer produces a new model, or does the tool have published standards for external models so that different software developers can implement models from other sources? In some cases, it is highly desirable to add user-specific models or at least to have the capability if required.

The bottom line is that the foundations of a radio-planning tool are built on the propagation models they incorporate. It is impossible for a good radio-planning tool to emerge from a system built on shaky propagation models in the same way that a robust building cannot be built on shaky foundations.

## 7.7 Modelling Functions of a Radio Network Design Tool

This section covers both the radio modelling and the radio prediction element of the design tool since as we will see they are closely linked. We will look at typical simulations that may be found in a planning tool and the technical information necessary for accurate simulation. We will look at tools for:

- Path-based predictions.
- Coverage-based predictions.
- Traffic analysis.
- Interference analysis.
- Frequency assignment.

We will also look at some other features that can boost productivity. We start with path-based predictions. For all coverage-based simulations, we have used synthetic images rather than ones taken directly from a planning tool. This is to allow us to bring out specific features in a manner that can be seen clearly in a monochrome media such as this book.

### *7.7.1 Path-Based Predictions*

As Figure 7.2 showed, path profiles predictions are the cornerstone of radio network design tools. Figure 7.9 shows the basic form of a path profile. The path is drawn between two known locations, and the terrain representation between the two locations (following the great circle path) is extracted from the digital representation of the terrin. Additionally, features from the clutter file on the same path are also extracted for inclusion in the radio prediction (trees and urbanisation is shown in Figure 7.9). This data is used in the selected propagation model to provide one of a number of metrics at the receive location. Although path profiles are more commonly used in fixed links, they are used in mobile network engineering to probe the cause of problem coverage areas, in order to determine the mechanisms that are causing the problem. Path profiles are also the only deterministic method of analysing mobile networks with no fixed infrastructure such as Combat Net Radio (CNR). Composites of path profiles can also be used to examine the performance of mobiles as they move through the network, as described later in this section.

There are a number of metrics that can be deduced, including:

- Path loss (dB).
- Received field strength (dBμV/m).

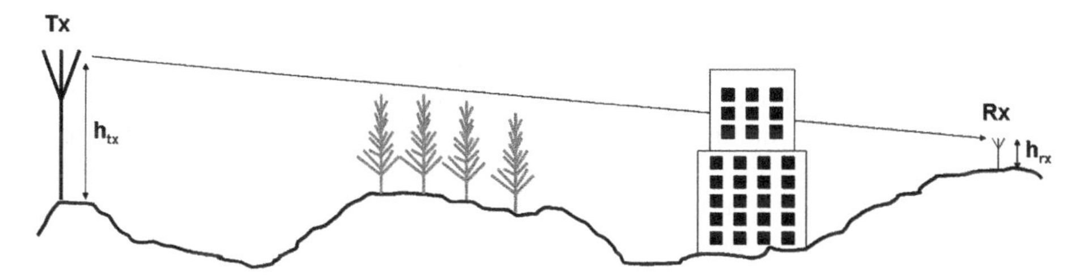


**Figure 7.9.** Path profile-based simulations.

- Equivalent power in the receiving antenna (dBm).
- Link margin (dB).
- Probability of Successful Operation (PSO) (Percentage).

The parameters required to perform these simulations are shown in Table 7.1. This is an illustration of the relationship between simulations and the technical information required, so that they can be accurately performed.

Besides being able to determine the parameters for individual links, modern tools should normally be capable of examining many paths at the same time. This can either be for the purpose of examining many links at the same time or because the aim is to examine the path between one or more base stations and a mobile subscriber moving through the network, as illustrated in Figure 7.10. In this figure, the mobile starts at one location and moves through the service area until the end of the survey. In each short sector the median field strength is recorded (see Chapter 15: this is not a simple exercise), and this can be compared to the predicted median values. As shown in Figure 7.10, this can be used to identify whether the link is acceptable or not in each measured location. The results analysis table only shows a small subset of the results generated.

When multiple paths are to be analysed, it is often convenient to summarise the link effectiveness without examining individual paths. This is illustrated in Table 7.2. In this case, the links are shown from a central team leader to individual patrols and in the reverse direction. Of course, when there are many paths, the tables will become very long. In such cases, the ability to focus on individual paths is important. The 'Warnings' column can act as a filter to allow only the problematic links to be displayed for further analysis.

Individual path analysis is appropriate for situations where the locations of both terminals are known. However, in most mobile networks the mobile subscriber will move throughout the service area and there will be far too many potential paths to examine them all in detail. Also, this degree of analysis is largely unnecessary in many cases. The principle mechanism for analysing mobile networks is based on coverage-based predictions rather than path profiles.

An additional feature in a radio-planning tool that can be useful is the ability to probe the network performance at locations specified by the operator. This can be used not only to probe for received field strength or equivalent power but also for the time of arrival of multiple base stations or for the delay spread of signal received from a single base station. The coordinates of such positions can be selected by coordinates or by cursor location. The results will depend on the ability of the propagation model to represent the effects of reflections and different paths between base stations and the receive location, and also on the environmental model used. Clearly, to determine reception from multiple paths it is necessary to be able to characterise and account for reflection points wherever they will influence the received signal. This becomes increasingly difficult for paths involving different structures and multiple reflection points. Potential methods for this include models based on specular or Lambertian reflections and are generically known as 'ray-tracing' models. These can be highly accurate but they do depend on precise, accurate environmental models, and there is always the question of how well a specific environmental model reflects reality; for example in a micro cell model for outdoor simulation, what about the effect of traffic as it moves through the area, and also the effects of people themselves.

**Table 7.1.** Parameters required for path predictions.

| Type of Prediction | Parameters Required | Results obtained |
|---|---|---|
| Path loss | Transmit antenna location Receiver antenna location Transmit antenna height (m)[1] Receive antenna height (m)[1] Frequency (MHz) | Path loss over path from transmitter and receiver (also valid for the return path). Value is expressed in dB. This can be compared to a maximum permissible path loss in a link budget to determine whether the link is workable or not. |
| Received field strength | Transmit antenna location Receiver antenna location Transmit antenna height (m)[1] Receive antenna height (m)[1] Frequency (MHz)ERP (W, dBW or dBm) Transmit antenna gain or loss in the direction of the receiver (relative to the ERP value) | Median field strength predicted for the short sector that contains the receive antenna. This is expressed in dBμV/m. This can be compared to a MMOL to determine whether the link provides an acceptable level of availability or not. |
| Equivalent power | Transmit antenna location Receiver antenna location Transmit antenna height (m)[1] Receive antenna height (m)[1] Frequency (MHz)ERP (W, dBW or dBm) Transmit antenna gain or loss in the direction of the receiver (relative to the ERP value) Receive antenna gain or loss in the direction of the transmitter (typically relative to a dipole antenna) | This is the value obtained by converting the median field strength into an equivalent antenna output power. This depends on the reference antenna and the input impedance of the antenna (normally 50Ω for mobile systems). The value is expressed in dBm. |
| Link margin | Transmit antenna location Receiver antenna location Transmit antenna height (m)[1] Receive antenna height (m)[1] Frequency (MHz)ERP (W, dBW or dBm) Transmit antenna gain or loss in the direction of the receiver[2] Receiver threshold term (see comments) (Optionally receive antenna gain or loss in the direction of the transmitter [2]) | The link margin is a value in dB above or below some reference point at the receiver. It can be the Minimum Median Operating Level (MMOL) at the antenna in dBμV/m or the MMOL at the radio input in dBm (in which case all gains and losses between the receiver antenna and the radio input must be accounted for). A positive value indicates that the link is predicted to be successful, and a negative value indicates that the link is predicted to be unsuccessful. See Chapter 5 for further information on link margins. |

*(Continued)*

**Table 7.1.** *(Continued)*

| Type of Prediction | Parameters Required | Results obtained |
|---|---|---|
| PSO | Transmit antenna location Receiver antenna location Transmit antenna height (m)(1) Receive antenna height (m)(1) Frequency (MHz) ERP (W, dBW or dBm) Transmit antenna gain or loss in the direction of the receiver(2) (Optionally receive antenna gain or loss in the direction of the transmitter(2) | The PSO is calculated by taking the link margin and using the appropriate statis tics for fast fading in the short sector. The link margin in dB is converted by the cumulative distribution function to the probability of exceeding the required threshold. Thus, PSO will be a value such as 90% or 99% for working links or less than 50% for links that have a link margin of less than 0 dB. |

Notes: (1) Antenna heights can be expressed in terms of Above Ground Level (AGL) or Above Sea Level (ASL).
(2) Typically reference a dipole antenna.

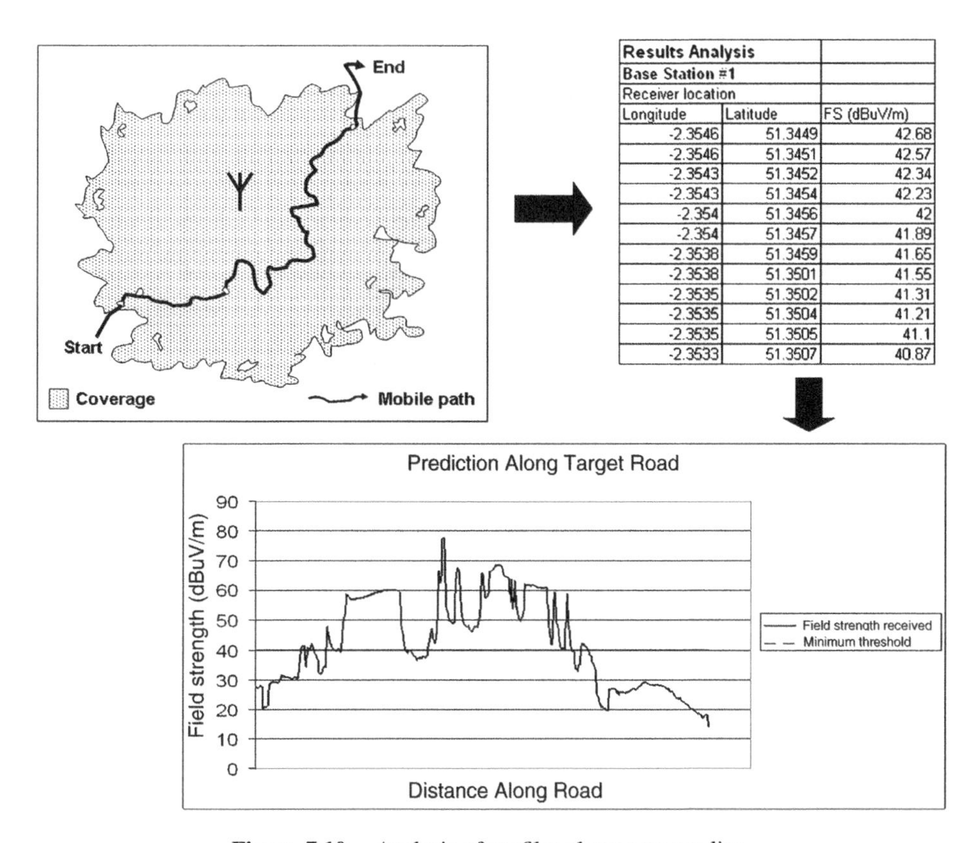


**Figure 7.10.** Analysis of profiles along a vector line.

## 7.7.2 Coverage-Based Predictions

Coverage predictions are effectively multiple path profile predictions, with the received field strength or equivalent power displayed in different colours or, as shown in Figure 7.11, as banded shading. This shows two levels of locations availability; 90% and 99% against a reference receive field strength. In general, planning tools will show levels against specific

**Table 7.2.** Multiple link analysis.

| Transmitter | Receiver | Link margin | Link effectiveness | Reverse link margin | Reverse link effectiveness | Warnings |
|---|---|---|---|---|---|---|
| Net Alpha | | | | | | |
| Team Leader | Bravo 10 | 10 dB | OK | 10 dB | OK | |
| Team Leader | Bravo 20 | 12.5 dB | GOOD | 11.5 dB | GOOD | |
| **Team Leader** | **Bravo 30** | **−3.2 dB** | **BAD** | **−3.5 dB** | **BAD** | ! |
| Net Bravo | | | | | | |
| ***Team Leader*** | ***Delta 10*** | *8.5 dB* | *MARGINAL* | *9.0 dB* | *MARGINAL* | *?* |
| ***Team Leader*** | ***Delta 20*** | *5.5 dB* | *MARGINAL* | *6.0 dB* | *MARGINAL* | *?* |
| **Team Leader** | **Delta 30** | **−0.5 dB** | **BAD** | **0 dB** | **BAD** | ! |

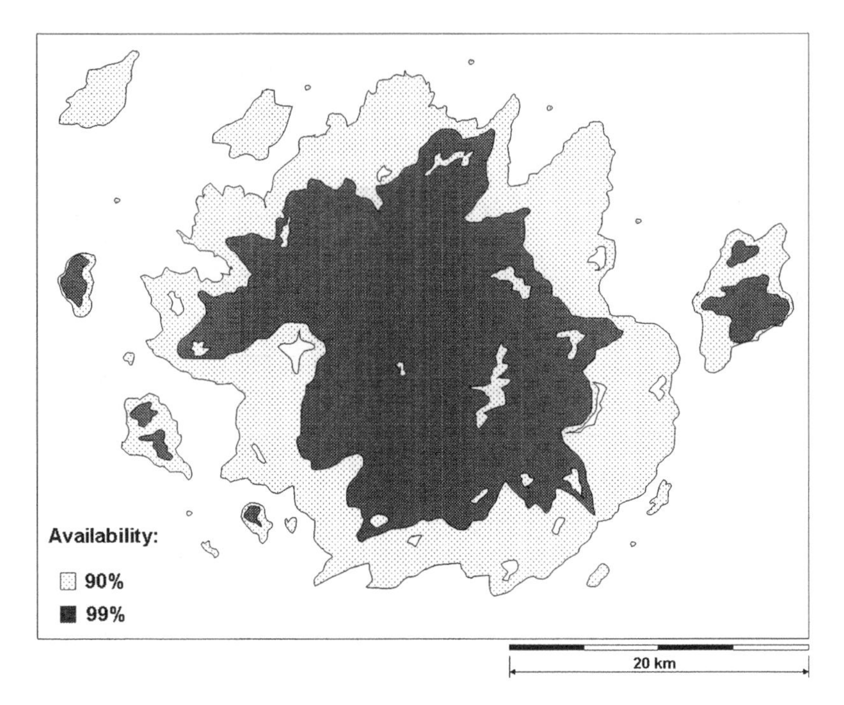


**Figure 7.11.** Coverage plot illustration (two-level display).

targets or against raw field strength in multiple banding levels. Typically, about 10–12 values are about the highest differentiation the eye can make out in different colours. It is also useful if the particular field strength underneath the cursor is displayed, thus allowing the operator to analyse the behaviour of the network in various areas. The meaning of the display must be consistent with the expected result (determined with respect to the link budget) and configured accordingly.

Coverage predictions require not only the parameters listed for the path-based predictions but also the base station antenna polar response, which identifies the amount of signal transmitted in each direction. For directional antennas, this will vary in different directions, depending on the azimuth and tilt of the main lobe, which is the direction of the strongest signal.

Coverage plots can be produced for individual base stations or for a number of them as a composite display. This is not however the extent of their usefulness; once raw coverage has been computed, it is possible to analyse the network in a number of other ways. This is illustrated in Figure 7.12, which shows an example of a 'best server' plot. A best server plot shows which base station produces the strongest signal at each point throughout the network. This can be useful for dimensioning the network for traffic and of course to identify which site provides the best coverage in particular parts of the service area. In Figure 7.12, there are a number of base stations. Callsign 1 shows an example of a base station with three separate 'sectors'. In this case, a single base station has three separate antenna systems, each providing coverage of 120 degrees. This is useful for providing more traffic capacity in areas of high demand. In the illustration, the sectors are split into zones described by a letter; A, B

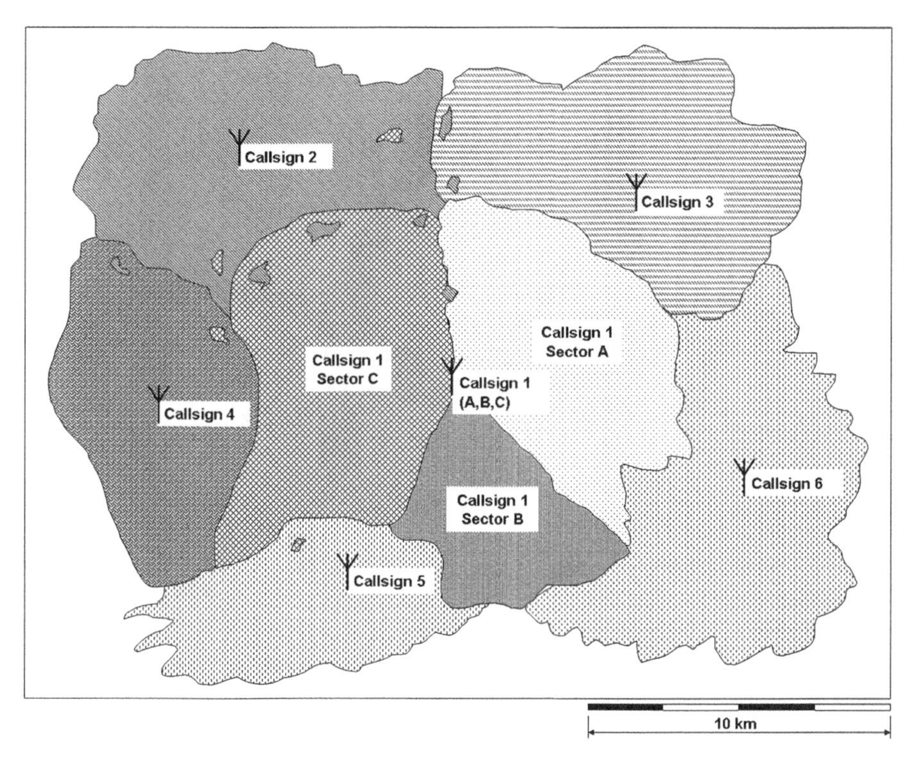


**Figure 7.12.** Best server plot.

and C. For sectored sites it is important to show each sector because they are in effect different base stations.

A number of other coverage-based predictions can be created to shed light on different aspects of the network. A 're-coverage' plot shows where a working signal is received from more than one base station. In many network designs, this can be a bad design feature since it potentially shows waste and may lead to interference. In other networks, redundant coverage can be highly desirable. This is particularly true for safety-of-life networks such as aeronautical and some emergency services networks. The re-coverage display can identify where this occurs throughout the network. In Figure 7.13, the re-coverage areas are shown in the darker shade. If this is a network requiring redundant coverage, then the design is poor. Otherwise the design may be good if coverage is only required from a single station.

A re-coverage display only shows where multiple coverage exists, but it does not show the degree of re-coverage. This can be seen via the simultaneous coverage analysis as seen in Figure 7.14. In this case, the shaded areas do not correspond with any individual site but rather with the number of sites providing a workable signal level at each location. In this example, there is a region where all six stations provide coverage. In general, this is not desirable but it will typically happen on the tops of high hills and therefore it may be acceptable (the tops of hills are not typically the most important areas of most mobile networks). For automatic direction finding systems, a degree of simultaneous coverage of three is normally required as a minimum and this is the type of scenario where this type of simulation is important.

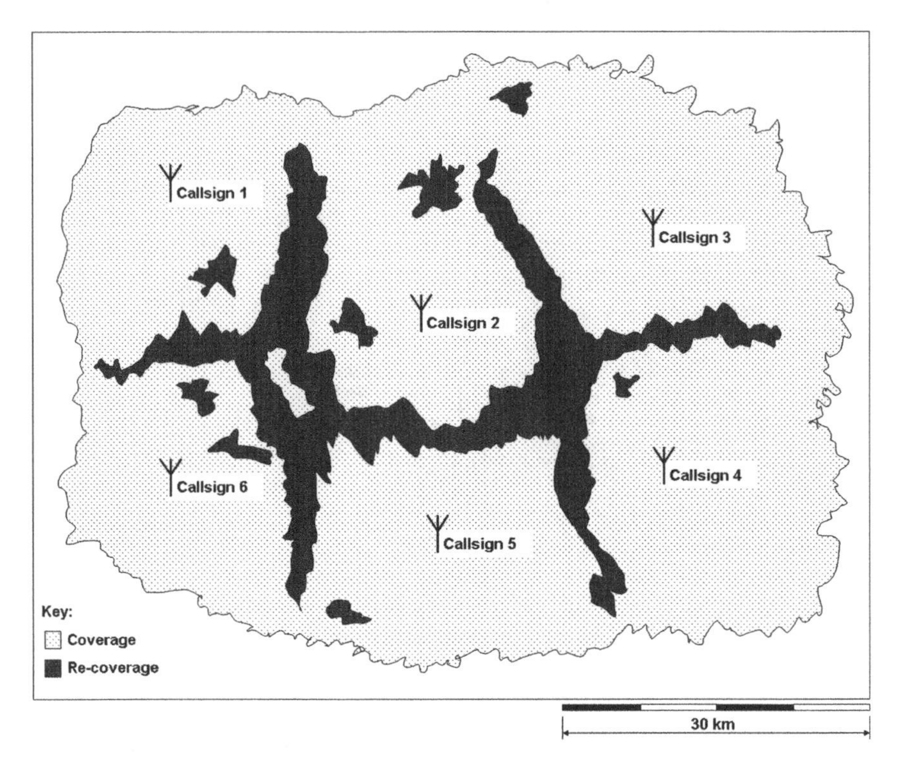


**Figure 7.13.** Re-coverage plot.

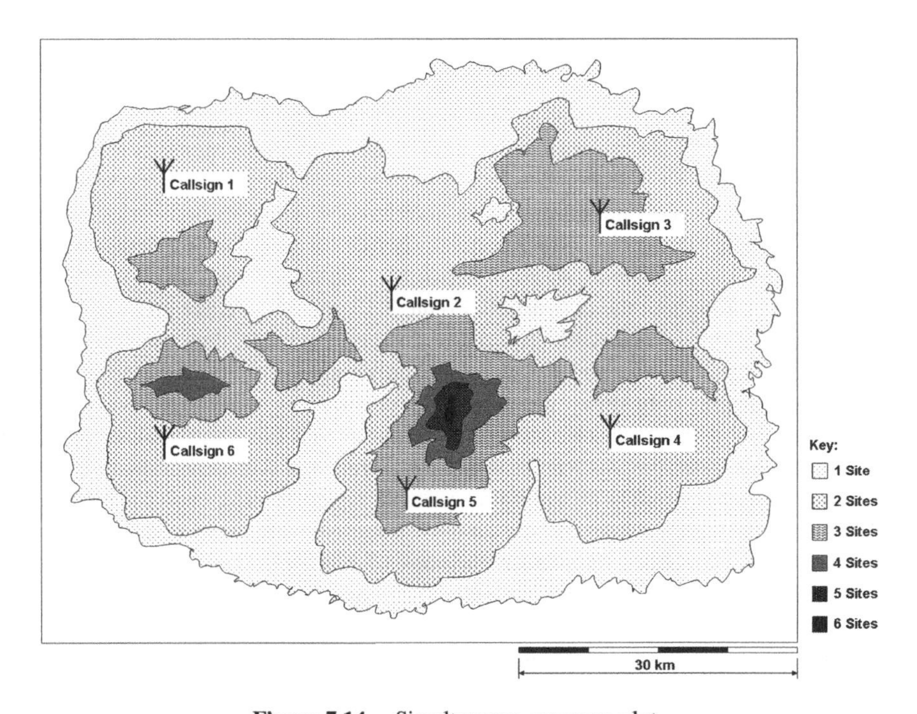


**Figure 7.14.** Simultaneous coverage plot.

Another important feature in radio network design is the regions where mobile subscribers may be passed from one base station to another. This is known as a handover region, and it is typically formed where the mobile leaves the service area of one base station and moves towards another. In order to maintain communications, the mobile may have to change to another frequency as it moves, although in some technologies there will be no frequency change but the control and traffic for the mobile subscriber will shift to another base station. The system has to recognise that the signal strength of the base station currently used is diminishing and that there are neighbouring base stations that can accommodate the mobile subscriber as it moves. This is not normally a straight comparison of field strength or equivalent power at the receiver since there is a hystereris component that means that the signal level of the neighbour is higher than the level of the base station the mobile is leaving. However, the analysis can be viewed as a coverage-type display. An example of a handover map is illustrated in Figure 7.15. The handover regions are shown as the darker areas.

There are a wide variety of ways of expressing mobile coverage in addition to those methods already discussed. For example, other potential coverage-based predictions include:

- Power sum, showing the total radio signal energy arriving from all active transmitters in the network.
- Margin display, showing the number of dB above or below an operator-defined threshold value the field strength is at a particular location.
- Availability display, showing the availability in percentage terms.

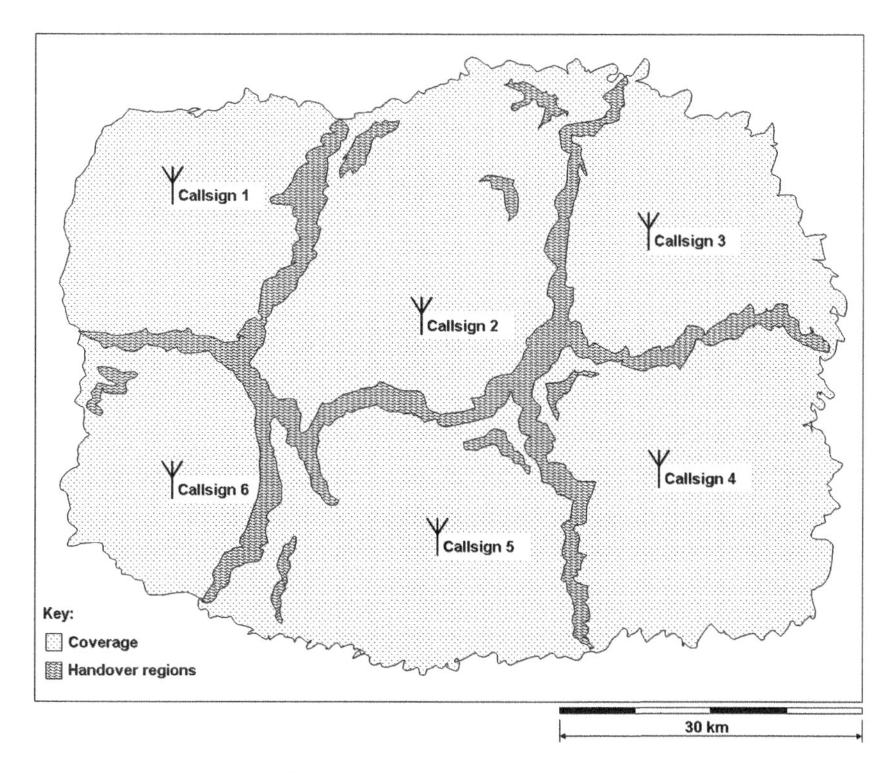


**Figure 7.15.** Handover region plot.

- Distribution display. This can show the probability of exceeding a particular threshold value at each location, given the standard deviation of the clutter environment present at that location.
- Pilot/sync channel coverage display for UMTS UTRA.

Each of the coverage-based predictions is a comparison of different field strength values for multiple base stations and they highlight the flexibility of this approach.

## *7.7.3 Traffic Predictions*

Traffic predictions can be based on coverage predictions or best server predictions. Coverage predictions are used to examine the traffic demand on single site systems, and best server predictions are used for systems in which the mobile will automatically choose the strongest signal when selecting which base station to communicate with. If the rules are more complex than this, then they have to be included in the analysis methodology so that the representation of traffic handling is realistic. We cover traffic design in Chapter 10, but in essence it is the business of trying to match traffic demand offered by the mobile subscribers in the network with the capacity of each base station to handle it. This is illustrated in Figure 7.16. The top of the diagram shows the best server coverage from an individual service, limited to a specific field strength that provides a service of the target availability. The lower part of the diagram is a completely different thing, and one we have not yet discussed. It is a map of the expected mobile subscriber density throughout the best server area, expressed as a given range of subscribers (abbreviated to 'Subs') per square kilometre. The range is relatively wide to reflect the fact that such descriptions are only estimates as described in Chapter 10. If we can estimate the density of subscribers, we can estimate the total traffic offered to each base station when we compare the area of coverage to the sum of subscribers in the covered area.

To illustrate this further consider Table 7.3. This shows the total number of subscribers estimated to be present anywhere within the service area of the base station under

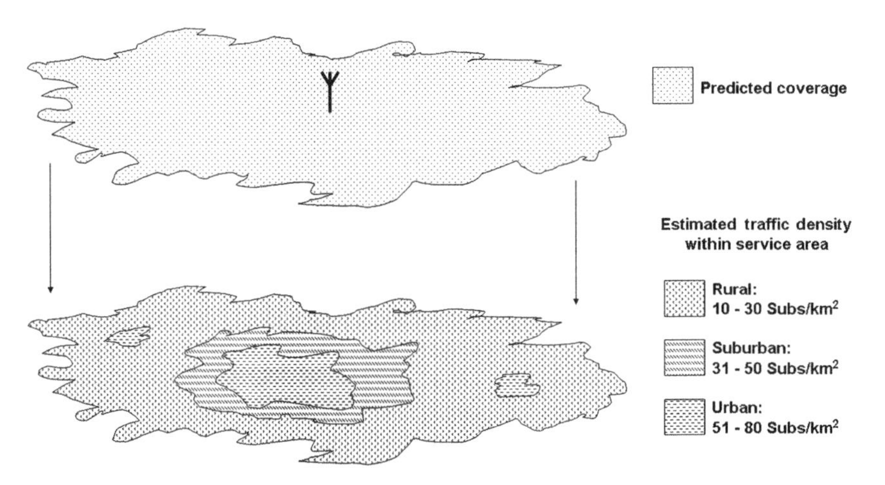


**Figure 7.16.** Traffic analysis, examining best server coverage area to estimated traffic demand.

**Table 7.3.** Traffic density estimate.

| Traffic polygon | Estimated subscriber density (subscriber per km$^2$) | Total area covered by base station (km$^2$) | Total subscribers covered |
|---|---|---|---|
| Rural | 20 | 100 | 2000 |
| Suburban | 40 | 50 | 2000 |
| Urban | 65 | 10 | 650 |
| **Total** | | | **4,650** |

consideration. By calculating the total area of each of the categories within the service area it is possible to estimate the total number of subscribers within it as shown in the table. If marketing or network usage statistics indicate that on average 1 % of subscribers may be expected to be using the network at the same time, then the base station must be able to handle a total of 46.5 (rounded up to 47) simultaneous calls. Once this figure has been derived, the information can be fed into the traffic capacity planning activity described in Chapter 10, and the decision can be made to split the service area into smaller sub-areas in which the traffic demand can be more easily managed, for example.

Of course, traffic demand will not only depend on the density of subscribers throughout the service area but will also vary according to time, as illustrated in Figure 7.17.

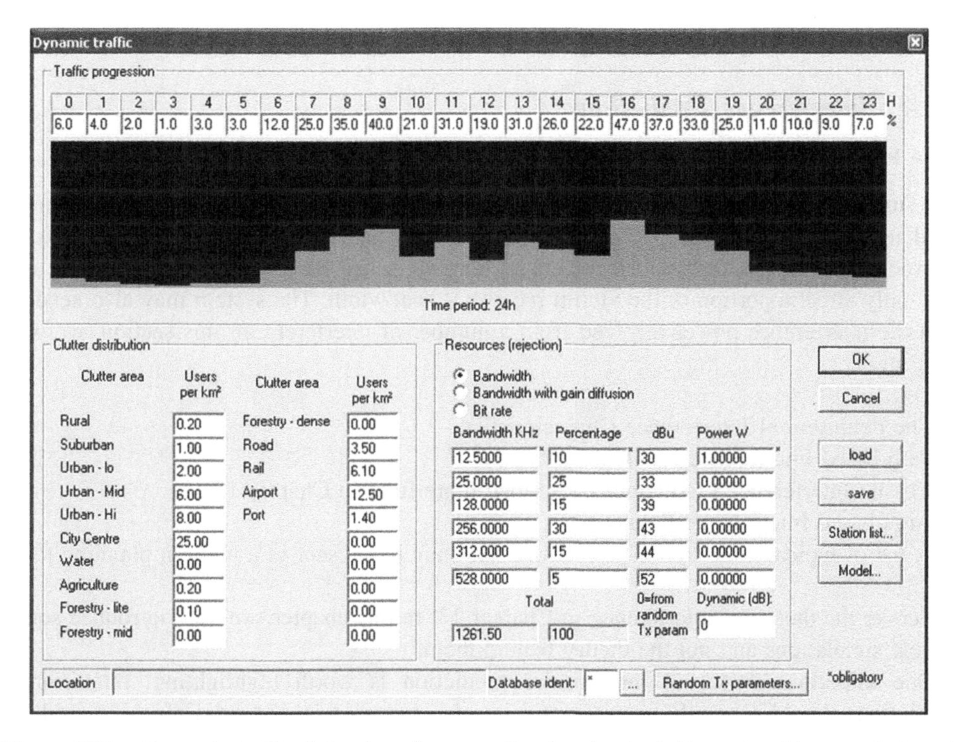


**Figure 7.17.** Dynamic traffic dialog box from a radio planning tool. Reproduced by permission of P & D Missud.

**Table 7.4.** Results of a typical traffic prediction.

| BS | Callsign | GOS (%) |
|---|---|---|
| 1 | Candidate 1 | 98.8297 |
| 2 | Candidate 2 | 99.9942 |
| 3 | Candidate 3 | 99.6982 |
| 4 | Candidate 4 | 98.5626 |
| 5 | Candidate 5 | 99.6919 |

For each expected level of traffic demand, the system must be able to compare the capacity of each serving station to handle the demand expected of it. This is described in detail in Chapter 10, but at the moment it is only necessary to identify that for traffic engineering, it is important that the planning tool is capable of expressing and using the number of calls that can be concurrently handled by each server in the network.

The result of a traffic prediction is often expressed in terms of the Grade Of Service (GOS) offered to mobile subscribers. The meaning of this is explained in Chapter 10, but Table 7.4 illustrates how the results of a typical traffic analysis may be represented. In some systems, the GOS term refers to the percentage of successful calls (as in our case) and in others, the GOS term reflects the percentage of failed calls (e.g. that method would give a figure of $1 - 98.8297 = 1.1703\,\%$ for the first line of Table 7.4). It is important to be clear on which representation is being used.

### *7.7.4 Interference Predictions*

#### 7.7.4.1 Introduction to Interference Predictions

The simulation of interference can be one of the most complex tasks that a planning tool may need to be capable of doing. The system must be capable of modelling the interactions between radio transmitters and receivers, potentially when the systems are dissimilar and may only affect a portion of the victim receiver's bandwidth. The system may also need to sum the interference power received from a number of interferers. In this section, we will look at:

- The definition of Interference Characteristics.
- Path-based Interference Predictions.
- Co-site Interference Predictions (described more fully in Chapter 13).
- Area-based Interference Predictions.
- A list of typical interference predictions that may be present in a modern planning tool.

We cover the theory of interference in Chapter 13. In this chapter, we only introduce some typical simulations and not the theory behind them.

One important aspect of interference prediction is worth highlighting; interference predictions should generally use specialist interference models, not radio planning models. This is particularly true for long-range interference predictions. For many interference calculations it will be necessary to perform the prediction of the wanted radio system using one (planning) model and to perform the prediction of interferers using another (interference) model. This is an often poorly understood point, but it is important. The planning tool must be capable of modelling in this way, and it must also be capable of considering different fading characteristics for the wanted and interfering systems (e.g. capable of modelling the performance of the faded wanted system against the unfaded interferer).

### 7.7.4.2 Definition of Interference Characteristics

Any simulations that need to model the interaction between radio systems will require mechanisms to describe the way each system relates to other networks that share the same environment. The details of this are discussed in Chapter 13, but at this point we merely identify some important characteristics. It may be useful to return to this section once you have read and understood Chapter 13. So in addition to the parameters already described, the system may include one, some or even all of the following characteristics:

- Transmitted spectrum occupancy. This may be a relatively simple description such as carrier frequency (or some other central point) plus bandwidth, or it may be more complex, for example describing the Power Spectral Density (PSD) over a sizeable percentage of the reference frequency.
- Receiver interference rejection characteristics. This may be a list of the Carrier-to-Interference (C/I) values in dB for co-channel, adjacent-channel and other channel offsets (+/−) for interferers of the same type as the victim system, or it may be dynamically calculated Interference Rejection Factor (IRF) values derived by the convolution of the interfering PSD with the victim IRF, which gives the Net Filter Discrimination (NFD). The particular form of wanted signal against unwanted signals and noise will depend on the particular system to be modelled. An example of both is shown in Figure 7.18.
- Spurious emissions that are emitted along with the wanted spectral components. This information is particularly important for co-site analysis and frequency assignment at a location.
- Receiver characteristics relevant to frequency assignment. This includes such characteristics as Intermediate Frequency (IF) and other design-related issues that will prevent assignment on particular channels.
- Time occupancy. For Time Division Multiple Access (TDMA) systems, the particular slots used by individual cells will determine whether interference will occur or not. It will also be necessary to be able to specify the maximum time delay in the RF path, because this will constrain the maximum range of an individual cell. For Frequency Hopping (FH) systems, the hopping sequence for each part of the network will have to be described to allow the degree of co-existence possible to be determined.
- Activity ratio and call statistics. For systems where interference must be determined statistically, or interference must be analysed under certain considerations (for example, an emergency services system in a disaster scenario), it may be necessary to examine the behaviour of users and be able to express this within the system.
- For Code Division Multiple Access (CDMA) systems, it will be necessary to determine the interaction of the codes used to differentiate between individual channels. In soft-limiting systems, a number of other characteristics will also be necessary.

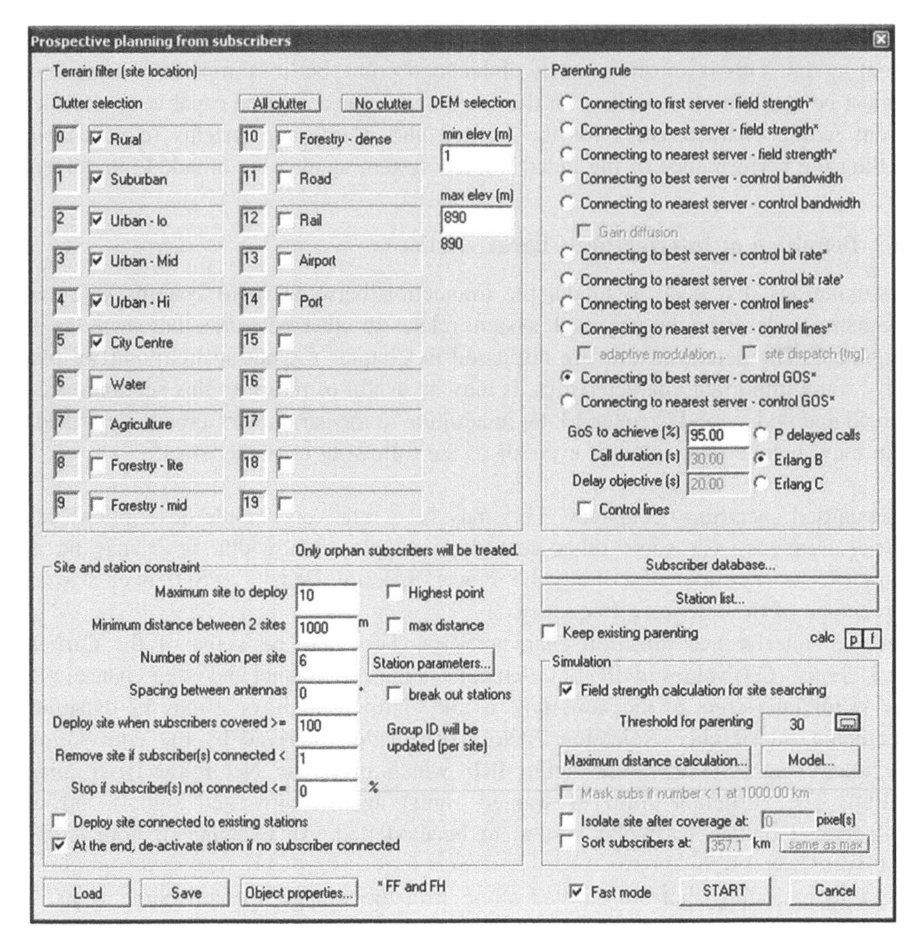


**Figure 7.18.** Automatic planning based on random subscriber generation. Reproduced by permission of P & D Missud.

### 7.7.4.3 Path-based Interference Predictions

It can be important to understand interference caused at the base station by other elements. This may be in terms of degradation of the system threshold due to interference (which may be co-sited or at a distance) or local noise. Table 7.5 shows a simple example where the power received from a number of fixed (i.e. not mobile) interferers is considered for each of the stations, together with the effect of reducing the receiver threshold at that station. Each potential victim is considered in turn and the other stations in the network that cause an interference problem are listed. In each case, the interfering power is shown in dBm, and the effective degradation in the receiver threshold is also shown. The threshold degradation shows how the receiver threshold is increased due to the presence of interference compared to the noise floor without interference (see Chapter 13). The level of threshold degradation experienced depends not only on the received interference power, but also the ability of the victim to reject interference from adjacent or offset channels.

**Table 7.5.** Typical results of point-to-point interference simulation.

| Victim | Frequency (MHz) | Bandwidth (kHz) | Interferer | Frequency (MHz) | Bandwidth (kHz) | Interferer (Power dBm) | Threshold Degradation (dB) |
|---|---|---|---|---|---|---|---|
| Callsign 1 | 150.075 | 25 | Callsign 2 | 150.1 | 25 | −105.28 | 0.6 |
| Callsign 1 | 150.075 | 25 | Callsign 3 | 150.05 | 25 | −81.89 | 15.24 |
| Callsign 1 | 150.075 | 25 | Callsign 4 | 150 | 25 | −105.9 | 0.53 |
| Callsign 1 | 150.075 | 25 | Callsign 5 | 150 | 25 | −129.79 | 0 |
| Callsign 2 | 150.1 | 25 | Callsign 1 | 150.075 | 25 | −105.28 | 0.6 |
| Callsign 2 | 150.1 | 25 | Callsign 3 | 150.05 | 25 | −84.21 | 13.01 |
| Callsign 2 | 150.1 | 25 | Callsign 4 | 150 | 25 | −120.9 | 0.02 |
| Callsign 2 | 150.1 | 25 | Callsign 5 | 150 | 25 | −133.67 | 0 |
| Callsign 3 | 150.05 | 25 | Callsign 1 | 150.075 | 25 | −81.89 | 15.24 |
| Callsign 3 | 150.05 | 25 | Callsign 2 | 150.1 | 25 | −84.22 | 13.01 |
| Callsign 3 | 150.05 | 25 | Callsign 4 | 150 | 25 | −79.81 | 17.27 |
| Callsign 3 | 150.05 | 25 | Callsign 5 | 150 | 25 | −89.18 | 8.48 |
| Callsign 4 | 150 | 25 | Callsign 1 | 150.075 | 25 | −105.8 | 0.54 |
| Callsign 4 | 150 | 25 | Callsign 2 | 150.1 | 25 | −120.91 | 0.02 |
| Callsign 4 | 150 | 25 | Callsign 3 | 150.05 | 25 | −79.81 | 17.27 |
| Callsign 4 | 150 | 25 | Callsign 5 | 150 | 25 | −86.12 | 11.22 |
| Callsign 5 | 150 | 25 | Callsign 1 | 150.075 | 25 | −129.8 | 0 |
| Callsign 5 | 150 | 25 | Callsign 2 | 150.1 | 25 | −133.67 | 0 |
| Callsign 5 | 150 | 25 | Callsign 3 | 150.05 | 25 | −89.18 | 8.48 |
| Callsign 5 | 150 | 25 | Callsign 4 | 150 | 25 | −86.12 | 11.22 |

Path-based interference predictions will typically result in a tabular output as shown. To analyse interference caused to mobile elements, a graphical or statistical display is more typical.

#### 7.7.4.4 Co-site Interference Analysis

Another special case of interference prediction occurs when several link terminals are located in the same place or within a short range of one another. This co-site condition applies further constraints due to the potential for high level transmission energy to be in the same place as low level energy used for radio reception. In such cases, the transmitter and receiver filters may be insufficient to prevent interference occurring, and there may well be other nonlinear effects such as inter-modulation. Co-site interference can be computationally difficult and requires a lot of information about the potential victims and interferers. This is described further in Chapter 14.

***Area-based Interference Predictions***

For modelling interference to mobile elements, a coverage-type display can be appropriate. An example of this is shown in Figure 7.19. In this example, areas where interference is predicted are highlighted in a dark shade. This particular diagram also shows the coverage of the network to allow comparison of the interference areas to the total coverage area, but this is an option. Interference can be shown on its own without any backdrop, over a map image or against a best server display, for example.

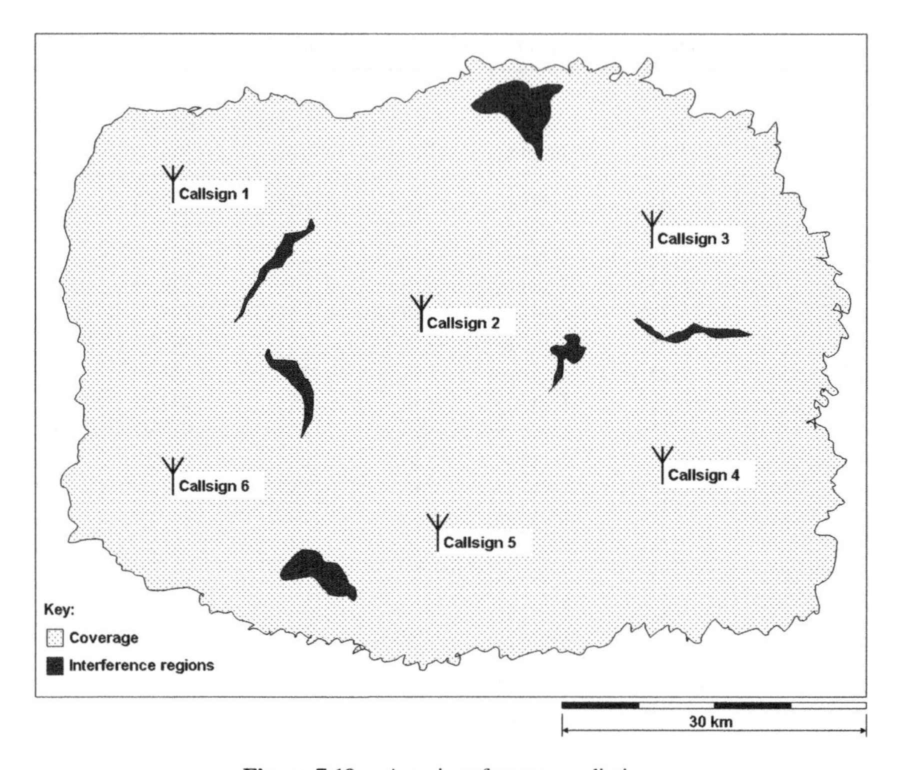


**Figure 7.19.** Area interference prediction.

It is also useful to be able to summarise the graphical information in a tabular form, as illustrated in Table 7.6, which shows the percentage of each service area (defined as where each base station is the best server) that is victim to interference from other elements in the network.

Interference is discussed in greater depth in Chapter 13, which examines different interference methods and the results that are produced.

### 7.7.4.5 A Typical List of Interference Prediction Capability

The tools offered by a candidate planning tool should be compared with the requirements dictated by the design methods for the particular technology used in a given network design.

**Table 7.6.** Results of an area-based interference prediction.

| Station | Frequency (MHz) | Interference(%) |
|---|---|---|
| Callsign 1 | 150.0750 | 3.50 |
| Callsign 2 | 150.1000 | 7.11 |
| Callsign 3 | 150.0500 | 8.07 |
| Callsign 4 | 150.0000 | 3.58 |
| Callsign 5 | 150.0000 | 3.87 |
| Callsign 6 | 150.0500 | 5.12 |

This yet again implies that before selecting a radio planning tool to assist in the design project, it is necessary that the engineers involved understand all the principles of design for each technology to be handled; radio planning tools do not replace engineers, they enhance their capability, but the fundamental knowledge must reside in the engineer, not the tool, unless the planning tool is an advanced expert system, where the engineering expertise has gone into the design of the tool rather than at the point of use.

Some typical types of interference predictions that might be present in a planning tool include:

- One-to-one interference (one victim and one interferer). The tool should allow any base station to be identified either as victim or interferer.
  - Interferer interference zone, which shows the areas of the victim's coverage interfered by the interferer.
  - C/I map (Coverage to Interference), which shows the relative level of signals between the victim and interfering system.
  - Interferer level in the victim service area.
- One-to-many or many-to-many interference (one-to-many is a subset of many-to-many)
  - Ability to compute power sum or Simplified Multiplication Method (SMM) nuisance fields.
  - Interference + best server, which shows interference from all other stations to the set of victim stations.
  - C/I map, showing the worst case situation of C/I for each location in best server areas.
  - Protection margin, showing the degree of protection to interferers at each location in a service area.
  - Protected field strength, which shows the field strength available to mobiles within the noninterfered service areas.
  - Quasi-synchronisation areas for single frequency networks.
- Point-to-point interference.
  - Threshold impairment at base station due to energy received from other base stations owing to increased noise.
  - Interference predictions for backhaul network (discussed in Chapter 13).

The interference predictions required will depend on the technology, and the interference functionality of candidate radio planning tools should be compared to determine which meets the interference needs best.

### *7.7.5 Frequency Assignment*

Frequency assignment capabilities provided by a planning tool can be fairly rudimentary or they can be highly advanced. For more advanced assignment functionality, the system may need to have the following information for the necessary calculations:

- For every station to be assigned:
  - Location (latitude, longitude or X, Y and altitude).
  - Transmit and receive link budget figures (less the frequency, which will be assigned).

  - A list of frequencies, bands of frequencies or minimum and maximum frequencies that can be assigned.
  - Assignment rules for the technology to be assigned, such as the number of channels required, duplex spacing for fixed duplex assignments or minimum duplex spacing for variable duplex assignments and so on.
  - Interference characteristics to and from the other stations to be included in the assignment process. Note that this should include all other stations of all types that are present in the assignment spectrum, not just those to be assigned. These stations may belong to other operators and may provide different services including noncommunications applications such as radars and TV broadcast, for example.
  - For stations listed as being co-site, this may include co-site characteristics (see Chapter 14).
- A switch to indicate whether the assignment should be based on fixed paths or based on coverage predictions. Fixed path assignments are suitable when the interferers and stations to be assigned have known locations, whereas coverage should be used for the situation where the mobile stations will move throughout the service area.
- The interference methodology to be used (see Chapter 13) to determine interference levels.
- The assignment methodology to be used. There are a number of methods of which some of the more common are:
  - Greedy sequential methods
  - Area sterilisation methods
  - Monte Carlo
  - Evolutionary algorithms
  - Simulated annealing
  - Taboo search

The description of the characteristics of these methods is beyond the scope of this book, but there are mathematical books available that describe them.

Frequency assignment for mobile networks is the subject of much academic research although many proposed methods have little real-life proof of their applicability beyond the academic principles. It is therefore appropriate to determine for a candidate the planning tool, what methods are used and how well-proven they are.

The outputs of a frequency assignment will be a list of assignments against the assigned stations and may also feature an interference analysis report to identify any potential problems with the solution offered.

Frequency assignment is covered in more depth in Chapter 14.

### 7.7.6 *Modelling Radios*

A single radio may be capable of supporting a number of modes and settings. If this is the case, then the technical parameters may vary according to mode. This is illustrated in Figure 7.20 for a hypothetical radio that supports analogue FM and digital transmission via FSK for voice and data, both of which can be in clear or encrypted. Each configuration must be properly selected with the relevant mode of the radio being simulated.

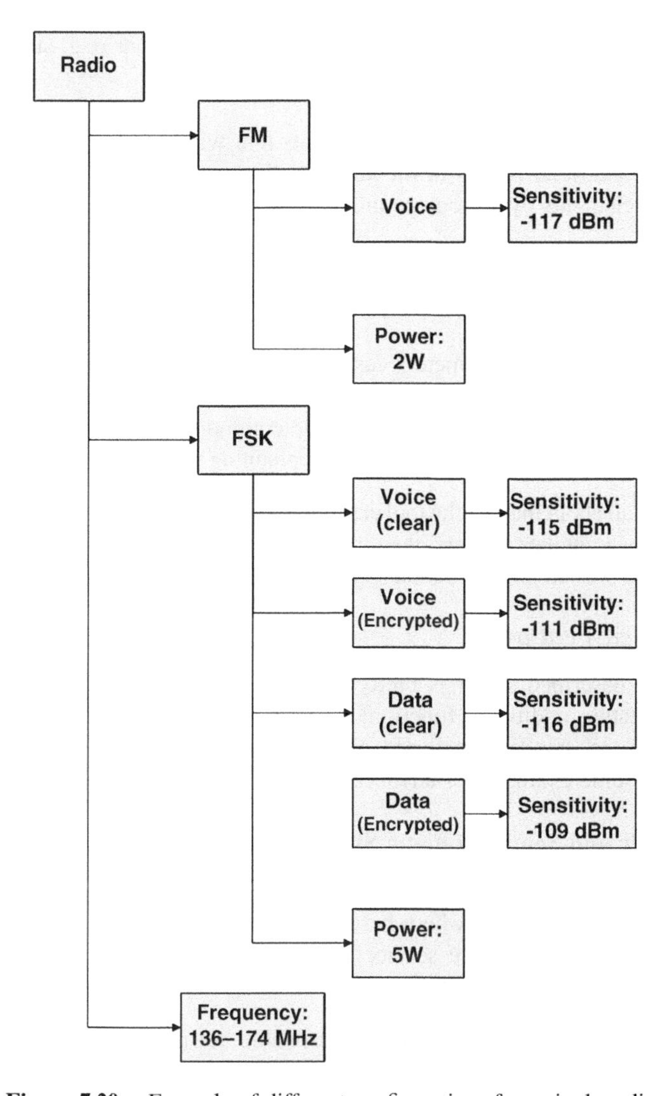


**Figure 7.20.** Example of different configurations for a single radio.

### *7.7.7 Ancillary Features*

Like other software systems, planning tools should have a variety of common functions that will be important to allow the best exploitation of results and continuity of work. These include the abilities to:

- Load previously created files and save current work for:
  - Network designs
  - System configuration files

  - Clutter and other methods of defining environment behaviour (e.g. subscriber density and location)
  - Vector files
- Import from and export to external data formats that will be used often.
- Print coverage and other results to file and directly to printers.
- Filter network elements by various parameters such as frequency, network ID, addresses and so on.
- Filter results by vector and environment type.
- Link to external databases, ideally to any Open DataBase Connectivity (ODBC) compatible database.
- Change network elements parameters easily, even when many sites have to be changed.
- Display network elements in a variety of formats to allow flexibility.
- Produce reports directly in formats that will be commonly required.
- User-accessible calculators for common radio planning conversions.

These features are all about making the tool easier to use on a day-to-day basis. If the tool is to be used intensively by a number of users, then it is vital that the tool is efficient to use, otherwise much effort will be wasted and the benefits of having a planning tool will be much reduced.

### *7.7.8 Advanced Features*

The functionality described so far is pretty standard for most radio planning tools. Some advanced tools feature additional functionality that can offer significant improvements in productivity over the standard methods. Such functionality will vary according to the specific tool, but some examples of advanced functionality are described in this section.

#### 7.7.8.1 Automatic and Semi-automatic Site Finding

This type of functionality works on the basis of identifying the wanted service area, and then working backwards to determine where base stations can be positioned in order to provide coverage to the largest part of the service area. This is illustrated in Figure 7.21, which shows a simple example. To generate this display, the given steps are followed:

- The basic link budget is compiled, using reasonable but nominal values.
- The height of the typical mobile antenna is determined.
- A nominal antenna height for the fixed base station is specified.
- The wanted service area is specified by the user drawing a polygon or importing the polygon from elsewhere.
- The simulation is run.

Figure 7.21 can be interpreted in the following way; the calculation tries a sample of the locations in the service area, selected at random. The more sites selected, the better the quality of the results but the longer the simulation takes to run. Each time a simulation is performed, the coverage to the specified availability is stored over the entire polygon, so in effect each one is a single-value coverage result (typically a value of one indicates success and zero indicates failure). After all the sample sites have been calculated, a composite score is calculated. This will be the sum of each successful coverage over each point in the polygon. The results are then scaled to a reference of 100 % and displayed, according to the

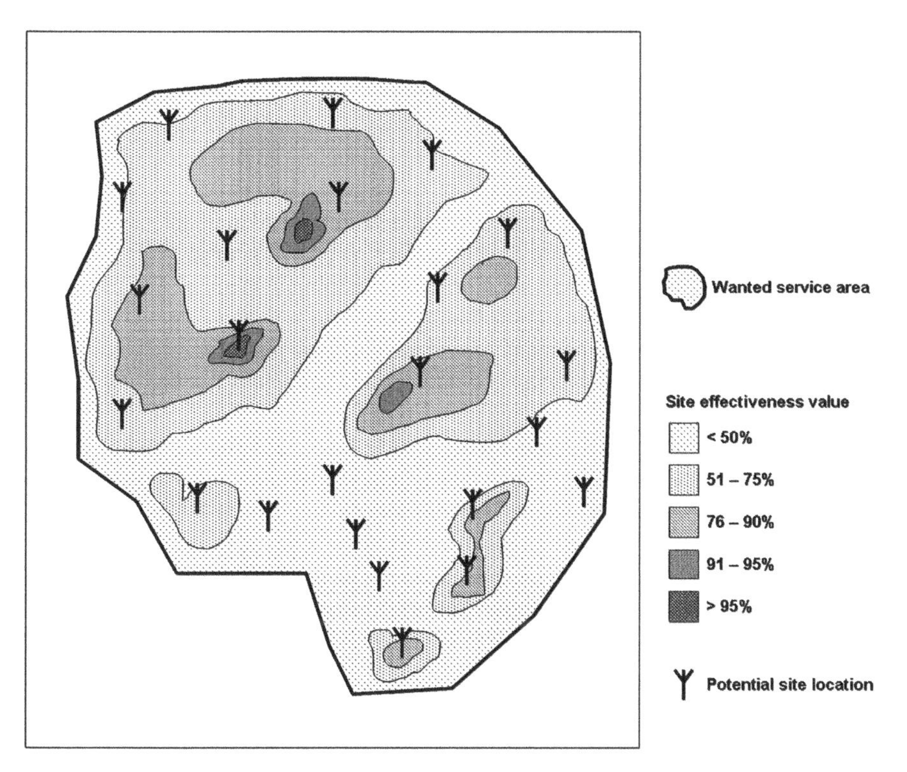


**Figure 7.21.** Automatic site finding functionality. Reproduced by permission of P & D Missud.

banding shown. The higher the value, the better the potential base station site location. For green field site projects, this information can be used directly to identify potential locations. If a database of potential sites exists (as shown in the figure), then the location of each base station can be compared with the results to see which is best. In this illustration, only one of the potential sites is within the highest banded area and is thus a prime candidate for a base station location.

This type of functionality is useful for complex site finding problems.

### 7.7.8.2 Real-time or Near Real-Time (NRT) Calculations

Some systems must provide answers in real time or near-real time in order to allow decisions to be taken quickly. Given modern processing time, this is only really possible for path profile based predictions or for very simple coverage predictions. In order to be able to do this, the system must receive updates of each of the elements to be modelled via electronic means and then must return the results for analysis as quickly as possible. Data transfer formats such as the US FBCB2 exchange format exist to promote the ability to exchange data for this type of functionality. If this type of functionality is required, it will be necessary to find tools with this functionality included, or to have existing tools modified to support it. This turns acquisition of a planning tool from a purchase into a project requiring significant effort.

#### 7.7.8.3 Communications Electronic Warfare

For systems designed for military environments, it may also be appropriate to have predictions methods for ES (Electronic warfare Support), EA (Electronic warfare Attack) and EP (Electronic warfare Protection). Typical functionality might include the following:

- Site finding, optimisation and analysis of radio detection systems.
- Site finding, optimisation and analysis of intercept receiver systems.
- Site finding, optimisation and analysis of direction finding systems.
- Link design for detection and direction finding networks.
- Site finding, optimisation and analysis of communications jammers.
- Optimisation to protect against detection, interception, direction finding and jamming.

All of this functionality should be provided for both static and mobile elements such as UAVs and aircraft. Also, using the same type of functionality described in the site finding section, the system must be capable of performing meaningful analysis even when the disposition and technical parameters of enemy equipment are not fully defined.

Because electronic warfare is such a niche activity, there are few systems commercially available, but there are a variety of differing levels of sophistication, and there are also bespoke systems that have been designed for specific applications which can be modified for similar requirements.

#### 7.7.8.4 Scenario Modelling

So far, we have modelled mobile subscribers as virtual elements, which for coverage predictions are placed anywhere, and for traffic analysis are distributed according to defined density values. On some occasions, however, it can be useful to examine network performance under an example of the expected network state. This can be achieved by producing a simulation of how individual subscribers may be distributed throughout the network. This equates to a 'snap-shot' of the network at a particular point in time. This snapshot must be developed using the best model of user behaviour that can be achieved. It might have to take into account the time of the simulation (perhaps during the rush hour, when many people are travelling, for example). These simulations can take some time to generate, however, automatic tools to generate random subscribers and call types can make the process considerably easier. An example is shown in Figure 7.22, which shows a simple method of randomly generating subscribers based on clutter distribution. To the subscribers generated in this way are added subscribers along specific roads, to give a composite picture.

In this figure, the number of subscribers has been enhanced to demonstrate the distribution according to the various categories such as urbanisation and roads. The idea of using such a detailed description of the distribution of subscribers is that advanced automatic methods of planning can be used against it. This can be repeated with different instances to ensure that the snapshots really do match the actual requirements. A typical automatic planning dialog box is shown in Figure 7.23.

In this figure, the types of environment to be considered are set in the top left hand frame. The rules for site deployment are set up in the bottom left frame. The rules for 'parenting' subscribers to base stations for calculating the grade of service are selected in the top right-hand frame. There are also options for selecting potential sites from a database rather than

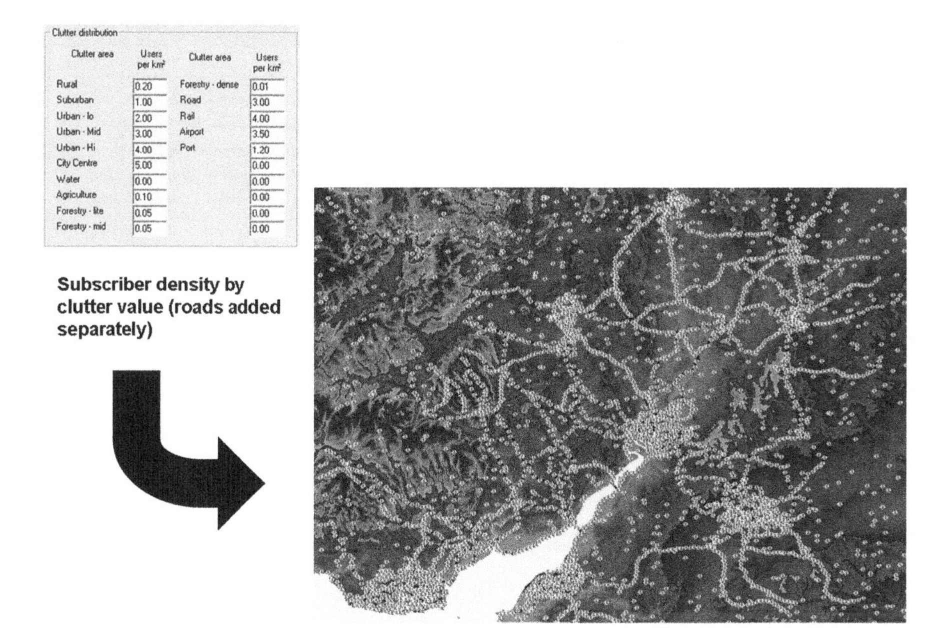


**Figure 7.22.** Random subscriber generation. Reproduced by permission of P & D Missud.

simply selecting Greenfield sites. It takes a little while to set up such a dialog box, but it saves time in the end since a network design is the result once the calculation is complete. This is likely to require some manual optimisation by the engineer, but it provides a good start.

### 7.7.8.5 3G Planning Functionality

3G technologies such as UMTS UTRA and CDMA-2000 differ substantially in that all subscribers share the same channel. Each subscriber is like noise to all the other subscribers, and more the subscribers there are in the same area as the wanted subscriber, the higher the noise level at that point. The performance of the system and the effective cell radius of each base station will constantly be in flux based on the number of subscribers currently using the system, and this means that the planning process must take cognisance of the fact that the system is essentially dynamic rather than static, as is the case for most other technologies. There are effectively many variables constantly in flux, and so where does the designer start?

Well, if one starts with the modelling of subscribers as described in the previous section, this isg a starting point for the design. The system must then be capable of iteratively determining the network performance as the link budget values change. A system capable of modelling 3G technologies must be capable of the following (all calculations refer to simulations from base stations to subscribers in the model):

- Allowing 3G-specific technical parameters to be entered and stored.
- Performing pilot coverage predictions.

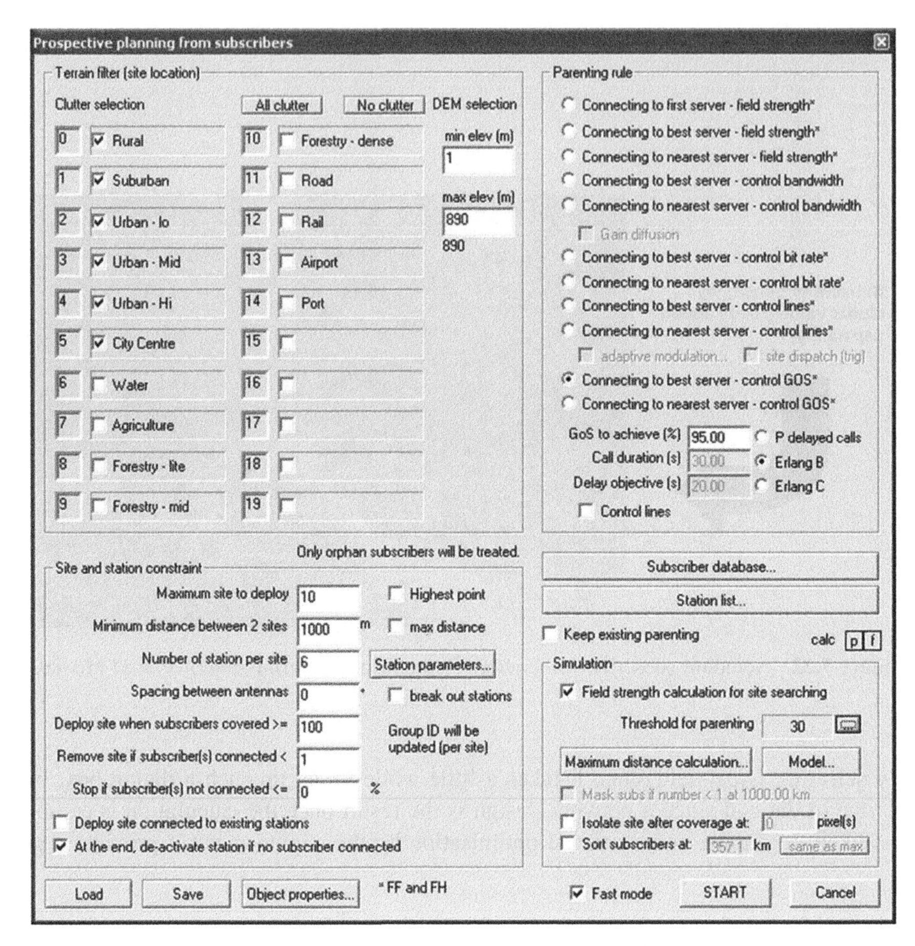


**Figure 7.23.** Automatic planning based on random subscriber generation. Reproduced by permission of P & D Missud.

- Performing $E_b/N_0$ reverse coverage predictions.
- Performing $E_c/I_0$ forward calculations.
- Performing $E_c/I_0$ pilot pollution.
- $E_b/N_0$ forward calculations.
- Perform traffic analysis, with mixed services.
- Perform noise calculations.
- Determine number of base stations connected to subscribers.
- Monte Carlo or other iterative simulation to test with different configurations of subscribers and to tune parameters.

There are a number of tools capable of modelling 3G systems, and their manufacturers would no doubt be happy to advise on their functionality. It has to be recognised that non-3G tools cannot be used for effective 3G planning.

### 7.7.9 *System Integration Features*

For large networks, site handling and manipulation of other user-defined data becomes a major part of the radio network design process, and the ease by which this is achieved with the radio planning tool will have a major influence on how efficient the design process can be. This will particularly be true for networked systems that have multiple operators sharing common data where data integrity must be maintained. Thus for radio tools to be used for large-scale design projects, it is important to consider the IT aspects of the system and also its scope. Critical factors include:

- What processes exist for the import of large amounts of external data? Large systems will require import of large amounts of data such as potential site lists, technical characteristics of radio equipment and antenna information. In major projects, this will constitute a very significant activity. If there are no automatic processes for data import, then this activity may come to dominate the entire project, whereas if there are suitable processes, then the import of even large amounts of data may be achieved relatively easily.
- How many radio sites can the system handle concurrently? This is not just a question of how many sites can be loaded in a given project but also a question of whether all simulations handle all of the stations loaded (especially considering simulations such as inter-connectivity, interference analysis and frequency assignment) that increase dramatically in complexity the more sites that need to be considered.
- Can the system operate as a shared system over a network, with access to communal data sources such as a central database, and if so is the data integrity maintained properly? Data integrity can be easily achieved in small to medium-sized projects, but requires very careful management in larger systems, using specially designed configuration management tools.
- Is the system flexible enough to be integrated into existing databases without major software engineering?
- Does the system provide for integration into other software systems? This may be of interest in terms of connectivity to spectrum management, radio monitoring, command information systems and communications electronic warfare systems. It is likely that there will be some requirement for development in this case, but if the system is designed in an open manner, it may be readily achievable.

Clearly, the larger the network that needs to be designed, the more important the planning tools to be used and therefore the more effort that needs to be put into the decision process when selecting a tool. It may be appropriate to bring in external consultants with specialist skills in radio planning tools or designing networks of the proposed architecture when putting together the requirements specification for a planning tool. Spending money on experts at this stage may well pay for itself many times over by avoiding expensive mistakes by cutting corners when choosing the right tool in the first place.

### 7.7.10 *Selecting the Right Tool for the Job*

We have discussed some of the features that radio-planning tools may have. It is important to recognise that more is not always necessarily best. The more comprehensive a tool is, the more complex it is to operate and the higher the training requirement. It is also likely to be more expensive than other systems, and in many circumstances a more expensive tool may not offer a significant advantage over others that may be significantly less expensive and easier to use. There will be occasions, however, when only the best will do. The key, as in most engineering, is to find the most cost-effective solution to the problem at hand.

The key thing is to match the tool functionality to the design demands. If the tool is to be used in support of a single type of network, then a tool that has only the functionality required for that type of network architecture should be selected. This will reduce the training requirements for the tool, since only the necessary functions need to be taught. It has to be borne in mind however that it will be necessary to enhance (if possible) or replace the tool whenever the design strays from the original requirements. For consultants offering radio planning and associated services, it would be best to select tools with the widest possible flexibility, despite the additional training effort required to acquire and maintain the skill necessary to use the tool to the best of its capability.

The planning tool must also be matched with the environmental data used in the project; it is not sensible to buy an expensive planning tool, and then use low-resolution, low-fidelity environmental data with it. In this case, the accuracy of the propagation model will be low, because of the poor data, and many of the advanced functions of the sophisticated planning tool will not be available or will also produce poor results. Conversely, is it also not a good idea to buy expensive, high-resolution terrain and clutter data at vast expense if the planning tool does not have features that are capable of exploiting that data effecively.

## 7.8 Environmental Data

### *7.8.1 Introduction to Environmental Data*

#### 7.8.1.1 Types of Environmental Data

Environmental data is that geographically coded data to be used with the radio planning tools. It typically consists of the following data types:

- Terrain elevation data, with or without additional obstructions such as buildings and vegetation.
- Radio clutter data, which is effectively a map of terrain usage insofar as it relates to the effect on radio propagation. This includes buildings, either individually represented or as belonging to a particular category such as typical sub-urban and different categories of urban. Note that these categories and their effects will often be different for different countries and architectural styles.
- Map or photographic images of service area, used to relate the calculated coverage to important social features such as roads and areas important to the end user community.
- Service area data, used to specify the service coverage area and elements within it. This may be split into different categories so as to allow user behaviour to be described in greater details. So, for example it may be possible to determine the proportion of users that may be in a specific part of the service area, or how many will be using the network while travelling along specified roads or while using particular train routes.

- For engineering the backhaul for the mobile network, it may be necessary to include rainfall rate maps to determine link reliability during the year. Rainfall rates will vary across the world and within individual countries and so a map makes the process more accurate than using nominal figures. Rainfall rates are only of importance at frequencies above about 6 GHz, so are not relevant to VHF or UHF mobile coverage.

We now go on to look at this data in more detail. Some additional guidance on environmental data for radio planning can be found in ITU-R P.1058: Digital Topographic Databases for Propagation Studies.

#### 7.8.1.2 Data Currency

An important issue in any environmental data is that of data currency. The question is how relevant the data is to the current situation, or which will prevail during the life of the network. For example, a map of social features such as towns and roads will go out of date depending on the rate of change. New towns, parts of towns and roads built since the map was created will not appear on the old map. If this out-of-date map is used for radio planning, then the network design will not take into account the requirements associated with these new features.

In general, Digital Terrain Maps (DTMs) have the longest currency, and it is only subject to change due to the largest activities; earthquakes and other natural events that change terrain features and massive engineering projects such as the building of dams or massive excavations. On a smaller scale, features such as embankments, cuttings and open-cast mining will also affect the terrain model and may well have local effects on coverage.

Radio clutter and map image data generally have less data currency, and they are linked (in fact they are often different ways of representing the same data). They will go out of data when the use of land changes, and the rate will clearly depend on the amount of developmental activity happening in the project area. In general, building and road data will change faster than vegetation such as tree heights, as trees tend to grow relatively slowly. However, bear in mind that vegetation will undergo seasonal changes which may or may not have an effect on coverage, depending on frequency of the network and the degree of vegetation involved.

It is important to recognise that the higher the level of precision of the data under consideration, the lesser is its data currency. For example, a Digital Elevation Model (DEM) that includes individual building heights will lose currency when any of the buildings in the model are demolished or new ones built. For urban areas under considerable re-development, the rate of change in this way can be very fast. This is not just an issue of getting the most recent data for the radio planning activity, it will also continue after the design has been completed and the network implemented. This means that if the design has been critically designed against high-resolution data, then the assumptions made during the designing will not hold during the life of the network. It may therefore be better to use planning methods that require lower resolution data and provides robustness through margins, rather than a method that relies on environmental conditions remaining static, such as is the case for link planning.

#### 7.8.1.3 Wavelengths and the Environment

Another issue of precision in the modelling of the environment relates to the wavelength of the radio network to be used. In general, effects that are affected by wavelength-sized perturbations (fast fading) are dealt with by margins derived from an appropriate statistical model (e.g. Rayleigh). In radio prediction, the intention is to model slow fading mechanisms which will be orders of magnitude larger than the wavelengths affected. The idea is that the terrain models capture salients in the environment without containing irrelevant data. It is difficult to determine a hard and fast relationship between wavelength and appropriate terrain model resolution because it may be possible to use large spaces between terrain data for relatively flat environments, but in other cases it will be necessary to have higher resolutions that capture all terrain variations. An often-used approximation is that the terrain step should be about 40–50$\lambda$ (the rationale for this is explained in Chapter 15).

#### 7.8.1.4 Geographic Projection

Confusion can often be caused by the issue of the geographic projection used by environmental data. This will modify how data is represented, how locations are expressed and the actual location of a specific point. There is no absolute standard for this, and a variety of systems have evolved over time. The characteristics of a projection system will vary according to:

- The underlying ellipsoid on which the data is based.
- The associated geoid on which the data is present.
- The geographical projection system used.
- The method of expressing the coordinates of a location.

The underlying ellipsoid is the mathematical construction used to represent the shape of the Earth, which is not a sphere but a slightly flattened ellipsoid ('biaxial ellipsoid'). Ellipsoids are used to provide a simplified mathematical representation of the shape of the Earth. A variety of ellipsoids have been generated to describe the shape of the Earth, and systems based on them will give different coordinates for the same location, perhaps differing by something of the order of more than 200 m or so. It is important to note that none of the systems are 'right', with the others being wrong – they are merely slightly different in their representation. What this does imply, however, is that it is necessary to know what ellipsoid has been used so that it can be compared to other representations and converted between one system and another as required. So why use different ellipsoids, and why not simply settle on one standard method used by all? Some ellipsoids are better for specific regions of the Earth, whereas other methods will work over the whole of the surface of the Earth. Those that are based on a region will tend to describe that region better than a generic system, but their applicability is limited to that region alone. In practice, the GRS80 (Geodetic Reference System 1980) is becoming somewhat of a global standard, because it is used for the GPS system. However, many data sets that the planner may encounter may not fit this standard but instead be based on another system; it is important to understand the distinctions between them and the implications of their use.

One fundamental point to consider here is that a definition of location does not describe a point on the surface of the Earth itself, but instead a position on the ellipsoid in use; this explains the apparent discrepancy between different ellipsoids used to describe the same location. This is also true for vertical expression of terrain altitude; it is the height of a point above (or below) the level of the ellipsoid (or, more technically, a geoid, which provides a reference altitude point), not the surface of the Earth. This is called the 'orthometric height' or simply the geoid height.

## *7.8.2 Digital Terrain Maps and Digital Elevation Models*

### 7.8.2.1 Introduction to DTMs and DEMs

Digital Terrain Models (DTMs) and Digital Elevation Models (DEMs) provide a representation of environmental data to a propagation model. A DTM is a model of the terrain of a region, not including ground features that may add to that, such as building heights. A DEM is a model of the environment including such effects. The decision of which of these to select will depend on how the propagation model works and how the subscriber behaviour is to be modelled. Both DTMs and DEMs can be of any resolution, precision and accuracy and the key thing in any mobile radio network planning tool is to identify which characteristics are necessary for a given project. It is important to recognise that more precise and accurate terrain and elevation models are not necessarily the most cost-effective; in fact, they may not even lead to the greatest correlation between performance predicted by a planning tool and those achieved in reality, as we shall shortly see.

### 7.8.2.2 Data Capture of DTMs and DEMs

There are a variety of methods of capturing the data that ends up in commercially available and indeed private DTMs and DEMs. In the early days of digital environment data, it was common for digitised maps to be created based on manual contour extractions from paper maps. The course of each contour would be traced by a digitisation pen or extracted using the colour characteristics of the contour line contrasted with the other features shown in the map. In both cases, these methods were relatively error prone even when using robust but commercially cost-effective quality control measures. In addition, the paper maps contained no data on terrain height between the contour lines and these needed to be interpolated in some mathematical fashion. There are a number of methods of achieving this ranging from straight linear interpolation through to more complex mathematical formulae, but in essence the intervening data is missing and thus efforts to approximate it cannot be very effective. More modern methods include capture of data by airborne radar and lasers. Both methods can typically capture DEMs and differentiate between forest canopy and ground level but in the case of buildings, the terrain data cannot be directly established in built up areas without processing the data to effectively remove the buildings. A commonly used radar method called LIDAR can be used to capture DEMs down to a vertical resolution of the order of 15 cm vertical accuracy. For mobile radio network planning, in general propagation models will not be able to utilise object information of a smaller scale than about 40–50$\lambda$, which is 40–50 m for 300 MHz, and 4–5 m for 3000 MHz. Arguably, ray tracing models can work with better data but given the amount of service area for typical mobile applications, this approach is computationally unfeasible and in any case it is not sensible to represent an environment in one set of circumstances when reality means that the real life situation will be at variance to the modelled version on many occasions (the movement of any reflecting, scattering, refracting or diffracting object will render the prediction inaccurate, and since this includes people and vehicles, it is somewhat difficult to prevent). Approaches such as airborne and space-borne radar interferometry can typically provide suitable data for various parts of a project at lower cost, while LIDAR and optical methods remain valid for point-to-point links in the microwave range.

#### 7.8.2.3 Accuracy and Precision in DTMs and DEMs

Accuracy and precision do not refer to the same attribute. Accuracy relates to how closely a data point in a model relates to the actual value in real life. Precision refers to the tightness of the specification of the data point. So, for example, horizontal precision may be quoted to 0.001 m, but the actual accuracy may be 100 m. This is clearly nonsense and is due to inappropriate use of precision, but unfortunately this does occur in real life either as a result of lack of understanding or even of dishonesty. For both DTMs and DEMs, accuracy is normally quoted in terms of RMS (Root Mean Square) error for both the horizontal and vertical range. Accuracy may vary over localised areas due to features present on the ground but not included in the model. This may include such important features as railway and road cuttings and embankments, open cast mines, areas of industrial spoil, landfill and areas of land reclamation. Such features may not be present either because the capture method did not account for them, or that have been created or changed since the original data was captured. Thus, in order to understand the accuracy of a DEM or DTM it is not only necessary to know the RMS accuracy but also how the data was obtained and processed, and when it was captured. This should be available from the data supplier.

#### 7.8.2.4 DTM or DEM?

The question arises of whether a DTM or a DEM should be used in a particular planning task will depend on the propagation model to be used, the resolution of the data available and the behaviour of the mobile subscribers. For example, for models that use street canyon approaches to refining propagation prediction, a DEM model will be most appropriate. For models that use radio clutter to represent urbanisation, a DTM should be used together with the clutter layer and a building model. It is important to ensure when combining data sets that relevant features are neither ignored nor double-counted, so a careful analysis and comparison of the data sets must be carried out. This underlines the requirement for radio planning designers and engineers to understand in depth how propagation models to be used actually work and the characteristics required for the data they need in order to produce accurate results. In all cases, it is important to refer back to the original model reference.

#### 7.8.2.5 Matching Data Resolution to Application and Project Phase

During a project, there will be different requirements for data at different times and for the different tasks to be carried out. Table 7.7 shows the type of considerations that may apply at certain times. Often in the early stages of a project, before commitment is made to progress with the project or to support a bid into an external customer, the amount of money available for the data will be a major constraint. Later in the project, once the business is won or full management commitment has been received, then the constraints imposed by performing the

**Table 7.7.** Data relationship with task.

| Project Phase | Task | Primary Data Constraints |
|---|---|---|
| Building the business case for bid/no bid activity or to obtain management buy-in to develop project costs. | Use estimation techniques to determine approximate project cost | Low acquisition cost Low processing/import cost Short processing timescales High level of accuracy not required |
| Producing the bid/refining project costs | Network dimensioning activities | Reasonably low acquisition cost Reasonably low processing and import Accuracy requirements will vary |
| Network design | Coverage planning Traffic dimensioning Intra-network interference Frequency assignment Inter-network interference | Appropriate accuracy to technology Cost dictated by required accuracy Processing and import time dictated by data Data must cover large geographic area Resolution requirements generally lower than for planning Accuracy requirements generally lower than for planning |
| International coordination | Proving stations meet coordination rules | Data must cover very large geographic area Low precision data acceptable Only major terrain features need to be modelled |

planning process to an acceptable level will be the dominant aspect. For specialised tasks such as proving interference to networks far away or for international cross-border coordination, the data must allow prediction over a very large area, and in which only major terrain features will affect the results.

#### 7.8.2.6 Sources of DTMs and DEMs

Few radio network planning organisations feature an organic capability to generate their own DTMs and DEMs, so in general it will be necessary to get data from other organisations that specialise in data collection and processing, or in distributing data bought in from data collectors.

Commercial organisations will normally have to buy in data from data supply companies, most of which can be found on the internet through search engines. These will normally supply data in a variety of formats, and it is important to ensure that the data can be imported into the radio network planning too (and in a suitable geographic projection) before purchasing it. In some countries, there may be a national data organisation that will generate, maintain and sell data (or supply it for free) to the public. An example of this is the UK's Ordnance Survey, as is the US Geological Service (USGS). These organisations will normally provide data in publicly available data formats.

One specific data source that should be mentioned is that derived from the Shuttle Radar Topographic Mission, an 11-day mission flown by the Space Shuttle Endeavour during

February 2000. This captured data for much of the world, which are available on the Internet. The website www2.jpl.gov/strm/ provides information about the dataset, its coverage area and its accuracy. It is important to check this if the data are to be used, because the original data were issued in a relatively raw form in which bodies of water were not well defined as they do not reflect radar very well. If these data are used, they must be processed to improve the accuracy of the data. A new version of the data with further processing is now also available and it corrects some of the original problems.

Military organisations or other governmental departments will often be able to obtain data generated by specialist military mapping departments. The military are large users of such data for command and control systems, navigation systems and for targeting, and thus there is a large volume of data available. However, again remember that geographic data are normally captured and processed for a particular application, which may have different demands than those of radio planning. For example, data captured for the purpose of aircraft and missile navigation must ensure that the highest points in each location are captured to avoid collision with the ground, so lower points within a particular cell may not be so important.

#### 7.8.2.7 Summary of Important Aspects of DTMs and DEMs

When looking for DTM and DEM data for use in radio planning, the following aspects should be examined:

- Cost.
- Availability (how long for delivery).
- Data file formats available, and how they can be imported into the radio planning tool (particular care must be taken to determine whether the data are in raster format (an array of data points with one data point for each value in the array) or vector format (where individual features are defined by type such as line, polyline, polygon etc, and where there will be no definition or where no feature exists).
- Delivery format (CD-ROM, download over the internet, DAT tape etc).
- Data capture date.
- Data capture methods.
- Processing methods used.
- RMS error, and variation of RMS error over the data coverage area.
- Exact extent of data to be provided (i.e. how is the boundary defined and is it completely unambiguous and understood by both parties).
- Whether manmade features such as embankments and cuttings are included or excluded from the processed data (this is true for DTMs as well as DEMs).
- Geographic projection.
- Whether there are any discontinuities in coverage area (e.g. due to combining different data sources, or that offshore islands are provided as separate data files).
- Licence conditions of use (e.g. can the data be provided to a third-party if required, can it be used publicly on line or in printed material, can it be re-sold etc).

This list is not exclusive, but should be a good starting point. It is important to take data provision to the project seriously, because it is all too easy to get wrong and the implications of an error can be significant.

## *7.8.3 Clutter Data*

### 7.8.3.1 Introduction to Clutter Data

The purpose of clutter data are to model ground based factors that will influence the performance of radio systems when they are located in particular environments. Simple models are based purely on background noise and may be defined by a low number of categories, such as business, residential, rural and quite rural. Others may contain considerable more categories, such as defining a variety of different urban types ranging from very dense city centres to out-of-town commercial parks for example. As with DEMs and DTMs already discussed, it is important to use a clutter model appropriate to the capabilities of the radio prediction systems being used. It is sometimes useful to use clutter models that contain more data than are used in radio prediction, if such data are to be used to identify service provision over those additional categories, such as streets, roads, railways, airport and so on. This type of analysis will be performed via the clutter data layer in some tools or via a completely separate layer in other tools, so it is important to ascertain how this is to be achieved before purchasing a specific set of clutter data.

### 7.8.3.2 Data Capture of Clutter Data

Clutter data can often be derived from layers contained within digitised map images. In other cases, the data can be obtained from ground usage imagery collected from space or air-borne earth observation systems. Such systems can produce data that can be automatically processed to provide clutter maps with multiple urban and rural categories.

A clutter file will internally relate a number code to each physical coordinate. The planning tool will then need a facility to convert ground clutter categories into figures such as attenuation values (or other values to be used in the treatment of clutter).

### 7.8.3.3 Accuracy and Precision of Clutter Maps

Once a DTM or DEM has been selected for a given application, an appropriate clutter map can be identified to work with this data. This will depend on the treatment of clutter in the radio-planning tool being used. The selected clutter data file must meet the requirements of the propagation model (in terms of the number of categories and their meaning), the terrain step interval (which in some tools much exactly match that of the DTM or DEM) and the geographic area covered (which again may need to match the area covered by the DTM or DEM).

The precision of the data points in the clutter file may be constrained by the planning tool, but the precision to which real life features are described can vary greatly. For example, consider a small residential part of a town, in which there will be streets, possibly parks, local shops and so forth. It would be possible to represent this as a single common value, defined in a polygon that matches the boundary of this specific area. It is also possible to resolve individual features such as each house, each street and so on down to the resolution of the clutter file itself. In this case, the area would be represented as a mixture of values rather than one specific value. Whether it is necessary to go to this level of precision will depend on whether such data will actually improve the prediction accuracy or not, and whether any improvement is worth the probable extra cost of the more refined data.

Of equal importance is the currency of clutter data. Clutter data are likely to be far more volatile than DEMs or DTMs since land use changes frequently. For urban areas, this will depend on the construction of new buildings or the clearing of areas. Agriculture and forestry will change throughout the year, and eventually will their effects on radio performance. Because of this, and the decision of the number of clutter categories used to represent each potential type of environment to be found in the network service area, it is very difficult to accurately quantify the accuracy of a clutter file, particularly since it is only valid for a short time. This also prompts the radio planner to ask another question: Should the clutter model be modified to include expected changes in environment throughout the life of the network, besides the design phase? This is not only more difficult data to obtain, there will be a risk that the proposed changes may never happen – often even advanced building plans fail before they achieve reality.

#### 7.8.3.4 Matching Clutter Data to Applications and Project Phases

In general, clutter maps with more categories, more tightly defined clutter areas and that are more regularly updated will be more expensive than data with fewer categories, more coarsely defined and that are not updated so regularly. Thus, just as in the case for DTMs and DEMs, the issue of cost will be important. This means that coarse maps will be appropriate for the early stages of the project when only approximate plans need to be made, with more refined maps possibly needed later depending on the influence clutter has relative to the other factors such as terrain and link radius. It is worth bearing in mind that clutter maps are more specifically linked with radio planning than DTMs or DEMs, which are used in a large variety of applications. This smaller demand necessarily results in smaller supply and in practice the designer or engineer may have to work with whatever data are available.

#### 7.8.3.5 Sources of Clutter Data

In general, clutter data can be obtained from the same types of sources as for DEMs and DTMs, but some of them may not also supply clutter data. The same methods for finding DTM and DEM data apply for clutter maps, as do most of the factors discussed in Section 7.8.2.7 which should also be considered, except that those aspects relevant to terrain accuracy are obviously not required and instead it is necessary to identify the number of categories available and the numerical codes that represent them. Also, it is essential to ensure that the data can be imported into the planning tool without undue problems. The issue of whether the clutter file needs to incorporate additional data layers for analysis of network performance (for example roads and rail links) also needs to be considered before purchasing a specific data set.

### *7.8.4 Building Data*

#### 7.8.4.1 Introduction

Building data are important when modelling indoor coverage, and also for penetration of outdoor networks into and beyond buildings. It has traditionally been difficult to obtain widespread areas of building data, beyond that of specific buildings, but it also possible to build maps of numerous buildings by making assumptions about them rather than knowing their exact composition and room layout; for example, an internal floor will takes up about 3–4 m vertically so by knowing the building height it is possible to estimate the number of floors. The converse is also true; when true building height is not known, it is possible to estimate the height by counting the number of floors.

Building data are not relevant to all mobile radio network engineering and many propagation models will not be able to exploit such data, but it is worth considering some aspects of building data.

#### 7.8.4.2 Data Capture of Building Data

Building data can be captured accurately from building design schematics, where these are available. However, in general such schematics are not publicly available so this does require having a source that can provide the data. Other than that, estimations of building structure can be made from very high-resolution maps, overhead imagery and also photography taken from ground level. The processing of such data are relatively intensive, and so it may be difficult to do accurately for large areas. In such case, the methods of approximating buildings already discussed may be more appropriate.

#### 7.8.4.3 Accuracy and Precision of Building Data

We have already discussed some issues of accuracy and precision in building data, but the one key aspect is that the higher the degree of accuracy and precision, the lower the data latency. Thus it may not be appropriate to spend too much time, effort and money on capturing data that will change quickly in any case (this may not be the case for one building, but when considered across all the buildings in a town, for example, there will be a continued state of flux).

### *7.8.5 Image Data*

#### 7.8.5.1 Introduction to Image Data

Some planning tools do not use images, since they are not strictly necessary for radio prediction. However, they do provide the ability to relate the results of simulations to social features such as towns and roads or other important features. For practical planning purposes, maps can also be used to identify whether green field sites are likely to be viable, or whether it is likely to be too difficult to find nearby power or access to the site. The most common image type used is a digitised map, but overhead imagery and site schematics are also occasionally used.

#### 7.8.5.2 Map Data

Map data are now commonly available in a digitised format at almost any scale from a variety of sources, easily found by searching the Internet. For radio network designers, it is useful to have access to maps at different scales, such as:

- Large scale maps for long range interference modelling or to display total network coverage for very large networks, typically in the region of 1:500 000 and 1:1 000 000 scale. These can also sometimes be used for marketing purposes to display the coverage of the network to potential customers.
- Maps for displaying coverage over relatively small areas, such as a regional network, or parts of a large network. Map scales in the region of about 1:250 000 have proved about right for many mobile networks in the VHF and UHF ranges.
- For site finding and for examining the network in more detail over smaller area, local maps in the region of 1:50 000 scale or less are useful.

Whatever map scales are used, it is important to ensure that the maps show the data required and that they can be used in the radio-planning tool. This may require that they are in the same geographic projection as the rest of the data being used, depending on the planning tool to be used.

### 7.8.5.3 Photographic Imagery

Imagery taken from aircraft or space platforms can be useful to augment map data. Although it can be more difficult to see features on photographic images if a large viewing area is required, imagery can be updated more easily than maps can be created. This thus allows consideration of features that have been added since the available map images were created. For military applications, it can also allow the network to be planned around mobile or semi-static structures that are tactically important. Photographic imagery is also available from many publicly available sources, and the resolution of current system is entirely adequate for mobile radio planning purposes.

### 7.8.5.4 Schematics

Schematics or site layout diagrams can be important for networks designed to cover a specific installation, such as a major airport, port or industrial site. As long as such schematics are described in terms of a valid geographic projection that can be used in the planning tool, then there should be no more difficulty in using such a schematic than is the case for any other image format. Of course, site schematics are normally not sold commercially but are produced for internal projects. Thus, special arrangements must be made to obtain these data from the project department responsible for producing such data.

# 8

# Starting the Project

## 8.1 Introduction

Starting the project is all about understanding the project context, the stakeholders involved and its scope. It will be necessary to arrange access to a suitable planning tool and appropriate environmental data. It will also be necessary to determine the interfaces between the various stakeholders and to set up the Method Statement incorporating all of the technical and nontechnical methods and rules to be used during the project. In this chapter we look at all of these issues in detail. It is important to consider the initial phases of a project in such depth because it is difficult to manage an effective project throughout its life without having set the system up correctly in the first place. Naturally, each project will be unique and so it will be with the start up process, but we can generalise the common features likely to be present in most projects. We will illustrate some of the principles by looking at a hypothetical project to design a new network.

We will look at this hypothetical network using a typical modern radio technology. For this example, we will assume that the network is to be designed for a noncommercial organisation requiring a digital mobile radio network, and that the customer has decided that for inter-operability reasons that the network will use the TETRA technology. Before going on to show how we would go about setting up the project, we will first try to capture the intentions of our hypothetical customer in a manner in which the requirements may be stated. The project is based on terrain found in the UK (since it is difficult to generate realistic terrain from scratch), but we have generated an entirely fictional social environment, with hypothetical towns, roads and forests. This area is illustrated in Figure 8.1. The desired service area is within the polygon shown.

For the purposes of illustration, we will look at the entire genesis of the project from the customer's perspective, long before the network design activity begins, so that we can examine the issues that may arise later in the project. Without understanding these early stages, their importance would not be clear.

We will also look at the start of the project documentation, outline project plan and risk register. As we go along, we will build a picture of the stakeholders involved.

Mobile Radio Network Design in the VHF and UHF Bands: A Practical Approach
*Adrian W. Graham, Nicholas C. Kirkman and Peter M. Paul* 

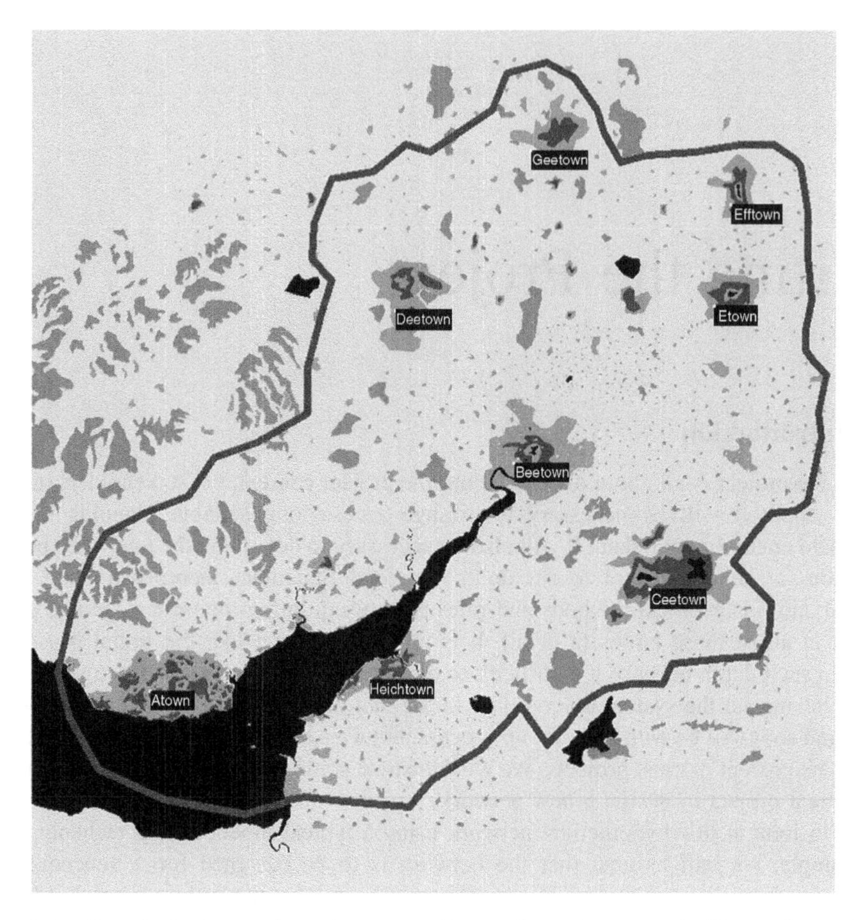


**Figure 8.1.** Hypothetical network service area, with a variety of towns and villages, forested areas and a large estuary. Towns are named alphabetically, and dense urban, urban and rural areas shown.

## 8.2 Project Requirements Statement

We will assume that the customer comes under the auspices of the emergency services and has access to a portion of spectrum reserved for such use. The customer is one of a number of such departments operating over the whole country, each of which is responsible for a specific region. We will call the hypothetical service area Countyshire, and the customer is the Countyshire Investigation Service (CIS). The CIS works alongside the police, but has a role in interviewing people at their homes and offices, and investigating particular activities. They do not however have a role in directly dealing with emergencies.

Before looking at the requirements of the project, we will identify the *perceived* need for the network. By this we mean, what does the customer want to achieve? It is perfectly possible to have a need without having a particular perception of need, and vice versa. In practice, the only reason such a project will progress will be because of the perceived rather than actual need. This may a sound strange way of putting it, but it enforces the idea that it is needs of the people – stakeholders – that drive the project, rather than any technical or nontechnical need, which may go unnoticed or may simply be ignored. Sometimes the role of the consultant is to separate the real need from the perceived need in order to go forward and recommend a design.

### *8.2.1 The Perceived Need for the Project*

As a background to the project, we will assume that the CIS has long been using a bespoke analogue radio network, but this network cannot meet current and future needs. In particular, the following deficiencies have been identified:

The network infrastructure is old. Elements are starting to fail more frequently, and it is becoming increasingly difficult to source replacement parts, as they are no longer manufactured. This means the cost of maintenance is high and the knock-on effects of network downtime means that the CIS has difficulty in meeting the performance their customers expect.

Since its inception, the environment has changed. Towns have spread and changed, and there are whole new areas of urbanisation that do not have service from the existing network. Since field operatives require continual contact back to base, it has become necessary for them to carry mobile phones. This is expensive and also the services offered by the carrier are limited and do not meet all of the needs of the CIS.

The old network was easy to intercept, since it used analogue FM without encryption. This meant that in many cases, operations were compromised and sensitive information was available for interception by anyone with a radio scanner. This is unacceptable, and it is deemed necessary to replace this with a new system that cannot be easily intercepted.

The CIS wants to be able to benefit from new digital services that will speed up their work by allowing operatives to interrogate databases and other services over the air interface. This will increase productivity, although it will lead to a quite a change in the working practices of staff and will require substantial training.

New working procedures mean that the CIS now has to work more closely with other organisations, such as the police, customs and excise, and other agencies. Thus, there is a need to allow communications between these organisations. Since these other organisations are already using a network based on TETRA, it has been decided that the new system will also be based on TETRA.

The reasons for investing in a new network are compelling, and the Chief Investigator of the CIS and the Board have decided that a project to replace the existing network is necessary. They have charged the Chief Radio Officer (CRO) and her department (the Radio Department, or RD) to investigate the possibilities, determine order-of-magnitude cost, and report back to the Board. We now have the first part of a stakeholder diagram, shown in Figure 8.2. In this diagram, we have enclosed all of the internal stakeholders by a larger grouping, which is that of the CIS organisation itself.

### *8.2.2 The Spectrum Environment*

Since the CIS is classed as an emergency services organisation, they have rights to spectrum allocated to emergency services use. However, this is centrally managed by a joint department

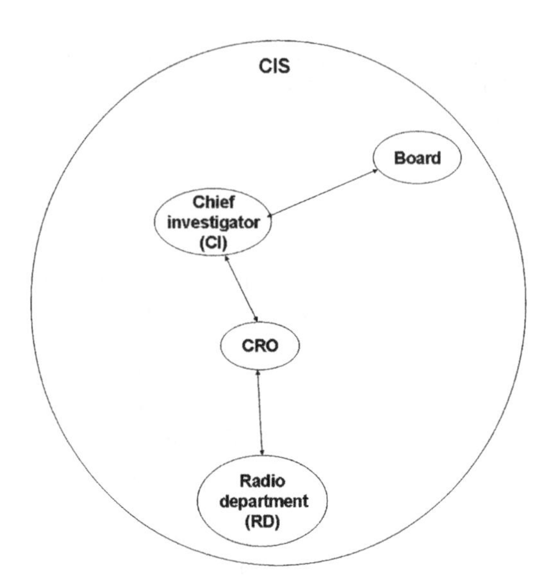


**Figure 8.2.** CIS internal stakeholders, shown encapsulated in the CIS stakeholder group.

of the national regulator and the Home Office (the government department responsible for the emergency services). We will call this the Emergency Services Spectrum Group (ESSG). Before obtaining the rights to use spectrum, it will be necessary to prove to this organisation that the use is justified, the design is efficient, and only the minimum amount of spectrum necessary is allotted to this application. They will also be responsible for de-conflicting the allotment with others in adjoining regions, already allotted to other regional Investigation Service departments.

This means that before progressing, the CIS will have to be able to prove that the spectrum is needed, and to be able to express exactly how much is required. If the case is accepted, then the ESSG will provide the allotment, which can then be used for frequency assignment in the network design.

The CRO identifies that it will be necessary to bring in external contractors to develop this case, since the department does not have the relevant skills and experience to determine the spectrum requirements of the job. To this end, the CRO recommends that a well-known consultant, Spectrum Services Ltd (SSL) be brought in to assist the customer build the case and also to provide technical expertise to assist in the selection of contractors to implement the network. SSL provides a specialist, Dr Grey, who will provide technical input to the remainder of the project on behalf of CIS. Dr Grey performs an assessment of the required spectrum and produces a report on behalf of CIS. This is submitted to the ESSG, who approve the reservation of most of the required spectrum, but do not offer quite as much as was asked for. The spectrum is only reserved for the application rather than approved, because the ESSG will need to see the final design before making a final decision. Dr Grey also puts together a cost estimate for the network and this goes before the Board. After some discussion, the CIS Board agree that the project should go ahead. The CRO is tasked with

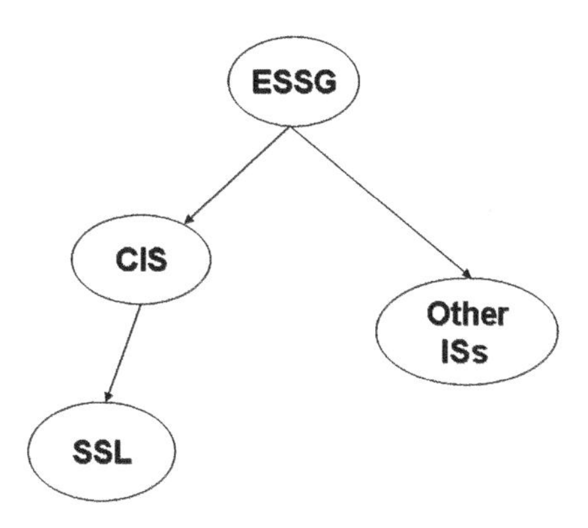


**Figure 8.3.** Spectrum stakeholders.

putting together an ITT, and to do that, she realises that it is necessary to define the requirements for the new network. She decides that the ITT should specify:

- The service area to be covered.
- The services that must be supported.
- The tasks that the contractor will be called upon to perform.

These are described in the next sections. At this stage, we can draw another stakeholder diagram, shown in Figure 8.3.

### *8.2.3 The Service Area*

The CRO originally believed that the service area is very well defined; the network must cover the entire operational region, which is Countyshire. However, after discussion with Dr Grey, she has realised that neither the budget nor the spectrum available will support anything approaching 100 % coverage. She has her own ideas about how the required service area can be prioritised, but rather than do so at this stage, she will instead ask bidders to describe how they believe the resources available should be employed. Because of this, she decides that the ITT will provide no further information about this aspect, and that bidders should recommend their solution.

### *8.2.4 The Required Services*

The required services are voice and limited data services for field Officers deployed within the service area. Voice is required for Officers in vehicles or on foot. The data services are only required within CIS vehicles, and there is no information about demand because this would be a new system and there are as yet no cars fitted with the equipment. Information about the exact data services required will be provided to the successful bidder, but all bidders should be aware that the service will be provided over the TETRA network.

## 8.2.5 *Contracted Tasks*

The CRO knows that her department, although highly experienced with their current network, do not have the necessary skills and knowledge necessary to be involved in the new network design in any major capacity. She intends that the contractor will take on all tasks to complete the network, within the budget and timescales set. This includes full design services, procurement of infrastructure and mobile equipment, and installation of all fixed infrastructure. The ITT will reflect all of this.

## 8.2.6 *The CIS Project*

Implementation of the new network is a major project for the CIS, and although they will not be carrying out the work themselves, it is still a major management activity requiring significant resources, and needing a dedicated team to ensure that the delivered network meets the need. To this end, a project team is to be set up consisting of the following capabilities:

- The CRO, to manage the whole activity.
- A representative of the finance department, to monitor project spend against the overall budget and spend profile.
- Two field Officers temporarily attached to the project to act as representatives of the end users.
- A senior member of the RD, to advise on the current network and to examine the differences between the coverage of the new service and the old service.
- Dr Grey, the consultant, to advise on technical matters relating to TETRA.
- Others may be brought in on an ad hoc basis to contribute as necessary.

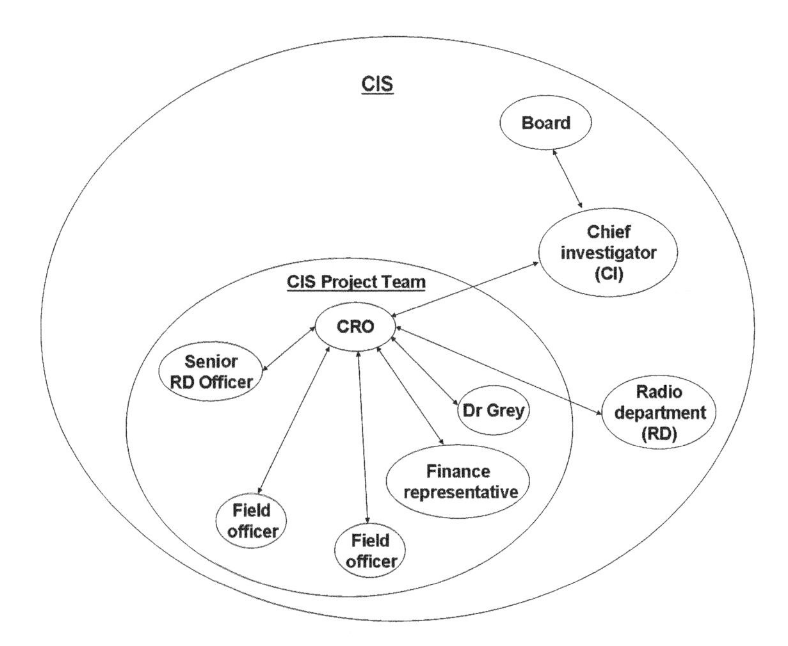


**Figure 8.4.** CIS internal stakeholders including project group.

We can revise our stakeholder diagram to include the new project team, as shown in Figure 8.4. Notice that the senior RD advisor is not shown as being part of the Radio Department – he is not, in the context of this project, and his views and opinions will probably differ from his colleagues in the RD that have not been assigned to the project. We have created a new, separate stakeholder within a new stakeholder entity; the project team.

With the team now established, and the top line requirements identified, it is now time to issue the ITT and await responses from interested parties.

We will now change our viewpoint from the customer to our design company, which we shall call Radio Network Design Services Ltd (RNDS). RNDS specialises in radio network design, but does not deal in hardware or installation.

## 8.3 RNDS Initial Actions

We will assume that the design company RNDS is very experienced, has carried out a number of similar projects in the past, and has built up a range of relationships with different companies in the network design and implementation field. On receipt of the ITT, RNDS realises that the scope of the project is beyond their capabilities, and if they are to be involved, they will need to be part of a consortium with a hardware supplier and installation teams. They contact a company that supplies TETRA equipment and that has installation teams (we will call them Radio Infrastructure Services (RIS)). This company has worked with RNDS often in the past, and usually uses them for network design activities. Both parties agree to look into the ITT. After initial examination, and following bid/no-bid process in both organisations it is decided to go ahead with the project. RNDS will provide network design services and specialist experience relating to the design aspects. RIS will provide the necessary infrastructure hardware and installation services, and will sub-contract the mobile radios.

Despite repeated requests to the CIS, no additional information is provided other than that provided in the previous sections. The consortium recognises that the relatively sparse information provided is a project risk, and does not provide sufficient information for a fixed and firm bid. Instead, a consultative bid is submitted instead with defined phases, with breakpoints for re-pricing, and the option for the CIS to re-compete the project should the consortium pricing being considered too high. This has the drawback for the CIS in that it means that there may be more effort required before the network is delivered, and there is the risk that the project will overrun, and it has the drawback for the consortium that they may not get all of the business. However, it has the principle benefit that it minimises risk for both parties, and this is a major consideration. It is agreed that it is the best approach for the project, and the consortium is given the job for the first phase, which is the detailed design activity to determine the exact resources required for the network, from which detailed cost figures can be derived. If these values are acceptable to the CIS, then Phase 2 will commence.

## 8.4 The RNDS Project Team and the Stakeholders

It is worth reviewing the stakeholders present in this project. We will then look at the structure of the RNDS project team and how they will relate to the other stakeholders. So far we have identified the following organisations as being involved.

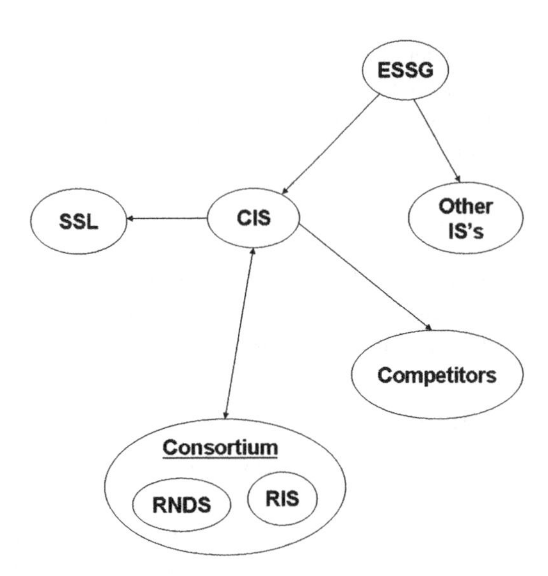


**Figure 8.5.** Project top-level stakeholder diagram.

- The Countyshire Investigation Services (CIS), who need the new network.
- The Emergency Services Spectrum Group (ESSG), who have control over the spectrum.
- Spectrum Services Ltd (SSL), the external consultants.
- Radio Infrastructure Services (RIS), the equipment suppliers and installers.
- Radio Network Design Services (RNDS), the network designers.
- The top-level stakeholder diagram (from the point of view of the consortium) now looks like Figure 8.5. In addition to the stakeholders listed above competitors are also there to be considered.

From this simple stakeholder diagram, it is possible to identify some top-level aims and risks associated with each, such as:

- We want to convince CIS that the consortium will offer the best solution. To achieve this, we may want to promote the idea that:
  - The consortium offers the best technical solution.
  - It offers an attractive price which, even if it is not the cheapest, offers the mot cost-effective solution.
  - It is fully committed to the project at every level, from senior management of both companies, down.
  - It offers stability for supply and support, on the basis of the standing of the individual partier.
- We will want to satisfy any technical issues raised by Dr Grey.
- We want to satisfy ESSG that the solution will be effective, and that the other ISs will not suffer unacceptable interference from our solution.
- In most cases, it is poor practice to set out to undermine the position. This is often seen as underhand by the customer and is therefore counter-productive. We will however want to counter any such moves made by the competition, by satisfying the CIS that we

understand the issues better and that any negative comments made by them are unfounded. We will in any case want to keep as good an eye as we can (within reason) on what the competition are doing.

As part of the same analysis, we can also identify potential risks so that we can be prepared to mitigate them if they arise, such as:

- The risk that the CIS will believe our solution is over-priced. We must therefore be prepared to justify our pricing.
- The risk that the CIS will not believe our solution is technically sound. We must therefore be ready to defend it.
- The risk that the CIS will be courted strongly by one of our competitors. We must be prepared to put in the effort to keep up contact with the customer and to ensure our voice is heard.
- The risk that the consultant, Dr Grey, will be negative towards us and our solution. We must be capable of offering a convincing alternative view to the CIS to counter this.
- The risk that the ESSG will not be convinced of the spectral efficiency of our solution, or that the risk of unacceptable interference is low. We must be in a position to prove that this is not the case.

*Because we have started to identify the stakeholders involved, it makes it easier to identify the aims we must achieve at each stage in the project, and the risks that exist. We have effectively used engineering principles (break a large problem down into smaller ones) in order to make the nontechnical problem more tractable, and thus helped us find a way to identify and deal with all the issues.*

These organisations are stakeholders in themselves, but this is not the whole story. We must break each one down and consider the various factions that might exist within them. We will now go on to look at the individual stakeholder groups and try to second-guess their possible motives and concerns. This is something that should be carried out with great caution because it is very easy to get it wrong, but on the contrary, not considering these options can also be a mistake. The principle aim of such a study should never be to exploit weakness, but rather to maximise the probability of success for the whole network and as many of the stakeholders as possible. In all of the discussion that follows, please remember that this is a hypothetical example used for illustration, and we do not mean to tar all such groups with these characteristics!

### 8.4.1 Countyshire Investigation Services (CIS)

The CIS can be split into a number of stakeholder groups including, as a minimum:

- The Chief Investigator.
- The Board.
- The budget holders.
- The CRO.
- The Radio Department.
- The field Officers.

In any real organisation, there may well be other factions within these groupings, and there may well be additional stakeholders not mentioned here, but even restricting ourselves to this list we can uncover many potential concerns and motives.

The *Chief Investigator* is strongly linked to the project, and will be whole-heartedly in support of the project. The Chief Investigator will most likely be more concerned with providing the necessary services to the field Officers than the detail of design and technical implementation. He or she may take a hands-off approach to the network implementation, or be more fully involved, depending on management style and other commitments. In the case of dispute, the Chief Investigator will side with the CIS and in such circumstances, may be unwilling to listen to explanations from the consortium if it looks like the project is going badly. The Chief Investigator reports to the Board and to the public the CIS serves, and would therefore be sensitive to anything that reflected badly on him- or herself in the eyes of either the Board or the public. The largest risk would probably relate to an individual, high profile failure of the network, followed a close second by the network being delivered late and well over budget.

*The Board* are the highest authority within the CIS, but their position will require public support. To this end, the Board will also be wary of the same potential failings as the Chief Investigator. They also would far prefer to be covered in glory for a successful network, and not tarred by a spectacular failure.

The *budget holders* are at a lower level of authority than either the Chief Investigator or the Board. In theory, they exist to support the activities of the organisation and should offer no resistance to the network. Sometimes, however, this does not always work in practice quite as well. If they do not appreciate the importance of the network to the aims of the CIS, financial matters relating to the project may be relegated to a lower priority than other activities. While this may not be intentional or even a conscious decision, it may cause delays in the project because of the slow release of funds to progress the project. While the actions of budget holders and their staff may appear at first sight to be only peripheral to the project, their cumulative actions may potentially cause significant delays if they are not committed to the project. The authors have seen many examples of this in practice, sometimes causing significant detriment to the project.

The *CRO* will probably be enthusiastic for the project, but may have some reservations. The new network represents change, and possibly quite a large change, to the day-to-day working life of the CRO and her department. Most people are wary of change, particularly if their current circumstances are comfortable. Hence, we may have a desire not only to bring into fruition a new, enhanced capability, but also to bring background wariness about what the changes may involve for her. She will also be committed to getting the absolute maximum out of the contractors within the fixed price; she may be a prime mover in any mission creep that may occur.

The staff of the *Radio Department* may well be less committed to the new network. They are technical staff who do not go on field operations and therefore will not benefit from the new network. In fact, it is often the case in these circumstances that the new network will primarily represent a threat to them and their department. At the back of their minds may be the thought that the purpose of the new network is, at least in part, designed to make them redundant. Even if this is not to be the case, they are still in a position that they will be forced to learn new technology and processes, and this also may be far from welcome. This can in some circumstances lead to reticence regarding the project and either tacit or even overt resistance.

The *field Officers* may also be wary of new technology. They may be aware of the limitations of their current system, but set against that is the threat that the new network may be worse. They may have heard stories of failure in the early stages of other TETRA networks, and even stories about possible health risks from the TETRA portables. Thus, even the most enthusiastic will still be somewhat reserved about what the new network will do for them, and they will require proof of the benefits (as well as having any health fears answered) before they accept the network as an improvement. This is likely not to occur until the network has been in service for some time.

It should be clear that even within the CIS, the organisation that is the prime mover in getting the new network implemented, commitment to the project and its success will be by no means shared by all. This is fairly typical in this type of project, and although the designer will not be able to influence all of the stakeholders, it is still important to know that these undercurrents may exist. The key thing is that the designers are sensitive to the individual concerns and motives of each faction within the CIS. This will only come by listening to them, appreciating what is being said and understanding what lies behind them.

A risk register can be drawn up to identify any potential risks in the project, and this can also be applied to risks generated by stakeholders. The register should be a living document embedded within or linked to the method statement and it should be regularly updated as the project progresses.

An example of a risk register we can put together based on the above might be as illustrated in Table 8.1.

When assessing the risks, we can multiply the severity by the likelihood to give a simple metric of project impact. This allows us to prioritise risks and the resources applied to mitigate them. Whereas technical risks will normally be dealt with via technical means, social risks such as those identified can normally only be solved by communicating with the people concerned.

### *8.4.2 Emergency Services Spectrum Group (ESSG)*

The ESSG are responsible for managing emergency services in this example (they, too, are an imaginary organisation). Like most organisations of their type, they will probably be very reticent in approving frequency requests, which is right and proper. This means they will need to be convinced of the need before they release any spectrum, and that consequently the CIS case must be strong, and the logic must be irreproachable. The CIS, their consultants and the designers must all work to ensure that the case for spectrum is as strong as it can be. The ESSG involvement is primarily as a potential block to network deployment, and one that must be overcome. The designers must work to provide information to the CIS for submission to the ESSG that best supports the CIS case.

### *8.4.3 Spectrum Services Ltd (SSL)*

SSL are the consultants supporting the CIS. Their specialist, Dr Grey, will provide technical expertise to help guide the CIS through the project. The role of the consultant will vary between projects. In all cases, the consultant will be dedicated to providing the best support to the customer, but they may also have other agendas such as protecting their reputation and also promoting themselves and their company so that they gain more business from the customer. This is not underhand, but rather the way that consultants typically work.

**Table 8.1.** Partial risk register.

| Risk | L | S | Potential cause | Mitigation |
|---|---|---|---|---|
| CI loses faith in the project, leading to withdrawal of support | 1 | 5 | CI | Maintain close contact with CI and ensure good communications maintained. Referany adverse comments with Project Manager. |
| Board loses faith in the project, leading to withdrawal of support | 2 | 5 | Board | Produce progress reports regularly aimed at the Board and other senior management in an easily digestible format. |
| Budget holders hold project as low priority | 3 | 3 | Budget holders | Try to engage budget holders and influence positively by explaining the importance of the project whenever possible. |
| CRO has misgivings about project, leading to less than whole-hearted support | 3 | 4 | CRO | Identify misgivings and counter them by demonstrating why they will not arise or how the new network will overcome them. |
| Radio department drag heels when providing vital information required for project progression. | 3 | 4 | Radio department staff | Try to engage enthusiasm of radio department staff. Monitor performance and raise any issues with CRO early to minimise impact and ideally prevent developing into a trend. |
| Field officers reluctant to buy into new network and offer resistance to it. | 4 | 3 | Field officers | Try to promote the new network at every level. This includes RNDS engineers, not just project management. Talk up the new features and allay fears about performance, ease-of-use, health issues or any other issues identified as negative. Raise any widely held negative views with project management. |

*Note*: L: Likelihood (1 – 5); S: Severity (1 – 5)

Depending on the personality of the consultant, they may choose to work constructively with the designers to get the best result for the customer, or they may choose to be confrontational in order to raise their profile in the project and in the eyes of their customer. A confrontational approach is not necessarily to the detriment of the project; the designers may benefit from searching questions, and a better network can result. Conversely, the situation may be negative if the consultant proves obstructive; the authors have seen this situation on occasion. In such circumstances, the worst thing to do is to reciprocate; there are better ways of responding to this situation. The best situation in this case is that direct communication between the consultant and the designers is minimised, but instead is carried out through the CIS. Assuming that the customer is reasonable, it should become clear to all who is being reasonable and who is not. This situation is undesirable and everything should be done to avoid it.

### *8.4.4 Radio Infrastructure Services (RIS)*

RIS are the equipment suppliers and installers. Compared to the rest of the project, the cost of equipment and installation will be huge. RIS therefore have most to win from the project. They will recognise the need for proper design, but may not fully appreciate all of the aspects of network design. This means that there may be a need to ensure they are apprised of all of the issues involved, so that they do not unintentionally undermine the case for putting sufficient effort into the network activity.

### *8.4.5 Radio Network Design Services (RNDS)*

RNDS should be keen to put together the best network design within the resources available for the project, which will as ever be set by the budget available. Ideally, the whole company should be working for the project's success, but in reality the project is likely to be one of many, all competing for limited resources. It is therefore incumbent on the project manager to fight his or her corner internally in order to procure – and keep – the resources necessary to fulfil the requirements of the project. Although it would be nice to think that this would never happen, but in reality it is rare that it does not. Any project is always in competition with other uses the company's resources. Bearing in mind our earlier comments about stakeholders, which we defined as being groups or people who have an influence over the project, we must recognise that internal and external competitors are also, in effect, stakeholders in the project – but not necessarily ones dedicated to the success of this particular project!

### *8.4.6 Interacting with Stakeholders*

We have seen that the range of stakeholders and the influence they may have are many and varied, and interacting with them is a potential minefield. There are various mechanisms for managing interactions to reduce the inherent risks, but none are more important than experience and common sense. However, some risk mitigation to avoid potentially dangerous results such as unplanned mission creep or increasing the customer's expectations beyond what is contracted include the following:

- Define a mechanism for interaction. It should not be a free-for-all, because this will lead to confusion and blurring of responsibilities. This means that it is important to differentiate between formal communications, which is always recorded in a written manner. This means that where face-to-face meetings are carried out, minutes should always be taken and each attendee should make their own notes. At the start of the project, it should be made clear and recorded that only formal communications are binding. This gets round the problem where one member of staff speaks out of mind and offers something not formally agreed – although this should be avoided even if there is a contractual method of back-tracking, it will not prevent bad feeling between the parties.

- It should be made clear between the parties who is authorised to offer or accept changes to the existing plans and who is not. There may be many people working on a project and it must be clear which ones have management responsibility for the overall project and which are there to perform work tasks without having authority to act on behalf of the company.
- There should neither be too much scope for formal communication, nor too little. Too much will result in the designers fire-fighting mission creep, and too little will mean that the customer's genuine requirements are not captured. This is of course a difficult balance to strike, and it depends largely on experience.
- There is a difference between contractual negotiations and high-level meetings, and the day-to-day working of engineers from both parties. It will in many projects be essential for the design engineers to work closely with the customer's engineers. This is right and proper and should happen without management hindrance. Engineers do however need to be briefed in advance of their responsibilities in regard to managing the customer's expectations.
- Any issues that appear to be arising should be reported to the Project Manager at the earliest possible opportunity, whether this is heard officially or not. Often, customer dissatisfaction may be first evident to design engineers on site well before they reach management ears; forewarned is forearmed, and getting a feel for the customer's mood will hopefully help to avoid situations escalating.

Interacting with other stakeholders is a crucial part of getting a project to run smoothly to its conclusion, which also explains why it is necessary to spend some time identifying the stakeholders in a given project. Once we have identified stakeholders and their concerns, we can work to build confidence, persuade, remove obstructions and counter arguments, all of which are essential in order that the project proceeds according to plan. This is not engineering in the narrow sense of applying technology to a problem, but it is engineering in the broader sense of applying effort in order to successfully complete an engineering project. As has been said elsewhere in the book, more networks fail because of problems with stakeholders than ever fail due to technology problems; and if a network design project is not completed, then the task to engineer that network has necessarily failed.

## 8.5 Project Activities

By the time the project has been quoted for, we should have a good understanding of all the project activities required (otherwise the price offered for the project will be wrong).

In this case, the project activities will consist of:

- Setting up the project.
  - Starting the Method Statement
  - Obtaining resources
  - Elucidating requirements to produce the Functional Specification components.
  - Determining the contents of the Detailed Design Document
  - Creating a Test Specification
  - Creating a project plan
  - Creating a quality plan

  - Creating a risk register
  - Producing a project baseline
  - Outline planning and dimensioning
  - Calibration surveys
- Designing for coverage.
- Designing for traffic.
- Designing the backhaul network.
- Frequency assignment.
- Interference analysis and mitigation.
- Network verification and acceptance.

In the remainder of this chapter, we focus on the aspects of setting the project up. In the following chapters, we will look at the other aspects, with a brief interlude to look at some network rollout strategies and scenario modelling.

## 8.6 Setting Up the Project

If we now assume that the project has been won by the RIS/RNDS consortium, the next step will be to set up the project. RIS and RNDS will have to agree how the project is to be carried out, and who will lead on certain activities. In some cases, the designers may well work at a remove from the customer, and only gain access via the prime contractor. However, since radio network design has a strong consultative component, this should in most cases be resisted. The designers must be able to talk to the customers in order to ensure their requirements are properly captured.

Apart from the normal administrative activities associated with setting up a new project (obtaining a works number, arranging secure storage of project documentation etc), the first steps will be to start the Method Statement and to start identifying the necessary project resources, and to go about procuring them. This of course may be different in your own organisation, which may have its own way of proceeding, but we present this method as one that works and has been adopted by a number of consultancy organisations.

### 8.6.1 *Starting the Method Statement*

In our approach, the Method Statement (MS) is the living, breathing project document. It grows and changes as the project progresses and throughout it is the principle document that describes how the project will be done and what deliverables will be produced. Since the document is designed to be used throughout the project, it is important to ensure that the design of the document will support these changes and enhancements as they are made without requiring large-scale modification of existing material later on. Whether the MS is produced purely to the details of the contract or whether the MS extends and enhances (are therefore in part replaces) the contract will depend on how the customer wishes to proceed, and will vary from contract to contract. In our example, we will take the view that since the ITT is sparse in information and does not specify the requirements in sufficient detail, we will use the MS as the principle document. The customer will need contractual protection in this case, to ensure that the MS does not vary what was offered in the RNDS response to the

ITT. This will be achieved by the customer having veto over the contents of the MS, and the project will not proceed without the customer agreeing to the contents of the MS and signing it off to prove the fact.

The specific contents and layout of the MS will depend on each project's specific requirements, but we will define a typical layout for this project. Initially, many of the sections will not contain much information, but these will be filled in later as the project progresses. Our structure will be put together as follows (in this case we elect to incorporate the contents of the FS, DDD, TS, project plan and quality plan directly into the MS):

## METHOD STATEMENT FOR CIS PROJECT

## CONTENTS

1. **Introduction and Background to Project**
   This will put the project into context. It will describe how the project came about, who is involved and the high-level aims of the project. This will be used to ensure that all parties have the same understanding of what the project is about, and will also serve as a useful introduction to anyone else joining the project at a later date.
2. **Project Reference Documents**
   This will detail the commercial documents and their precedence, so that there is no doubt in this regard. This will not include technical references, which will be listed elsewhere (normally at the end of the document).
3. **Tools and Resources Used**
   This will precisely document the radio planning tools and associated data to be used. The data is particularly important, since it will come to be the definitive description of the service area and areas of interest within it (e.g. roads and areas of urbanisation).
4. **Outline Process**
   This will detail what is to be done. Even if it cannot be fully documented as yet because the precise methods have not been established, it will be possible to document in detail how the process is to be derived; in other words, the process to derive the design processes themselves should be identified.
   This section will provide an overview in the final document, with the details in the next section.
5. **Detailed Methods**
   This will be a sizeable section and will detail all of the tasks involved in the project. Examples may include:
   - The processes to import and convert all external data
   - The propagation models to be used, including all settings within the planning tool
   - The design processes to be used, detailing again all simulations and settings
   - The link budgets to be used (may be put in an annex and referenced instead)
   - Figures to be used for clutter attenuation and any other characteristics used.
   - The testing to be carried out during the project (TS). This is the definitive list of the criteria that constitutes success and therefore completion of the project.
6. **Project Plan**
   Normal project plan, including scheduling and resources applied.

7. **Quality Plan**
   Details of the quality procedures to be applied to the project. This will normally be laid down in company procedures.
8. **Deliverables**
   A complete, definitive list of all deliverables to be provided as part of the project. This is likely to consist of:
   - Candidate site lists
   - Configuration data for each candidate site, including equipment, antennas etc
   - Coverage plots
   - Interference plots
   - Frequency lists
9. **References**
   References are likely to be important, and in this case it may be appropriate to put then in their own section rather than at the end of another section.
   **Annexes**
   **Appendices**
   (As required)

The risk register, which is an internal document for RNDS may not be included in the MS since it is not necessarily for public viewing. However, in some cases the risk register for a project will be declared as a public document, and in this case all parties must be frank in adding to the register. If there is a desire to recognise risks publicly, but there are some risks that cannot be shared, then there may be a project risk register shared with the customer and another one for internal use only.

Once the draft MS document has been completed as much as it can be at the start of the project, this should then be presented to the customer. If necessary, the customer should be taken through the detail of the document until they are happy that it is a true representation of what they want to occur. After that, the customer should sign the document before work commences. It then becomes the principle design document.

### *8.6.2 Project Resources*

Project resources must be obtained in a timely manner for the project to proceed according to the plan. Some resources, such as environmental data are likely to be obtained from external sources such as dedicated data provision companies. This may take a while to be delivered, particularly in regard to data that has to be captured first. It is therefore essential that early thought is given to obtaining such resources in a timely manner; in fact, it would be normal to enquire about the availability, price and delivery timescales of such data at the quote stage, so that the figures and impact can be incorporated into the bid. Project resources required may fall into the following categories:

- Project leader.
- Consultant or activity leader (often the project leader as well).
- Senior engineer(s) to direct sub-tasks.
- Junior engineer(s) to perform sub-tasks under supervision of senior engineers.
- Technical specialist(s) to provide in-depth technical assistance to the project.

- Access to suitable planning tools for the required number of staff to be used.
- Suitable environmental data for the project:
  - DTM at suitable resolution (typically 50 m for TETRA), imported into the correct format for the planning tool
  - Clutter map at a resolution to suit the DTM, again imported into the correct format for the planning tool
  - Map images (if the planning tool uses them): These should be at a suitable scale to allow design decisions to be made. For TETRA, images of 1:250 000 for overview and 1:50 000 are reasonably typical. The 1:250 000 scale maps allow coverage plots to be interpreted in terms of the service area; the 1:50 000 allows detailed analysis of candidate base stations such as access and nearby structures. If the planning tool does not accept map images as an overlay, it will probably be necessary to buy paper maps, or digital maps for a companion GIS system.
  - A description of the service area and all areas of interest within it. This may have to be generated as part of the project, but in any case it will be necessary for such data to provide the following information:
    - The definitive outline of the service area
    - A definitive outline of each category of ground use within the project area. This is likely to include roads (split into the number of categories to be used in the project), categories of urbanisation and areas of special interest (airports, ports, golf courses, and known trouble spots – anything that requires individual attention within the overall design). It is vital that each category of data can be individually isolated from the others and analysed by itself.
  - Population statistics, based on the most recent census or some other representation of population distribution. This will allow the total population actually served within the service area to be determined.

When identifying the environmental data to be used in the project, it is important that both the customer agree on the *precise data set* to be used, not just the type of data. In other words, it would not be acceptable to say, 'the project will use 50 m resolution DTM data covering the whole of mainland UK'. Instead, it must be more precisely described to an individual, unambiguous data set, e.g. 'the project will use the Ordnance Survey 50 m Panorama database, 2001 edition'. This is not ambiguous; anyone can refer to the data documentation to discover the exact extent of the data, how the data was produced and what the mean errors and caveats of the data are. This is not true in a less specific description, and in theory this means that a less accurate dataset that may not cover the same geographic area may be substituted instead. This may be to the detriment of the project and therefore it is important to close the loophole.

The Project Manager must be responsible for obtaining the resources necessary for the project, and must be prepared to fight their corner in the face of potential conflicts with other projects. The degree to which this is required will depend on the philosophy and structure of the company, and also on how busy the organisation is. Although in a perfect environment, this should never be necessary, in practice there is usually some element of this in every organisation. As with handling of stakeholders, not being able to procure and keep the required resources will cause the project to fail. All project resources required should be reflected in the Method Statement (in our example, in Section 3).

## 8.7 Elucidating Project Requirements

One of the most important activities for the network designer or consultant is the translation of user aspirations into measurable metrics that the network can be designed against. This activity is normally large in scope and takes considerable skill and experience. In this short section, we can only scratch the surface and illustrate the principles by a few simple examples.

If we consider the original requirements as stated by the CRO earlier in the chapter, we can explore how this activity works. These are the statements for the service area and the services required:

*The CRO originally believed that the service area is very well defined; the network must cover the entire operational region, which is Countyshire. However, after discussion with Dr Grey, she has realised that neither the budget nor the spectrum available will support anything approaching 100 % coverage. She has her own ideas about how the required service area can be prioritised, but rather than do so at this stage, she will instead ask bidders to describe how they believe the resources available should be employed. Because of this, she decides that the ITT will provide no further information this aspect, and that bidders should recommend their solution.*

*The required services required are voice and limited data services for field Officers deployed within the service area. Voice is required for Officers in vehicles or on foot. The data services are only required within CIS vehicles, and there is no information about demand because this would be a new system and there are as yet no cars fitted with the equipment. Information about the exact data services required will be provided to the successful bidder, but all bidders should be aware that the service will be provided over the TETRA network.*

We can start at the beginning by looking at the required service area and the coverage required within it. The extent of the service area is defined as the boundaries of Countyshire. This should not be too difficult to agree, but there are two aspects that require agreement between customer and designer. Firstly, what is the reference for the regional boundary; where is the unambiguous reference that both parties can agree on? Maps may not always agree precisely – which one is to be used, or is the boundary defined as a set of coordinates, and if so, where is the definitive reference to be used? Also, are there any plans to change the boundaries of Countyshire during the life of the network? Boundary changes are far from rare. In many cases, these questions are easily resolved, which then brings us on to the second question; given an agreement on what constitutes the boundary, how should this be represented in the planning tool (which will be used to plan the network)? This question revolves around the issue of the selection of a data set that represents this area. If the data exists in an unambiguous format and can be imported into the tool while retaining sufficient precision and accuracy, then all to the good; if not, then it might be necessary to trace boundaries manually. Will this be acceptable to both sides? This may all sound like pedantry, but it is an issue that needs to be resolved before any design work is done against the representation of the service area. In most cases it is straightforward and can be achieved in minutes, but where it is a contentious or complex issue, it can become a major issue that requires sufficient discussion to be resolved unambiguously.

Whereas obtaining agreement of the boundary definition may be resolved relatively easily, other issues may well be more difficult to pin down exactly. We need to consider the requirement to prioritise the areas within the service area, and this can be cut many ways.

First, it may be appropriate to ask the CIS what their commitments are in respect to serving the public in their service area. There may be some requirements stated in their charter, mission statement or operating procedures that need to be reflected in the network design. Also, it may be necessary to look at how the CIS interacts with the public. We can contrast with a normal police service; this must operate anywhere within its service area, but usually it will be focussed on where people live, work or play, and on how they travel between these places. If the design is based against these areas, then it will best meet the pragmatic requirements of the network. But what about the CIS? Does it fit this model, or does its behaviour differ in some way? This can only be established by asking the right people in the CIS – they are the ones that know their business best. The designer should not seek to second-guess the requirements; this would be a mistake. When the best informed people are available to input their knowledge, use them – do not make less well informed suppositions on their behalf.

Let us assume that an interview with the CRO and the field Officer representatives has come up with the following observations:

- The CIS is primarily concerned with long term investigations, and most of its work involves interviewing people. This would normally occur at home or, less often, where they work.
- There are occasions when CIS Officers need to follow individuals in their cars. This is only done when they move within the boundaries of Countyshire; the CIS has no mandate to follow them beyond these boundaries.
- In most cases, CIS Officers will make calls when suitable rather than needing to make a call at a specific moment in time, as a police Officer might. This means that if it is not possible to make a call from a specific location, this is not normally a major problem.
- If a CIS Officer is to visit a suspect individual, or a particularly dangerous area, they will travel with a police officer, who will have their radio with them.
- If the CIS are following an individual in a car, they will need to be able to communicate the suspect vehicle's position and heading continuously. This is an identified problem for the current system, because there are coverage gaps along some major roads in the area.
- The CIS has aspirations to provide data services to field Officers, but in fact the infrastructure is not actually ready yet, and it is envisioned that the capability will be provided in Phase II of the project, in about two year's time.

It is possible to derive some ground rules from this information. The designer comes away from the interview with the following assertions:

- This is not actually a safety-of-life network. Calls are not made to call for assistance but rather to speed up normal work. If a CIS Officer is put in a potentially dangerous situation, they will be assisted by a police Officer, who will have their own radio.
- The best availability must be provided along major roads; this is where calls are time-critical. Urban areas must be covered, but this may be at a lower level of availability if necessary. We will need to define the levels of availability for each category.
- Data is actually an aspiration, rather than a requirement, at this point. The network must be capable of accepting these new services in the future, but they are not required at present.

The designer has started to separate the real needs from the perceived needs, and also to identify priorities for the service provided by the network. The revised requirements are however still far from sufficient to allow planning to begin. The task to convert these requirements into measurable metrics will typically be relatively long and complex, but we will summarise some key decisions made, and show how they are expressed in metrics that can be used for design and verification.

Given the guidance provided by the CIS, the designer proposes the following design metrics:

- The overall service area is defined by the county boundary map produces by National Mapping Agency 1:250 000 county boundary data set, Issue 1 of 2006.
- Coverage (which is defined further in the next bullet point) achieved by the design will need to meet the following metrics:
  - All cities, towns and villages of population greater than 5000 must receive 'urban coverage'. The list of such towns is drawn from the National Mapping Agency list of Cities, Towns and Villages, Issue 2 of 2005.
  - All regions with a population density of greater than 150 inhabitants per square kilometre must receive 'rural coverage'. These areas are derived by using the National Mapping agency Postcode and Population data set, Issue 4 of 2004. Linear distribution over postcodes is assumed.
  - All major roads, defined in National Mapping Agency RouteMaster data set, Issue 1 of 2006 will receive 'road coverage'.
  - At least 75 % of minor routes, defined in National Mapping Agency RouteMaster data set, Issue 1 of 2006 will receive 'road coverage'. The percentage is defined in terms of overall road length.
  - At least 75 % of the total service area (as defined by the county boundary map, described above) will receive 'rural coverage'.
  - All specific locations notified by CIS and agreed with RNDS will receive 'urban coverage'.
- 'Urban coverage' is defined as receiving a median field strength of at least 62 dB$\mu$V/m as measured at 1.5 metres above ground height, in the street, for 90 % of measured locations (outside the short sector). There must be no coverage gaps of more than 500 m in diameter.
- 'Rural coverage' is defined as receiving a median field strength of at least 47 dB$\mu$V/m as measured at 1.5 m above ground level, in the road, for more than 90 % of measured locations (outside the short sector). There must be no coverage gaps of more than 1000 m in diameter.
- 'Road coverage' is defined as receiving a median field strength of at least 36 dB$\mu$V/m as measured at 2 m above road level, in the road, for more than 95 % of measured locations. There must be no coverage gaps in excess of 1000 m along any road.

These requirements examine raw coverage only, and it is likely that there would be other requirements to ensure call quality, to traffic metrics and many other things. The important point to realise is that we can conduct a survey to determine whether each of these metrics are individually met. We have defined the measurement method in the street for each case, to allow rapid survey by vehicle. We have provided single, unambiguous values to determine whether the metric is met; 90 %, so that 89.9 % fails, and 90.0 % passes, on the basis of a field strength that can be directly measured by a suitable configured and calibrated measurement receiver. We have avoided metrics that would be difficult to measure. For example, we could have chosen the following examples, with attendant problems:

- 'The network will provide a working signal in 90 % of locations'. What constitutes a 'working signal'?
- 'The network will provide 47 dB$\mu$V/m inside 90 % of buildings'. Difficult to measure, because it would require access to many buildings, significant effort to manually measure on foot, and it is almost impossible to design against, given the variety of building types to be found.
- 'The network will provide coverage to 75 % of rural areas within the service area'. Get the walking boots out, to tramp across kilometres and kilometres of countryside with a measurement receiver and GPS. We have carefully designed the metric to be measurable by car on the road, on the assumption that this is representative of the rest of the rural terrain.

Once these metrics have been proposed, they will still need to be agreed by the customer. This is likely to lead to significant discussion, and it is important that the rationale for each metric is well explained, and the problems that would arise from alternative methods are equally well explained. In most cases, a reasonable compromise will be achieved.

The outcome of this whole activity is the generation of a *de facto* FS, which again in this case is incorporated into the MS. This is likely to be detailed and long, covering every important aspect of the network design, but it is ideal for designing the network against and also for generating the test specification. There will also be many additional requirements for the complete network infrastructure, such as network reliability, but these will be dealt with by RIS, the infrastructure suppliers.

## 8.8 Detailed Design Document

The detailed design document is effectively the network design, and will continue to evolve until the project is complete. As such it will be the last document completed. It can either be produced as a stand alone document, or it can be incorporated into the MS such that by the end of the project the single document identifies exactly what was done in order to achieve the design obtained (this is the main benefit of combining them). This means that at the end of the project, the MS should be renamed as Project Documentation or other meaningful name. Note that for large projects, the MS and associated documents may run into several volumes, and thus the DDD can be stand alone as a separate volume. In any case, it must be easy for readers to separate the how (how the project was completed) from the what (what the eventual design was).

At this stage we can however detail the contents of the DDD. In this case, we can propose the following contents:

- A list of each proposed base station locations, defined in National Grid Reference (NGR) 12-figure coordinates.

- For each antenna on each site:
  - The height above ground in metres.
  - Orientation in azimuth and tilt in degrees for both.
  - The power to be emitted by the antenna in dBm, reference a dipole.
  - The frequencies for transmission and reception in MHz.
- The frequency plan for the entire network, provided as a list of base station and frequency.
- Coverage plots for 'urban coverage' in A2 and A3 paper sizes, as colour plots.
  - Total coverage area of the network.
  - Coverage per base station (treating each base station individually)
  - Plots of areas with coverage holes of 150 m diameter or more.
- Coverage statistics for 'urban coverage'.
  - Total area covered in square km over whole service area.
  - Percentage of total area covered.
  - Total area covered in cities, towns and villages with population over 5000 or population density of more than 150 inhabitants per square km.
  - Percentage of cities, towns and villages with population over 5000 or population density of more than 150 inhabitants per square km covered.
- Coverage plots for 'rural coverage' in A2 and A3 paper sizes, as colour plots.
  - Total coverage area of the network.
  - Coverage per base station (treating each base station individually)
  - Plots of areas with coverage holes of 500 m diameter or more.
- Coverage statistics for 'rural coverage'.
  - Total area covered in square km over whole service area.
  - Percentage of total area covered.
- Coverage plots for 'road coverage' in A2 and A3 paper sizes, as colour plots.
  - Total coverage area of the network, filtered along roads.
  - Coverage per base station (treating each base station individually)
  - Plots of areas with coverage holes of 200 m length or more.
- Coverage statistics for 'road coverage'.
  - Total length of roads covered, in km.
  - Percentage of all roads covered.
  - Total length of major roads covered, in km.
  - Percentage of major roads covered.
  - Total length of minor roads covered, in km.
  - Percentage of minor roads covered.
- Coverage statistics for each specific point of interest defined by CIS and agreed by RNDS.
- Best server plots.
  - Best server plots for whole service area.
  - Best server plots for each city, town or village with population of more than 5000 inhabitants or population density of more than 150 inhabitants per square km.
- Interference plots.
  - Interference plots for whole service area.
  - Interference plots for each city, town or village with population of more than 5000 inhabitants or population density of more than 150 inhabitants per square km.
- Interference statistics.
  - Interference statistics for whole service area.

  - Interference statistics for each city, town or village with population of more than 5000 inhabitants or population density of more than 150 inhabitants per square km.

This actually is quite a short list and in a real project there be list would be more comprehensive, but at least it illustrates the principles of the DDD.

## 8.9 Test Specification

The test specification is used to determine whether the eventual network design (based on analysis of coverage obtained in the simulations) and the network implementation (based on physical surveys) meet each of the criteria laid down. Test specifications can be very complex to produce if sufficient thought has not gone into their creation. They are also classic areas of disagreement if they are developed late in the project, so it is very important to get them agreed as early as possible. Fortunately, in our design, we have factored this in when developing the FS. In fact, all we need to do is take each element of the FS and turn it into a test specification element.

In the FS, we derived the following elements for coverage:

- Coverage requirements.
  - All cities, towns and villages of population greater than 5000 must receive 'urban coverage'. The list of such towns is drawn from the National Mapping Agency list of Cities, Towns and Villages, Issue 2 of 2005.
  - All regions with a population density of greater than 150 inhabitants per square kilometre must receive 'rural coverage'. These areas are derived by using the National Mapping agency Postcode and Population data set, Issue 4 of 2004. Linear distribution over postcodes is assumed.
  - All major roads, defined in National Mapping Agency RouteMaster data set, Issue 1 of 2006 will receive 'road coverage'.
  - At least 75 % of minor routes, defined in National Mapping Agency RouteMaster data set, Issue 1 of 2006 will receive 'road coverage'. The percentage is defined in terms of overall road length.
  - At least 75 % of the total service area (as defined by the county boundary map, described above) will receive 'rural coverage'.
  - All specific locations notified by CIS and agreed with RNDS will receive 'urban coverage'.
- 'Urban coverage' is defined as receiving a median field strength of at least 62 dB$\mu$V/m as measured at 1.5 m above ground height, in the street, for 90 % of measured locations (outside the short sector). There must be no coverage gaps of more than 500 m in diameter.
- 'Rural coverage' is defined as receiving a median field strength of at least 47 dB$\mu$V/m as measured at 1.5 m above ground level, in the road, for more than 90 % of measured locations (outside the short sector). There must be no coverage gaps of more than 1000 m in diameter.
- 'Road coverage' is defined as receiving a median field strength of at least 36 dB$\mu$V/m as measured at 2 m above road level, in the road, for more than 95 % of measured locations. There must be no coverage gaps in excess of 1000 m along any road.

These can be easily turned into part of a test specification by putting them in a table, long with a record of whether they are met by prediction and subsequently by survey. The

**Table 8.2.** Test specification extract.

| No. | Test | Met by prediction (Y/N) | Met by survey (Y/N) |
|---|---|---|---|
| 1 | All cities, towns and villages of population greater than 5,000 must receive 'urban coverage'. [Notes 1 & 3] | | |
| 2 | All regions with a population density of greater than 150 inhabitants per square kilometre must receive 'rural coverage'. [Notes 2 & 4] | | |
| 3 | All major roads will receive 'road coverage'. [Notes 5 & 7] | | |
| 4 | At least 75% of minor routes, will receive 'road coverage'. The percentage is defined in terms of overall road length. [Notes 5 & 8] | | |
| 5 | At least 75% of the total service area will receive 'rural coverage'. | | |

*Notes*:

1. The list of such towns is drawn from the National Mapping Agency list of cities, towns and villages, Issue 2 of 2005.
2. These areas are derived by using the National Mapping agency Postcode and Population data set, Issue 4 of 2004. Linear distribution over postcodes is assumed.
3. 'Urban coverage' is defined as receiving a median field strength of at least 62 dB$\mu$V/m as measured at 1.5 metres above ground height, in the street, for 90% of measured locations (outside the short sector). There must be no coverage gaps of more than 500 metres in diameter.
4. 'Rural coverage' is defined as receiving a median field strength of at least 47 dB$\mu$V/m as measured at 1.5 metres above ground level, in the road, for more than 90% of measured locations (outside the short sector). There must be no coverage gaps of more than 1,000 metres in diameter.
5. 'Road coverage' is defined as receiving a median field strength of at least 36 dB$\mu$V/m as measured at 2 metres above road level, in the road, for more than 95% of measured locations. There must be no coverage gaps in excess of 1,000 metres along any road.
6. Major roads are defined in National Mapping Agency RouteMaster data set, Issue 1 of 2006.
7. Minor roads are defined in National Mapping Agency RouteMaster data set, Issue 1 of 2006.
8. The overall service area is defined by the county boundary map produces by National Mapping Agency 1:250,000 county boundary data set, issue 1 of 2006.

prediction and survey methods should be capable of providing a Yes/No answer to each of the elements in turn without requiring further interpretation.

Thus, we see that the TS can be created at the same time as the FS, and using the same metrics. This means that the network is designed to and tested against the same metrics throughout the project. An example is shown in Figure 8.2.

## 8.10 Project Plan

Project plans are part of any normal business activity and we do not propose to teach their use here. Of course, the project plan should be part of the project documentation and in this project we have chosen it to be included in the MS, but it can also be a standalone document.

## 8.11 Quality Plan

Any engineering project should have an associated quality plan. This is to ensure that errors are minimised and, if created, are not propagated. Most reputable consultancy and engineering companies will hold ISO 9001:2000 accreditation, and the company processes developed to meet this standard will be used to generate the quality plan for a given project. Again, this can be part of the MS of held as a standalone document.

## 8.12 Risk Register

We identified the need for a risk register when we looked at stakeholders earlier in the project. This should cover not only issues relating to stakeholders, but to any other identified risks in the project. Pro-active effort should be put into identifying technical and project risks, and maintaining the risk register as a live document throughout the project as part of the risk management activity. Each element can be identified by:

- A unique reference number.
- A text description.
- A text description of the problems it may cause.
- A value for the likelihood of its occurrence.
- A value for the severity of problem likely to arise from its occurrence.
- An importance value based on its likelihood times its severity.
- Mitigating action that can be taken to prevent it.

An example might be (on the basis of the above items):

- Risk number 01234.
- 'List of points of interest not provided by CIS in a timely fashion'.
- 'Design process may be delayed while waiting for data to be delivered, leading to overall project delays'.
- Likelihood value $= 4$.
- Severity value $= 3$.
- Importance value $= 4 \times 3 = 12$.
- Maintain pressure on CIS to provide the data. Remind them of the impact of delay and that we cannot be held responsible for any subsequent delays should this occur. Project manager to speak to CIS CRO or CI if problem persists.

Typically, the risk register is held on a spreadsheet or, for larger projects, on a dedicated risk management tool. Again, it can form part of the MS or be held separately. Of course, some of the risks identified may be sensitive in the context of the project, so great consideration must be put into determining whether the risk register should be an internal document or an open document that the customer can see.

## 8.13 Producing a Baseline

The new network will provide enhanced services compared to the old analogue network it will replace. Existing subscribers of the old network will not only expect that the new network will provide new functionality, but will also expect that the new network will at least provide service over everywhere that the old system does – even if this is not explicitly specified. Since subscribers will typically be at least partially resistant to change, it is important that the positive aspects of the new system are not undermined by clear deficiencies compared to the old network; a common cause of this is that the new system does not work where the old one does. This can often cause such resentment that it becomes increasingly difficult to get the subscribers to accept the new system, to the detriment of the entire project.

In order for designers to understand the issues and to allow them to manage expectation as well as to minimise the potential for this problem arising, it is usually a good idea to produce a baseline from the current system before starting the new design. This would typically involve producing coverage and interference plots based on the characteristics of the old network, and may even involve calibration surveys to prove the model and tweak the propagation model settings to fit the application and environment. To ensure that comparisons between the new system and the old system are entirely comparable, it is essential to ensure that both are measured against the same description of the requirements. This means that the method of modelling the entire set of requirements for the service area but be derived before the baseline can be produced.

The definition of the service area is shown in Figure 8.6. The county boundary of the hypothetical area of Countyshire is shown bounded by the polygon. Towns are shaded,

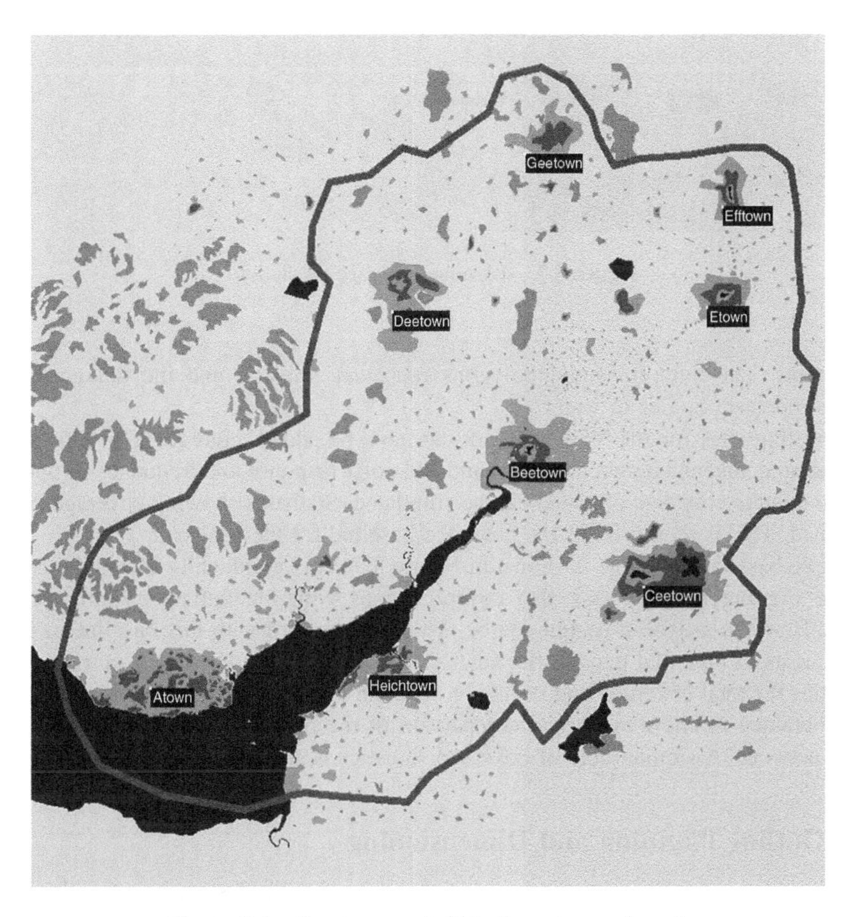


**Figure 8.6.** Service area (within the polygon shown).

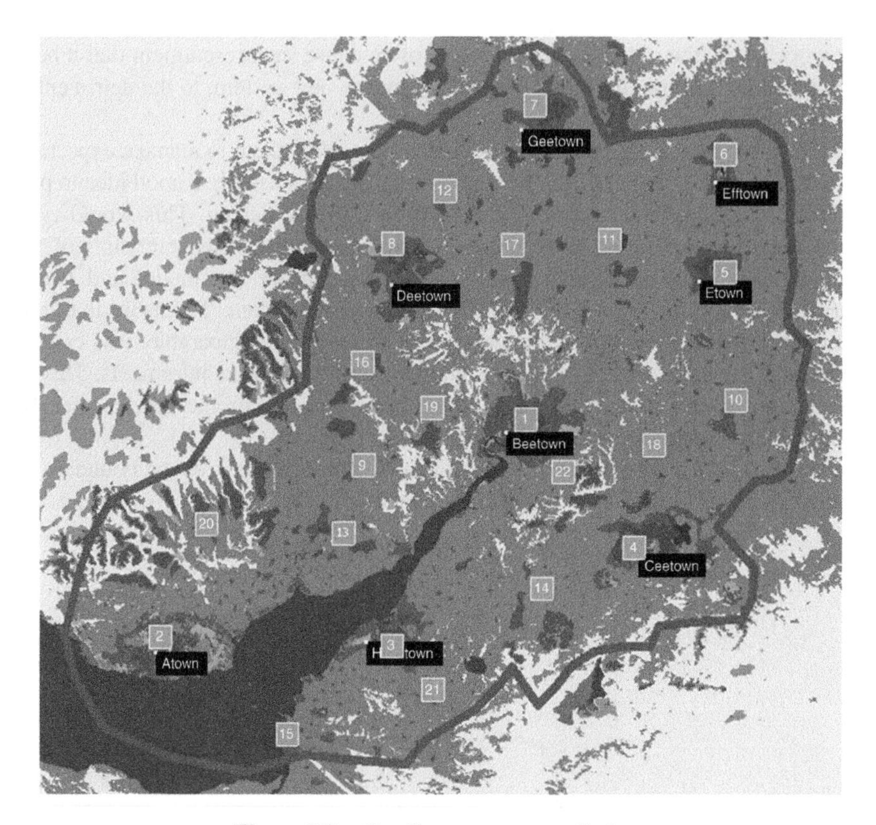


**Figure 8.7.** Baseline coverage prediction.

named areas. The light unnamed polygons represent forestry, and the darker unnamed polygons represent bodies of water.

The existing sites for the VHF network can now be added, configured to represent the power radiated, antenna height above ground and polar response and frequency of operation, and from this the coverage obtained can be simulated. Statistics showing coverage can then be obtained. This is then used as the coverage baseline for the project.

A simple baseline coverage is shown in Figure 8.7. This graphically shows the coverage from the existing network and also coverage gaps with the current system. Predicted coverage from the proposed design can be visually compared with this baseline to identify improvements or potential gaps in the new system that do not exist in the current system.

Statistics for total coverage and coverage of urban areas and major and minor roads can also be obtained to allow numerical comparison of the proposed network design with the existing network. A simple table of coverage statistics is shown in Table 8.3.

## 8.14 Outline Planning and Dimensioning

After generating a baseline, on the basis of the current radio system being replaced, it is often also useful to perform some outline planning and dimensioning. The purpose of this

**Table 8.3.** Coverage statistics for baseline.

| Coverage category | Total area covered ($km^2$) |
|---|---|
| Total service area | 89.9% |
| City, town and village with population > 5000 or population density > 150 inhabitant per $km^2$ | 81.0% |
| Rural | 90.0% |

activity is to quickly identify a rough idea of how many base stations are likely to be required and to identify any potentially risky areas to cover at an early stage. Like the more detailed planning to follow, this activity will use the same methods and tools described in the next chapter. The purpose, beyond identifying risk, is akin to performing an order-of-magnitude calculation before performing a detailed calculation on a calculator; it allows the result to be compared to a common-sense solution so that gross errors in the detailed process can be identified.

The rules used for outline planning will be a lot less stringent than for the detailed planning activity. For example, the following simple rules might be followed:

- Use a nominal antenna height of 15 m above local ground.
- Use a nominal power of 25 W.
- Assume omni-directional antennas.
- Antennas can be positioned anywhere reasonable, not just where commercial radio masts exist.
- Assume that expected number of sites will be 4/3 of number calculated using above rules.

The outline planning activity can also help to focus site acquisition teams on the most important areas within the network area, and can help refine infrastructure cost figures.

## 8.15 Calibration Surveys

If possible, and certainly for any sizeable project, it is highly recommended that a calibration survey be carried out, to allow propagation models to be tuned to the precise service area environment. Such a survey would be carried out according to the principles outlined in Chapter 15.

## 8.16 Continuing with the Project

Now that the groundwork has been done and the project has been well set up, the main activity starts. In the next few chapters we will be looking at detailed design methods for coverage, traffic, different rollout scenarios, noise and interference analysis and frequency assignment.

# 9

# Mobile Coverage Design

## 9.1 Introduction

In this section, and for the remaining chapters of the book, we look at some methods for engineering a network. A network should be designed against a clearly defined set of objectives, taking into account all of the various technical, regulation, commercial, financial and social constraints. The objectives should be as detailed and well-defined as possible to allow a critical design methodology to be adopted; by this we mean that the design can be made to just meet the requirements rather than being over- or under-engineered. Likely constraints will include the technical capability of the selected radio technology, regulations imposed by the spectrum administrator, commercial and other business-specific constraints that will vary among different projects. Cost will also normally be an issue, as will the ability to place base stations in particular parts of the service area. Local planning rules may have a major impact in this regard.

In terms of the geographic coverage of a network, often the limit of coverage will be determined by one of the following limiting factors:

- Noise limited, in which the environmental noise and the minimum signal level needed to provide the desired level of availability will be the limiting factor.
- Interference limited, in which the presence of other radio systems will increase the level of signal that will be required to achieve the necessary level of performance.
- Traffic limited, where the traffic demand will be the crucial issue in providing network performance.

In many networks, some of these elements will be present in different parts of the network. In each case, the limiting factor will modify the required service area for each base station within the network. The design of the base station will then be performed against the wanted service area, and the design of the whole network is the composite of all the required base station designs. For complicated networks, there will be more than one definition of wanted service area. For example the network design objectives might be

Mobile Radio Network Design in the VHF and UHF Bands: A Practical Approach
*Adrian W. Graham, Nicholas C. Kirkman and Peter M. Paul* © 2007 John Wiley & Sons, Ltd

the composite of the following:

- Coverage for vehicle-mounted antennas over 90 % of the major roads in the entire country.
- Coverage for helicopter-mounted antennas over 98.5 % of the entire country.
- Coverage for handheld antennas over 94 % of the composite land coverage of towns with a population of over 5000 and areas with a population density of more than 150 people per square kilometres.

As we discussed in Chapter 8 and will look at in more detail shortly, even these definitions would probably need to be further refined to be usable. In terms of defining service areas, it may be possible to produce composite service area descriptions by combining some of these individual targets, otherwise they should be kept separate and a candidate network design tested against each set of requirements. Thus, network design process can involve meeting a composite of coverage design requirements.

In this chapter, we will look at the following issues:

- Practical expression of performance criteria.
- Initial design approaches.
- Configuring, performing and interpreting coverage simulations.
- Establishing nominal characteristics for the network design activity.
- Examination of the effects of modifying base station characteristics.
- Deliverables from the design activity.

We will start by looking at practical expression of performance criteria.

## 9.2 Practical Expression of Performance Criteria

### *9.2.1 Introduction*

The design parameters will vary from project to project, and should be given a great deal of consideration. In this section, we look at how the requirements for mobile coverage are defined in a practical manner. Although this is of course vitally important for the network specification, it is equally important from a contractual point of view. In major projects, this can be a highly complex task, but it is not one that can be ignored or carried out in a haphazard manner. The key point is that each design requirement must be clearly defined in an unambiguous way, with a binary success criterion. By this, we mean that there can only be two outcomes of a test to measure whether the parameter is achieved or not; either a clear pass or a clear fail, with no room for uncertainty. Each requirement must be consistent with all of the others, and the sum of all requirements must fully express all of the objectives.

### *9.2.2 Practical Definition of Service Requirements*

The link budgets generated during the design activity will be used to plan the network and also have to reflect the commercial performance requirements. Some of the important aspects of link budget design that should be considered include the following:

- If an availability figure is used as a performance metric, this must be expressed both for location and for time. This is because at both VHF and UHF, radio propagation can change due to atmospheric conditions, particularly for long links. Thus, a definition that only expresses performance in terms of availability is not sufficient. A more tightly constrained definition might be, for example '95 % availability for 50 % of the time'. As we will see in Chapter 15, it is also necessary to carefully consider how this will be measured during surveys once the network has been built to confirm performance. For the expression above to be more complete, it would also have to include a term to describe the percentage of the service area that benefits from this availability level; such as '96 % of short sectors within the service area must benefit from 95 % locations availability for 50 % of the time, with no individual coverage gap exceeding 1 $km^2$'.

- For safety-of-life networks, availability may be an insufficient metric for network performance. An alternative definition is the metric of confidence rather than availability. This is the probability that a single call will result in a successful link, and it is a far more rigorous metric than availability. An example of this type of performance metric might be '95 % probability of a successful call for 90 % of the time in 90 % of locations'. Using a metric of this type will result in a more robust network than a simple expression of availability, but the network cost and infrastructure requirements will be considerably higher. This is not justified for most non safety-of-life applications.
- For in-building coverage, it will often be necessary to identify the margin to be applied for building penetration. This may be a single figure, as 'in-building coverage will be assumed to require 22 dB above the street value'. As part of the specification, it will also be necessary to identify how the percentage of buildings benefiting from the specified level of coverage is expressed. This could be as simple as expressing that, say, 90 % of an urban area must benefit from the required signal strength, but this definition is somewhat lax unless it is more closely defined; for example does 'urban area' include streets, parks and other open areas? There is little point in these locations benefiting from the 22 dB uplift; ideally there should be an expression that identifies that in-building coverage is required for buildings within the urban environment, based on a given environmental data set that differentiates between buildings and other land used in the urban environment.
- For road coverage, again the definition of the required coverage must be specified closely. It is not sufficient to state that 'the network must provide coverage over major roads'. A more detailed and less ambiguous definition would be better, such as 'the network must provide coverage over at least 95 % of major roads, defined as category A and motorways in the service area by the Highways Agency, and there must be no coverage gaps of greater than 500 m in length'. This defines the roads to be covered, how they are defined by unambiguous data and also provides two metrics that are measurable by survey (total coverage of 95 %, no gaps greater than 500 m).
- For area coverage, the same type of coverage requirement expression can be made as for roads, such as 'the network must provide coverage over 90 % of the service area, with no more than 10 coverage gaps, and none of more than 10 000 $m^2$. This type of definition can also be made far more specific by breaking the service area down into sub-sections, and also by clutter category. It is important with this type of definition to ensure that each sub-category does not overlap; otherwise it becomes more difficult to determine whether the overall specification is consistent.
- Besides defining general coverage areas, it may also be appropriate to identify specific locations that must be covered individually. These may be expressed as single point

locations or by small polygons. Typically, such individually defined locations might include customer premises, ports, airports, golf courses, known trouble-spots, popular tourist areas and so forth, depending on where network subscribers can be expected to be. For large networks, individually identified locations may run into the hundreds or thousands. The crucial aspect is that the locations that are important to the customer are not omitted by a design that generally meets the overall coverage figures for the service area.

It may appear that defining the network requirements to such a degree of detail is pedantic. However, in order to avoid possible contractual conflict later in the project, and also to allow the designers to focus on the customer requirements, it is important to go to this level of specification. Otherwise, it is highly likely that the final design will not fully meet the customer requirements, and the designers will not be able to prove that the requirements have been met.

### *9.2.3 Definition of Geographic Service Areas*

We have identified that coverage objectives may require a number of different levels of definition, and that the overall objectives are the composite of these definitions. An example is shown in Figure 9.1. Each of the areas specified will be present due to the meeting of one or more of the criteria in the network functional specification.

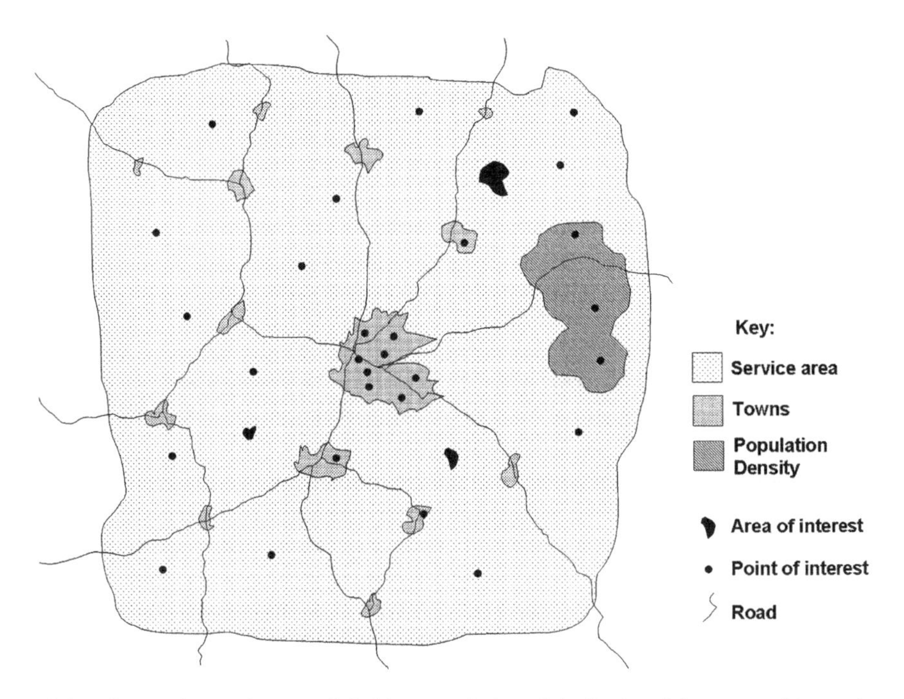


**Figure 9.1.** Composite service area definition, consisting of the limits of the geographic service area and sub-categories within the service area.

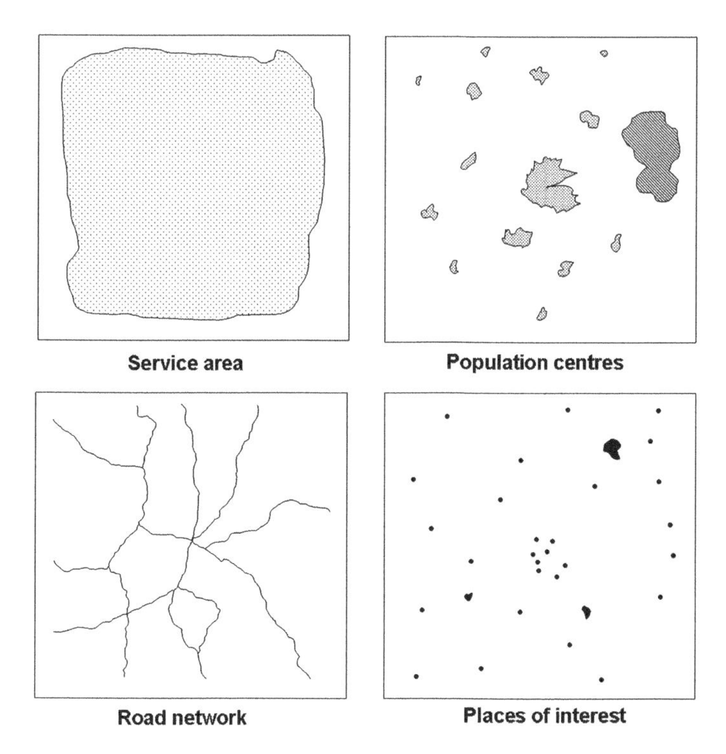


**Figure 9.2.** Service area definition elements.

The composite definition can be broken down into individual elements for further analysis as shown in Figure 9.2 and to be the basis of ensuring that individual objectives are met.

The specification of service area is not just an issue of proving the network design meets the objectives, it is also fundamental in the design process itself. Each element can be broken down into smaller areas to help to define coverage areas for individual base stations or small groupings of base stations. We will see more of this later, but first we look at one of the most difficult parts of any engineering activity; the approach to take at the start. After all, network design is a complex activity involving many different variables. Given such a complex set of design factors, what approaches can be taken to come up with an initial design so that the rest of the design can be built on some form of foundation? We will look at some simple examples next.

## 9.3 Initial Design Approaches

Before we go on to examine how to perform coverage predictions and to set up the parameters to be used in the design, we will look at some possible approaches to initially identifying potential areas to search for suitable base station locations. This may form the seed for the rest of the design process, or it may be the largest part of the base station selection process depending on how much can be achieved during this activity. This will typically depend on the data available and the approach taken. The following examples are not the only approaches that can be taken; an important part of the network designer's mental toolkit is the ability to determine the best site selection strategy for any given project.

### 9.3.1 Grid Style Approach

Probably, the least sophisticated approach to initial base station site finding is to use an approach based on a grid or cellular structure. This is the approach that was initially proposed for traffic-limited cellular coverage when planning tools did not offer any better solutions, and to some extent it still persists within the cellular network planning community, even though better methods may now exist. The basic approach is illustrated in Figure 9.3. The points indicate the grid, the spacing of which is calculated according to the expected distance between base stations expected for the final design. The service area is also shown. The idea is that initial searches for potential base station locations will be performed in the area around each point. Once this has been performed, coverage predictions can be carried out to determine how well each potential site contributes to the overall coverage of the network.

The seed point to generate the grid can be random or can fit to some basic criteria. For example the grid in Figure 9.3 is based on placing as many points as far as possible from the edge of the service area. The basic approach can be enhanced by using overlaid grids with, for example smaller, denser grids used for urban areas and wider grid patterns for rural areas. Extra grid points can be added along features such as roads.

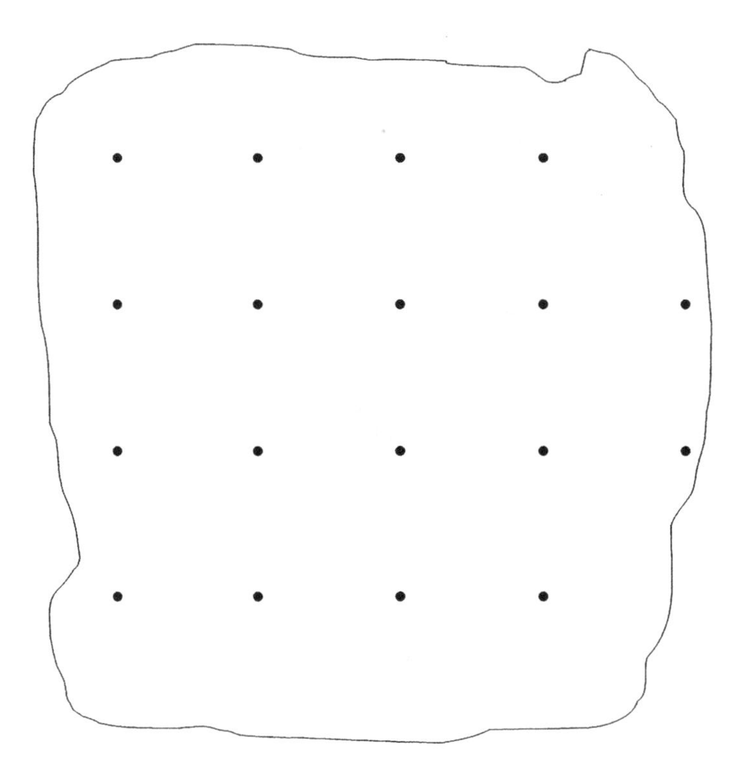

**Figure 9.3.** Grid style initial site layout approach.

This approach can be used when there is little or no information available and no better method presents itself, or if the expected design will fit the pattern (based on previous experience of similar networks), but in general it is a relatively unsophisticated method.

## *9.3.2 Selective Design Approach*

The selective design approach is illustrated in Figure 9.4. This diagram again shows the extent of the service area and all of the areas that need coverage due to the presence of population that meets the design criteria. Over and around each area is a patterned polygon, intended to represent the coverage achieved when an initial design activity has been carried out to provide coverage within each rural area. This is a slightly less complex activity than the overall design, and in many cases the actual locations for each urban base station may not greatly affect the coverage that extends out into rural areas. The white areas within the service area still need coverage, and the design can now continue until the white areas are covered. Thus, a simple activity has been performed and it then provides a seed for the rest of the design process.

The same type of approach can be taken for any specific aspect of the design, such as road coverage, coverage of places of interest and so on. If the network performance objectives are prioritised, then this should usually be chosen for this initial activity. Alternatively, if it is anticipated that meeting a particular objective will be most difficult to achieve, this is also a good candidate for the initial design process.

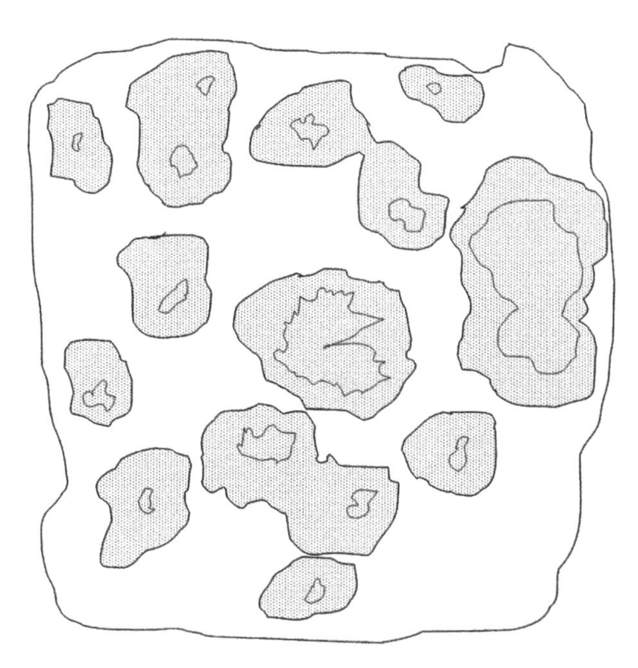

**Figure 9.4.** Selective design approach; providing coverage for a subset of the overall network requirements.

This approach can be used for new networks where there are no existing sites and it can be an applicable method to combine with the existing site approach discussed below. It is applicable to both coverage and traffic limited radio networks. For complex network design, it is often one of the best approaches. There is a risk with this approach that the 'islands' of coverage can leave irregular areas left to cover. There are some examples of this in Figure 9.4, where the islands almost, but not quite, meet and thus extra sites may be needed to fill in these small gaps. This leads to sub-optimal design. Again, the network designer should use their knowledge and experience to determine whether this approach is suitable for a given network design.

### *9.3.3 High Point Approach*

For networks where the aim is to provide the greatest geographical coverage using the minimum amount of fixed infrastructure, a simple approach is to identify high points that are likely to provide the largest coverage areas. The tops of hills are ideal, where access is available and facilities such as power can be provided. An illustration is shown in Figure 9.5. The inverted triangles show the high points of hills, and the shaded area around shows the coverage achieved.

When using this approach, the selected high points should be far enough away to avoid them in providing overlapping coverage.

The advantages of this approach are that it is simple and identifies sites that are likely to provide the best coverage over a wide area. It is not, however, appropriate for traffic limited

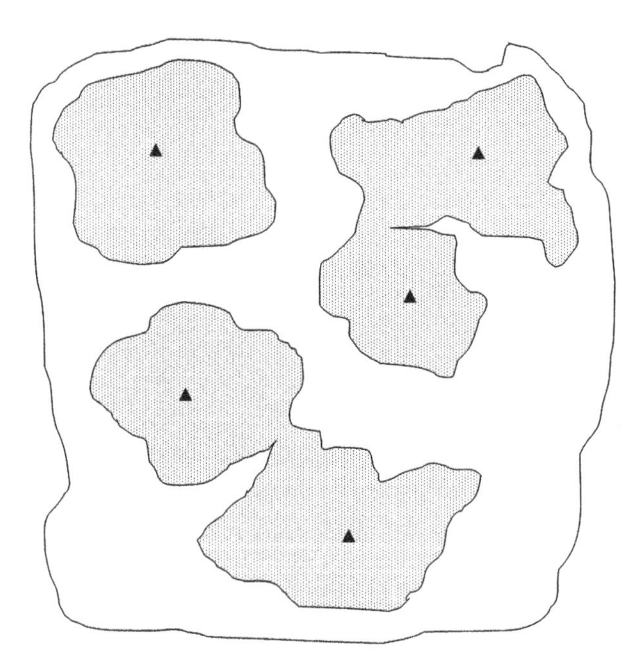

**Figure 9.5.** High point approach; choosing high points for base station locations and examining the resulting coverage.

radio networks, because wide area coverage is unlikely to be desirable. Also, this approach will tend to increase interference to and from other spectrum users.

### 9.3.4 Existing Site Approach

In practice, many networks will be constrained by the need to use existing sites, often stored in a database. This may be because the intention is to use commercial radio masts, masts already used by the business for other networks or the need to co-locate base stations with other infrastructure.

If this is the case, and the size of the database is not too constraining, then the best approach may well be to work in a different way. In this case, if simulation time will not be excessive, it may be best to perform coverage predictions for each potential location in the database as the first activity. It is important to recognise that the costs of buying several computers to allow the parallel processing of site coverage for the database may be cheaper than adopting other, less efficient approaches, so this should be borne in mind.

Once the coverage of every site has been predicted, tools such as simultaneous coverage can be used to selectively *remove* sites from the project until only the minimum number required remain.

Figure 9.6 shows a simple example of a simultaneous coverage display used in this process. The degree of re-coverage for each area is shown. The idea of the approach is to identify the areas with the highest degree of simultaneous coverage and start removing sites that contribute to the re-coverage. This is iteratively repeated until the minimum degree of re-coverage (usually one) is achieved across the network. This is an iterative process because

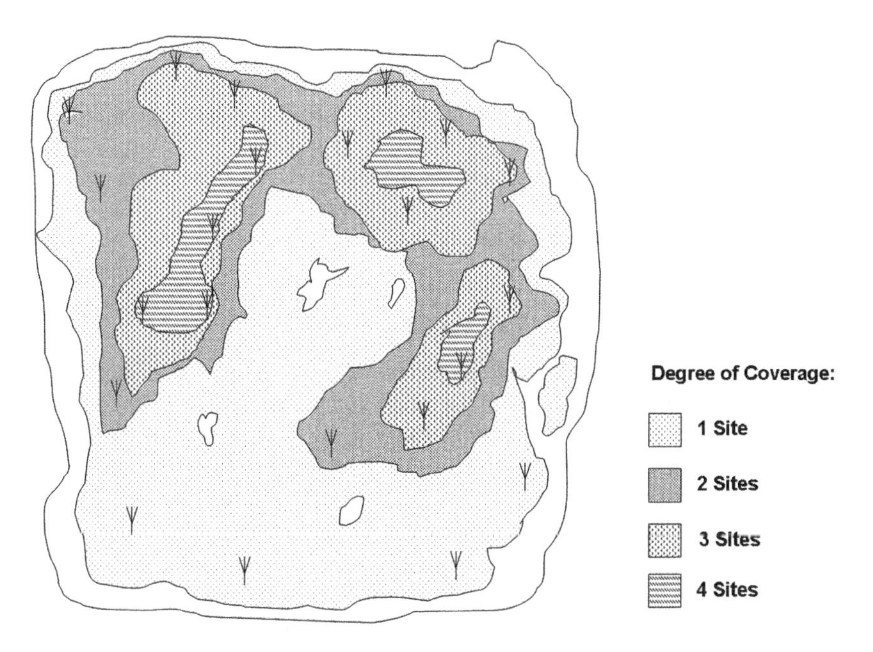


**Figure 9.6.** Existing site approach; using re-coverage method described in text to come up with a network design.

sometimes removing a site will lead to coverage gaps and thus it will need to be added back in again.

Although it may seem counter-intuitive to perform simulations for many sites that will not be used in the final design, in fact once the simulations have been completed, an experienced network design engineer can work out a good network design very quickly, so the overall design time is small. Also, since simulations can be set up on many PCs and left running unattended (given the right software and uninterruptible power supplies and so on), the design costs are typically much lower than many other techniques.

### *9.3.5 Automatic Site Finding Approach*

Many of the approaches discussed above can be automated in a sophisticated planning tool. If this is the case, then the higher cost of such a tool relative to less sophisticated systems is quickly offset by the reduction in design effort and time that results.

## 9.4 Configuring, Performing and Interpreting Coverage Simulations

### *9.4.1 Setting Up Coverage Predictions*

Modern planning tools suffer from the generic problem of all software systems, which is that they will process inputs automatically to produce outputs, whether the input data is sensible or not. In fact, this can be compounded by the fact that coverage plots can look very credible even if the data displayed is entirely wrong. The radio engineer therefore has to be careful to ensure that the tool is set up with the correct data before performing a coverage-based prediction. The data that needs to be validated before starting a prediction broadly falls into the following three categories:

- The propagation model and its configuration.
- The environmental data to be used.
- The parameters for the radio elements to be modelled.

Each of these is described in more detail next.

#### 9.4.1.1 The Propagation Model and its Configuration

If the propagation model and its configuration are not completed correctly for each simulation to be performed, then all of the results obtained are completely invalid and any decisions taken on this information will likewise be flawed. Radio network planning tools designed for only one technology are usually relatively easy to configure since the number of options are likely to be small. However, for sophisticated planning tools aimed at high-level consultants, the choices and options can be far more complex. This is illustrated in Figure 9.7, which shows the ICS Telecom propagation configuration box. Each core model has to be configured with appropriate additional factors.

This type of dialogue box is ideal for the highly trained consultant, but less appropriate for less experienced users. If dialogue boxes are complex, then there are mitigation methods that

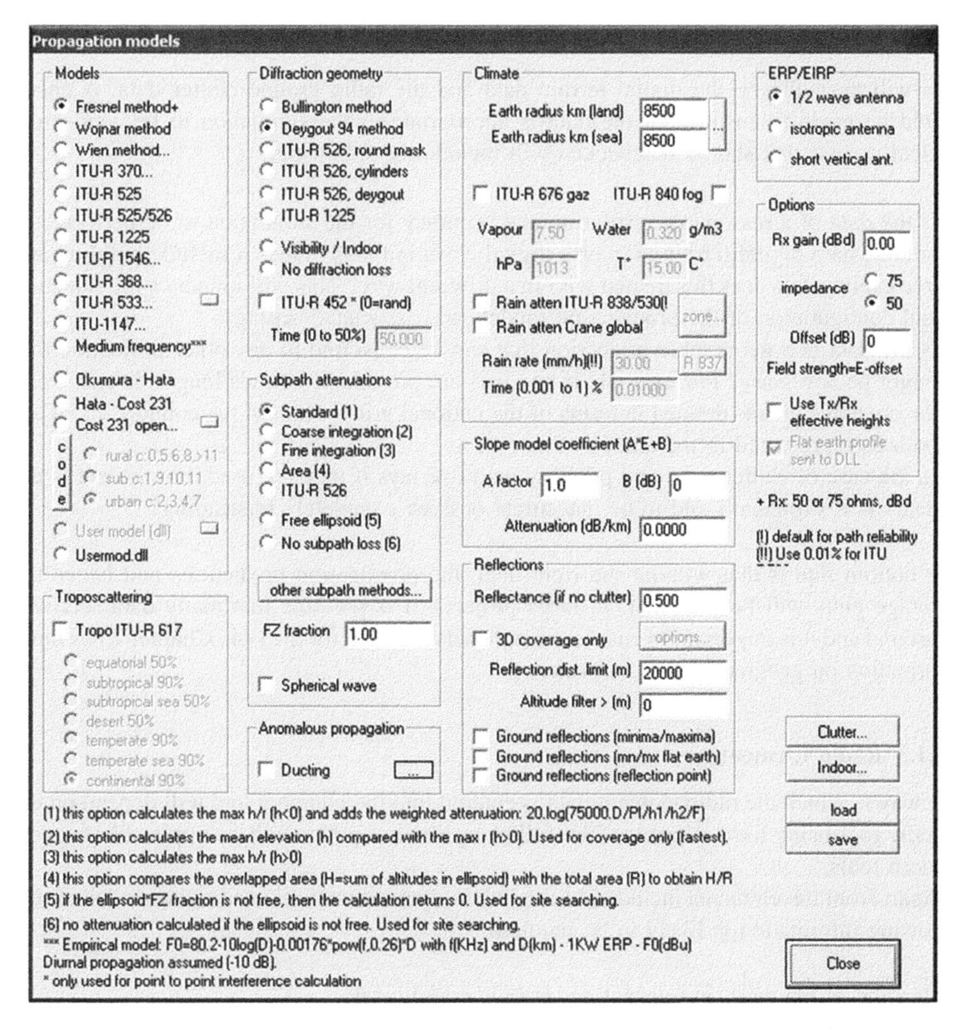


**Figure 9.7.** Sophisticated propagation configuration dialog box. Reproduced by permission of P & D Missud.

can be adopted. This includes provision of pre-configured propagation settings that the operator can select from a list, with the main dialogue box simply not available to less able operators.

It is also important to ensure that the configuration settings are recorded for auditing purposes later in the project. In this way, any errors can be detected whenever they have occurred and corrected.

The bottom line is that the engineer must ensure that they not only understand when, where and how specific models and associated configurations can be used, but how they are set up without errors in the planning tool used.

#### 9.4.1.2 Environmental Data

This will normally be the digital terrain data and the radio ground clutter data. A check should be made to ensure that the data is appropriate to the simulation to be conducted. Typical aspects that should be checked will include the following:

- Is the data of a reasonable resolution and accuracy for the simulation to be performed?
- Is the data a digital terrain map or a digital elevation map? Does it include building and tree canopy data or is this treated with in a different way? This will influence the selection and configuration of the propagation models and associated settings.
- Is the data in a geographic projection that can be converted to any other projection that might be necessary? For example if site lists are provide in latitude/longitude terms and the coverage area is defined in terms of the national grid system of the country, can these both be represented in the data?
- In the case of clutter data, and possibly map overlays if they are used, how recent is the data? Is it sufficiently old to be inaccurate or even completely misleading?

The bottom line is that without the right data, the propagation predictions and hence the coverage plots will be incorrect. In large projects, it is possible that many data sets will be around and it is important to ensure the right sets are used for the task. Chapter 7 has more information on geographic data information.

#### 9.4.1.3 Radio Elements

The way in which the radio element data is entered into the planning tool will depend on the specific tool being used. However, the following general points will be applicable to most modern tools.

Apart from the elements included in the link budget, which will need to be modelled, the following information is likely to be required:

- Position of the mast, in relevant geographic coordinates.
- Antenna characteristics, including the polar response for both the horizontal and vertical planes.
- Antenna height, expressed either in terms of above ground level (AGL) or above sea level (ASL).
- Antenna bearing (azimuth) relative to map grid or magnetic North.
- Antenna tilt in degrees.
- Administration information to allow the unique description of fixed base stations. This will be important for the management of large network planning activities, when there may be hundreds or thousands of candidate base station sites.

Information such as bandwidth, filter response and interference rejection curves may not need to be known for purely coverage predictions. For systems that report the results of a coverage prediction in terms of field strength, it will not be necessary to enter any parameters for the receiving system other than the height of the antenna. This will be AGL for terrestrial systems or ASL for aeronautical systems.

#### 9.4.1.4 Section Summary

All of this is an illustration that, as ever, the engineer must never lose sight of common sense even when using sophisticated tools. Without attention to this level of detail, the risk of producing misleading or entirely incorrect data is very high. For this purpose, it is essential that any project includes procedures, reviews and audits to prevent and detect any mistakes made during the project. This should never be viewed by engineers as unnecessarily intrusive or in some way showing a lack of trust. Engineering is a complex activity and even the best engineers can make mistakes. Proper project procedures minimise the probability that such errors may propagate through the project. These should be incorporated in the project quality plan and each engineer working on the project should be briefed on them.

### *9.4.2 Performing Coverage Predictions*

The more sophisticated propagation models are very mathematically intensive (or, in computer jargon, they are computationally costly). The time taken to complete simulations will typically depend on the processor speed, the RAM fitted, access time for data stored on disk as well as the efficiency of the software algorithm used. The factors that will affect the amount of time taken include the following:

- The resolution of the environmental data; the higher the resolution the more calculation points and the longer the calculation time.
- The propagation model and configuration settings selected. Although some models may be quicker than others, this should be used only as decision criteria if the faster model is as appropriate to use as slower models.
- The area over which the prediction is to be made.

Some systems will provide the operator with options to reduce simulation time, such as constraining the simulation to a target area, specifying a maximum range from each base station to calculate or constraining the simulation to a sector around each site. Although it may be attractive to minimise simulation time, it is necessary to consider the implications of doing so, such as:

- Is the prediction coverage being calculated to determine system performance or interference between systems? The level of interfering signals will be relevant at a far greater range than service limit, and so it will be essential to ensure that the interference prediction is carried out to the maximum possible interference limit, otherwise areas of potential interference at extreme range will not be accounted for.
- Does the technology have any timing constraints that may affect range? Most digital systems will have a limit based on the maximum transmission delay. In this case, when analysing system coverage it would be pointless to predict beyond the maximum limits imposed by the technology.

In order to maintain a balance between speed and reliable results, these points should be considered prior to the PC being set to run predictions.

### 9.4.3 Configuring Prediction Results

Coverage plots may be configured to show the results of predictions in a number of ways, and it is important to be able to display information that helps the engineer for the specific task being performed. Typical display configurations include the following:

- Path loss values in dB.
- Raw coverage prediction figures in dBμV/m.
- Equivalent received power in dBm for a specific receiver system.
- Link margin in dB for a specific receiving system.
- Availability figures in percentage terms, based on a specific receiving system and an appropriate representation of fading characteristics.

An illustration of a raw coverage display is shown in Figure 9.8. The field strength is shown in different field strengths received at each location, with shaded levels displayed in banded values. In most systems, the display will be in colour and it will be possible to differentiate between about 10 and 12 levels. This type of display shows the direct results of the propagation prediction algorithm. Although this is the true output, it is usually better to configure it to provide most system-specific information (for the system under analysis). Note that normally the coverage display shown would normally be displayed over a background map to allow interpretation of the performance of the system in terms of social

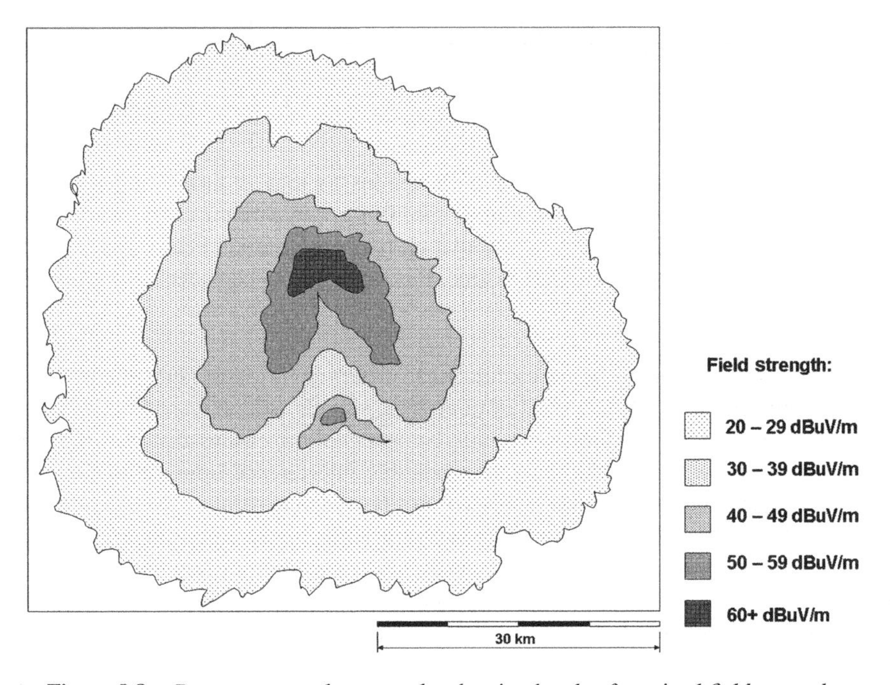


**Figure 9.8.** Raw coverage plot example, showing bends of received field strength.

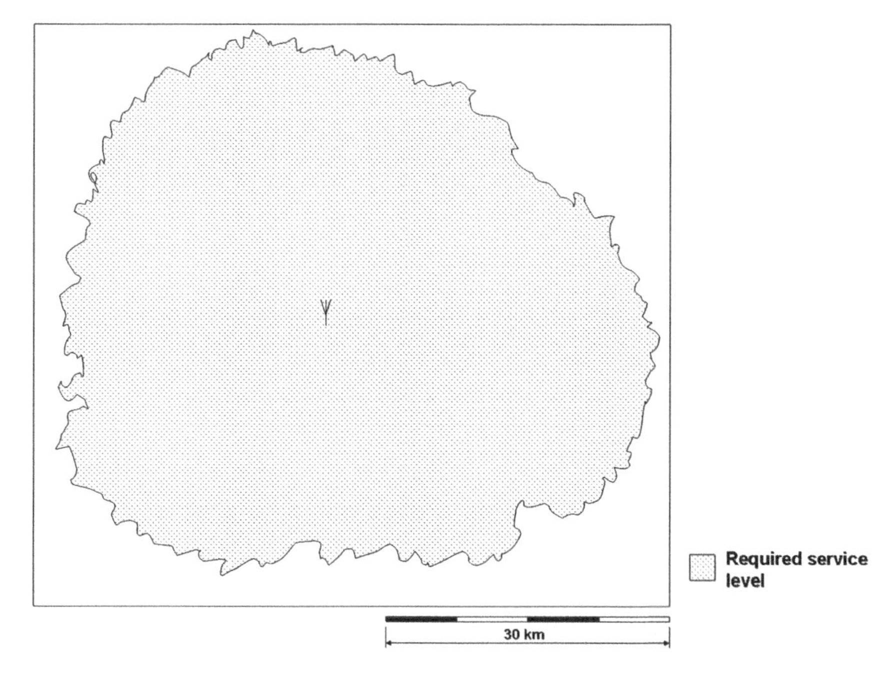


**Figure 9.9.** Prediction results configured to show required service level only to de-clutter display.

features such as towns and roads; for ease of interpretation in the monochrome medium, this has not been done for any of the diagrams in this section.

An example is illustrated in Figure 9.9. This shows part of the information of the previous display, but the system has been configured to show the areas where the system meets a 'required service level'. This would be used in the situation where the network performance criterion will be on the lines of 'the system must meet the required service level'; it is not important how much the performance exceeds the criterion only that it does. This means that providing extra layers of information is not only useless, but also counter-productive; it detracts from the only important piece of information.

It is also possible to configure prediction results to show more than one level of information. For example consider the situation where it is important to know where the network provides an availability figure of 90 % and where it provides 99 % availability.

It should be clear that where important information can be determined from a raw figure in field strength, it is possible to show it directly in a configured coverage plot.

In addition, it is worth considering an important feature of radio propagation, which is known as the reciprocity theory. This states that energy travelling from one antenna to another will suffer the same loss as energy travelling in the other direction. Thus the path loss from a base station to a mobile in the downlink direction is the same as that of the mobile to the base station in the uplink direction. If we consider that differences in the noise floor can be accounted for by the treatment of clutter, then we can relate the path loss figures in both directions to determine link performance in both the uplink and downlink directions in the same diagram. If the uplink and downlink link budgets are not balanced,

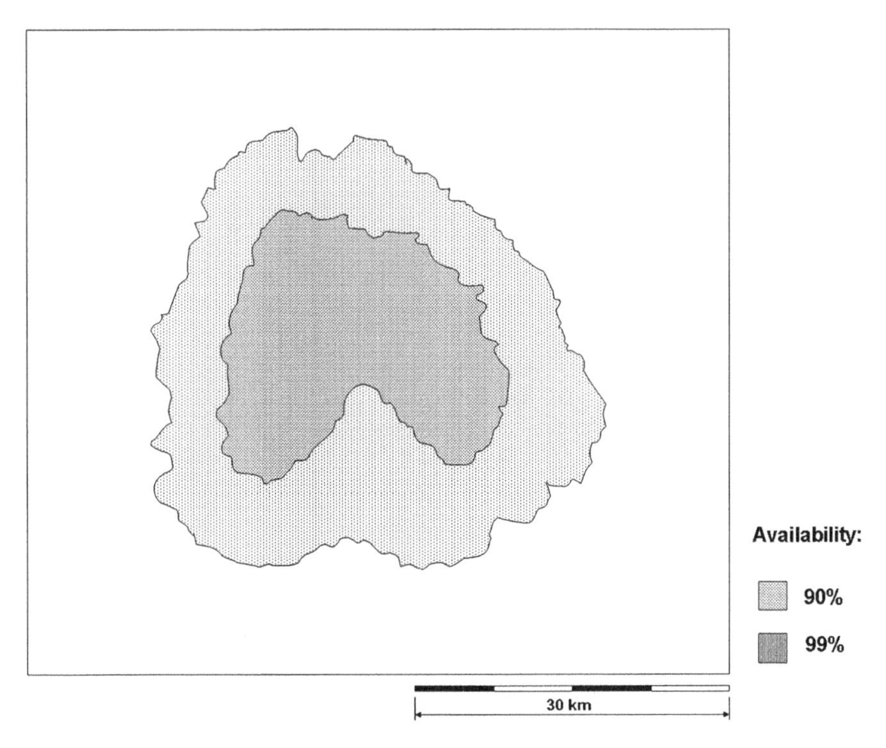


**Figure 9.10.** Results configured to show two levels of availability.

then it is possible to show both links on the same diagram as shown in Figure 9.11. This allows the engineer to probe potential areas where one link will not perform to the wanted standard. The engineer can use this to determine mitigating actions.

The basic principles described can be used to configure the display in a number of ways, such as:

- A direct comparison of a single mobile receiver type but with two antenna options that have different gain. This allows comparison of the performance achieved by the two antennas so that one can be selected.
- A direct comparison of two types of mobile receiver with different link budgets to examine the differences in performance achieved, with the additional service area achieved by the most sensitive radio shown in black.
- A direct comparison of mobile systems fitted to a vehicle with a hand-held radio, showing where one works but the other does not.
- The difference between a mobile element on the street or suffering excess loss by being within a building. This can be expressed as a single excess clutter loss, or a number to show the performance using multiple levels to indicate performance within different types of structure.
- A simultaneous display of the field strength required to achieve service of a given level and the field strength limit of interference that may cause degradation to other systems, which will occur at far lower field strengths.

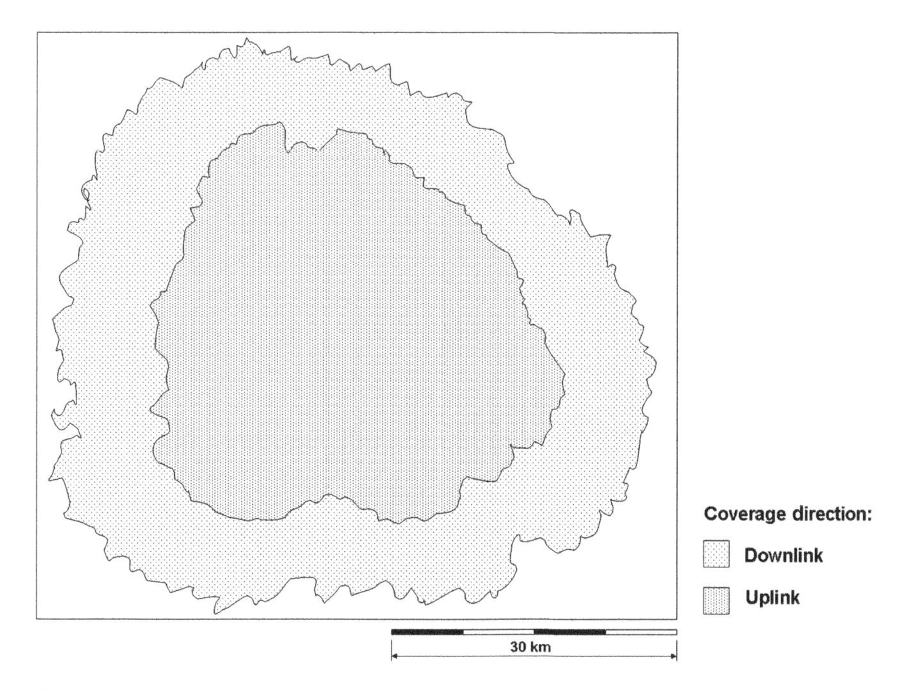


**Figure 9.11.** Coverage plot showing uplink and downlink directions on the same plot.

- A simultaneous display of communications range versus detection range possible by a hostile force, or vice versa. This may also be true for interception and direction finding and so is a useful analysis for electronic warfare purposes.

As illustrated by these examples, the simple coverage plot can be used as a powerful analysis tool far beyond simple display of service area. The engineer should endeavour to think of imaginative uses of such tools in order to maximise the benefits gained. Of course, having produced a coverage plot, it is necessary to be able to interpret what it actually means. This is what we will look in the following sections.

### *9.4.4 Interpreting Coverage Predictions*

Once a prediction has been performed and configured to provide the wanted results, the next task is to interpret the results and determine whether the results are good or bad and, if necessary, to identify the mechanism that is causing the features seen. We will look at the single site situation and also the composite coverage scenario.

If we assume that the shaded area in Figure 9.12 represents the limit of coverage to the wanted performance level, then the perfectly circular shape of the coverage area would tend to suggest that the performance is not limited by either terrain effects of clutter attenuation. If the radius of the coverage area is the same as that suggested by the link budget, then this represents the ideal coverage.

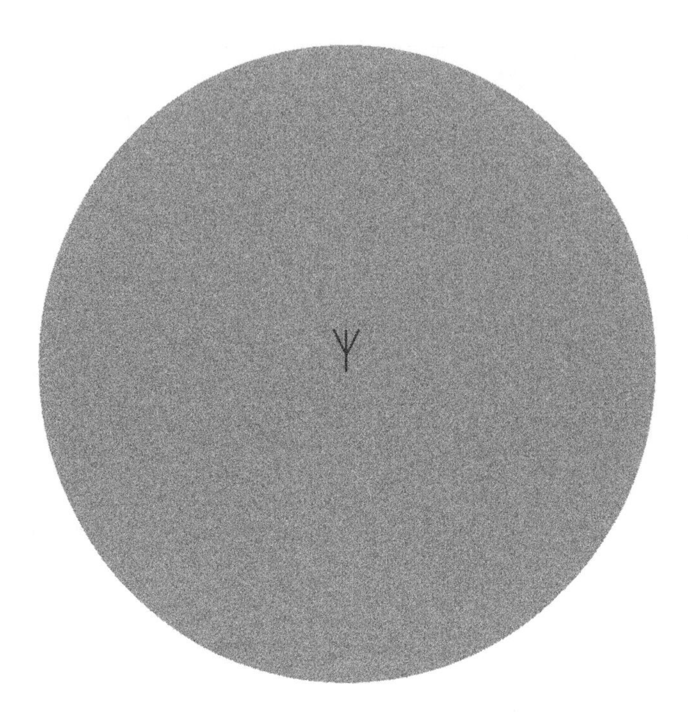

**Figure 9.12.** Coverage plot for single base station, limited by maximum link budget range.

More frequently, the coverage will not be circular but may form an irregular shape. A simple illustration of one irregularity is shown in Figure 9.13. To illustrate the principle, only a single irregular feature is shown. Examination of the coverage shown in Figure 9.13(a) indicates that the coverage is not circular, but it does not identify the reason for it. Figure 9.13(b) shows a very simple terrain model display, with a single hill identified. The radio shadow area caused by the hill is illustrated by the lightly shaded area. From this

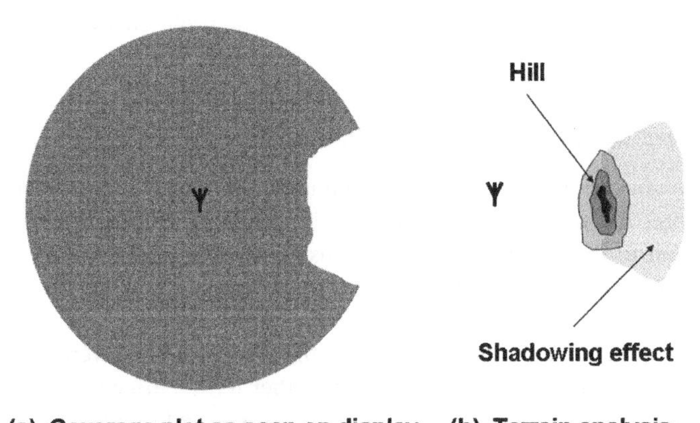


**Figure 9.13.** Coverage plot for single base station.

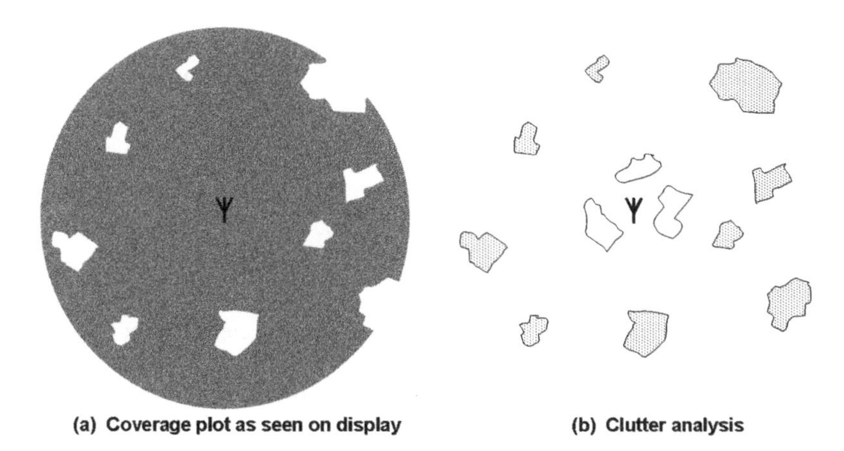


**Figure 9.14.** Coverage plot for single base station, and comparison with ground clutter used for the prediction. This comparison shows that clutter attenuation is the cause of the holes seen in the coverage prediction.

very simple analysis, it is clear that the hill is the cause of the coverage reduction in that area. Usually, situation will be more complex in practice, but the same principle of terrain analysis determine the reasons behind the coverage seen.

The same principles can be applied to any other form of effect that influences radio propagation. Another example is shown in Figure 9.14. The coverage plot is shown in Figure 9.14(a). It features a number of holes that could be due to terrain effects such as valleys, but in fact a comparison with the clutter map used, shown in Figure 9.14(b), shows similar shapes towards the edge of coverage. These are villages. This indicates that the clutter attenuation is causing the coverage pattern shape. Note there are other clutter features nearer the transmitter, but these do not cause the network performance to drop below the wanted threshold and thus are not seen on the coverage plot.

Naturally, both of these examples are somewhat simplified and contrived, however, they do illustrate the principles that should be used when examining more complex, real world effects.

## 9.5 Nominal Characteristics for the Network Design Activity

### *9.5.1 Introduction*

We have identified that the network design activity can start with a description of the wanted coverage service area, and that we can adopt an approach by initially identifying potential base stations. We also need to start thinking about how we go about identifying radio characteristics for the initial simulations to define network coverage. By the end of the design activity, each site will have been configured for frequencies, antennas, antenna orientation and height and so on, but at the beginning of the process these have not been fully established. So how do we go about starting the process? In practice, it is normal to set nominal characteristics for the radio characteristics that are as close as we can get to the final values used. These are then used for all potential base stations in the initial design.

The nominal characteristics used may fall into a number of categories so that, for example, there are nominal characteristics for rural sites which differ from those of urban sites, and so on. In this section, we look at some factors to take into account when setting these parameters.

## 9.5.2 *Nominal Base Station Parameters*

### 9.5.2.1 Nominal Power

The nominal power used in the link budget may have been selected straight from the values in the data sheet of the intended radio equipment, or an ERP value may be based on the maximum power usable under the terms of the radio network licence. In some cases, it may be appropriate to select a value a few dB under the value quoted on the manufacturer's data sheet to allow for performance drop over time.

### 9.5.2.2 Nominal Antenna Height

The height of the mobile antenna will be determined by how the mobile subscriber will use the network, but the height of the base station antenna can vary significantly. In general, the higher the base station antenna height above local ground then the further the coverage will extend. This is beneficial in low-density networks that have to cover wide geographical areas, but is not desirable for networks that are limited by traffic demands and the need to avoid interference. Because of this, it is important to consider the ground rules to be used when estimating antenna height above ground before the actual values are known. For example it might be appropriate to use a nominal value of 30 m for rural sites, and 5–15 m for urban sites. The selected values should represent what is achievable in practice and also what is allowed under local planning rules.

### 9.5.2.3 Nominal Frequencies

At the beginning of the design process, frequency assignment will not have been performed. For coverage prediction, it is not important whether sites interfere with one another or not; the engineering for this will be carried out later. For most applications, a single frequency can be used; either at the beginning, middle or top of the band. In general, propagation distance reduces with increase in frequency, so selecting the highest frequency is not unreasonable. If the network is to be dual-band, then it may be appropriate to design the network twice, using the highest frequency in each band. The situation is more complex for networks that will cover a wide range of frequencies; in this case, a set of rules governing the selection of frequencies used in the initial simulations will have to be set up, typically based on the expected distance of each link.

### 9.5.2.4 Receiver Sensitivity

The receiver sensitivity of the equipment to be used can be easily derived from the technical data sheets for that radio. However, it will also be necessary to determine suitable loss figures to be used for outline planning, when the actual feeder losses are not known. It may also be appropriate to add in some additional margin for man-made noise, particularly if the receiver will be housed with equipment that will add to the noise floor. Of course, adding too much margin will result in a poorly designed network that exceeds the link demands required.

For networks based on an open specification standard such as GSM or TETRA, the sensitivity offered by a particular manufacturer may be better than that prescribed by the standard. In this case, the decision has to be made whether to use the manufacturer's figures in order to plan a more efficient network, or whether to stick to the values specified in the standard. If it can be assured that radios with the improved sensitivity will always be available in the future, then there is no problem, but if the manufacturer stops producing radios to that specification and there is no further source, then when the radios are replaced in the future, network performance will suffer and any remedial re-design work will be expensive.

## *9.5.3 Mobile Parameters*

### 9.5.3.1 Different Types of Subscriber

It is important to recognise that mobile subscribers may use the network in very different ways. Although it may be the same subscriber, their behaviour may vary in terms of how they use the network and this must be reflected in the design. As an extreme example, we will consider a single network subscriber who belongs to a specialist police unit responsible for performing highly risky tasks such as surveillance and raiding suspect locations as well as normal police roles. This subscriber may use the network in the following circumstances:

- As a mobile subscriber in a normal police vehicle with an integral radio system.
- As a mobile subscriber within an unmarked car fitted with a covert antenna.
- As a mobile subscriber in a specialist undercover vehicle (perhaps a van or lorry).
- As a portable subscriber in a police vehicle, using a handheld radio.
- As a portable subscriber in an unmarked vehicle using a handheld radio.
- As a portable subscriber hidden in the back of a van.
- As a portable subscriber on foot, in a street using a standard handheld radio.
- As a portable subscriber on foot, in a street using a covert radio and antenna.
- As a portable subscriber on foot, inside a building (or underground) using a standard portable radio.
- As a portable subscriber on foot inside a building using a covert radio and antenna system.

This is only a short list of the possible alternatives, and each of the mobile options will also need to be considered for the condition where the vehicle is in motion at various speeds. It can easily be seen from this that modelling subscribers can be no easy feat, because in each of the cases highlighted the link budgets will be different. In practice, designing for each configuration is not normally performed, but it is normal to specify reference configurations, often drawn from the scenarios for subscriber use. Again, these rules need to be specified before any planning is performed.

#### 9.5.3.2 Antenna Height Above Ground

Not only will the link budgets be different for each category of mobile subscriber, it might also be necessary to use a different height above ground for each type of subscriber. This is clearly the case for networks that may have subscribers on aircraft as well as the ground as is often the case with military and aeronautical networks. However, if for example a network has a mixed set of subscribers that might be using chest high antennas (around 1.5 m), others with antennas mounted on the top of normal family-type car (possibly 2 m), another set with antennas mounted on top of large vehicles such as lorries (of the order 4 m) and even deployable systems that might be temporarily be mounted on the side of buildings (assume say 10 m), then the planning will have to take account of coverage at each of these heights and the design will effectively be carried out separately for each antenna height. This can take some time, because it will require re-calculation of the coverage for each different type. Often, a good approach is to define for the worst case (the lowest antenna), and then determine whether coverage can be achieved for the other combinations using the same sites or a subset of them (it is likely that the number of circuits required to handle traffic for the higher mobile subscribers may well only require a subset of the total site count).

#### 9.5.3.3 Nominal Subscriber Equipment Parameters

As with base stations, there will be choices as to what parameters to use for mobile equipment as well. For publicly specified networks such as GSM, TETRA and UMTS, minimum equipment performance, such as receiver sensitivity, will be specified. Individual equipment producers will, however, try to compete by improving the specification of their equipment above that of the minimum. This can have an advantage in network design, because it may be possible to use the improved performance to produce a design with fewer base stations than would be possible using the minimum specification. However, what happens if the manufacturer subsequently decides to change the equipment components for cheaper versions, and thus the specification is reduced (but still to the minimum specification), or if the manufacturer goes bust and no other equipment suppliers provide equipment of comparable specification? At this point the network may no longer meet its performance specification, and this will be a difficult issue to resolve. This whole issue is therefore a risky part of the project and must be properly considered at the project management level, rather than simply as a technical issue.

#### 9.5.3.4 Subscriber Movement Considerations

Mobile subscribers will, by definition, move and will use the network as they move. Some subscribers may well be moving slowly and others may be moving at speed. This causes two main issues; the first is that the sensitivity of the receiver under moving conditions will be different from the receiver at rest, and this will particularly influence those parts of the network designed with the fast moving subscriber in mind, such as roads and rail. The second issue is that if the subscriber is moving too fast, then there may be problems in network handover because the process will not be completed in time to avoid dropping the call. This means that design parameters for subscriber speed must be included in the specification, and the design must be built to meet it.

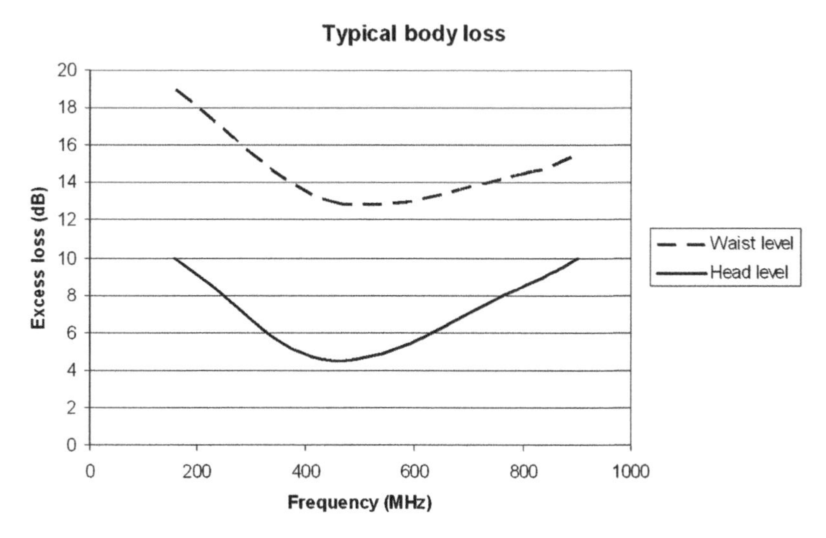


**Figure 9.15.** Typical body loss figures.

### 9.5.3.5 Subscriber Attenuation

Subscribers using portable systems will influence the efficiency of the antenna due to body loss. The amount of body loss will depend on the frequency and the position of the antenna on or near the body. In practice, each subscriber will have a different effect based on their size and the way they tend to orient the antenna. It will therefore be necessary to define nominal figures to be used as an approximation. Typical body loss figures are illustrated in Figure 9.15 for the condition where the antenna is used at head height and when it is waist mounted.

Another aspect that should be considered is that the mobile subscriber may not wear the antenna in the same location at the start of a call originated by another user. The subscriber may always bring the radio to their head to make a call and keep it there during the call, but may have the radio attached to the belt when they are not using it. However, calls to the subscriber still need to get through to meet the performance criteria. In this case, the design must be designed to provide the required performance when the antenna is belt mounted, or else subscribers must be trained to use the system in an acceptable way.

### 9.5.3.6 Building Attenuation

Building attenuation is difficult to specify because the variation in attenuation varies enormously according to building materials, thickness of external and internal walls, number of floors and even how far underground the building extends. Estimates of nominal building attenuation have been produced using building material, wall thickness and attenuation per floor. These can be used to estimate the losses to be expected at different parts within a building, but then in any large mobile system the service area will include a large number of buildings each of a slightly different design and orientation. Thus, in practical terms, it might be necessary to use a nominal figure and design the network to this level. This may mean that buildings that have losses higher than the nominal value may not experience an acceptable level of service, so this figure must be chosen carefully.

This particular factor is important in network specification, and it can cause friction between designers and customers, because the customers may demand that acceptable coverage is present within buildings no matter their construction, whereas the designer cannot accede to this as a design parameter. If the designer were to give in on this point, then the customer would quite likely be able to point to a specific building or buildings where coverage is not achieved, and then reject the whole design based on this. The key thing in contract negotiations about this point is to ensure the customer understands the issues and to manage the customer's expectations so that a reasonable compromise that both parties can accept can be achieved.

## 9.6 Base Station Design and Optimisation

The process of base station design and optimisation can be drawn out as a simple flow chart. The actual form of the flow chart will depend on the requirements of the design, technical characteristics of the radio technology used and commercial preferences (such as, do we choose a new location first rather than change the antenna type, or the other way round). An example of a typical process is shown in Figure 9.16. This ties together individual processes into an over design process.

The figure shows a list of changeable parameters and a flow structure that places certain more preferable changes ahead of others in the design process. So, in this example, a new candidate site is only considered when all other parameter combinations have been exhausted. The process also includes testing for each base station and for the whole network. The design process is likely to be iterative at the single base station level and also at higher levels.

This type of diagram is not only useful for prototyping the design process itself but also as the basis for work instructions; it is far clearer and easy to follow than a textual description. Where necessary, amplifying text can also be attached, referenced to the specific part of the diagram it refers to.

For example it might be suitable to create a table of actions for an engineer to follow when considering each of the four parameters shown in line across the diagram. A simple example is shown in Table 9.1.

**Table 9.1.** Design action notes.

| Design Action | Comments | Limitations |
|---|---|---|
| Adjust antenna height | Increase height for greater coverage area, reduce height to reduce coverage area | Maximum antenna location on mast |
| Adjust transmit power | Reduce power from licence maximum to reduce coverage area | Maximum ERP at any station is 25W |
| Adjust antenna azimuth | For directional antennas only; adjust to achieve maximum coverage in desired direction | Nil |
| Adjust antenna tilt | Adjust towards radio horizon for maximum coverage range; tilt towards ground to reduce coverage range | +/−10 degrees |

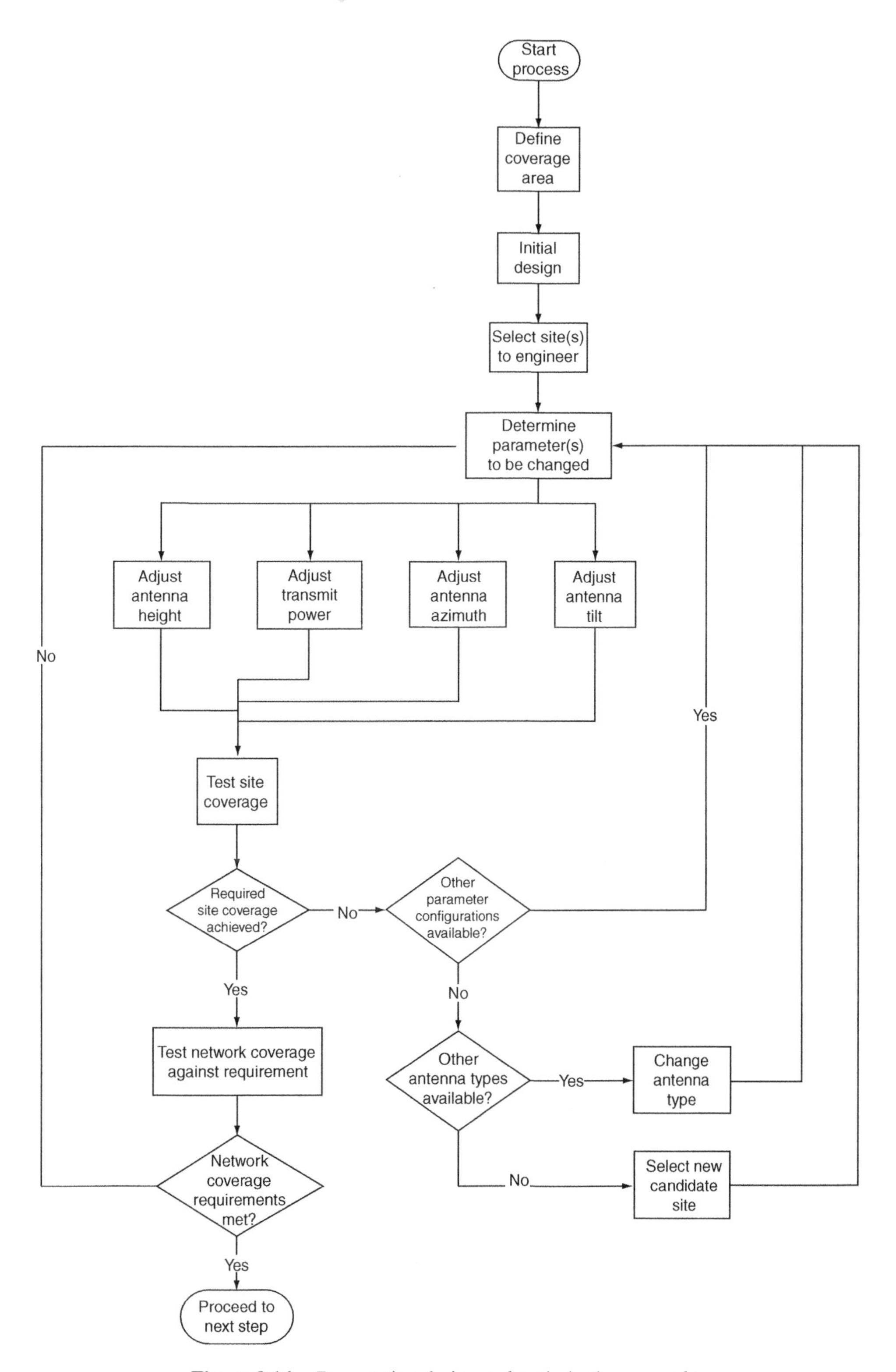


**Figure 9.16.** Base station design and optimisation example.

It should be noted that the design process diagram must be created afresh each time; although a diagram such as that shown in Figure 9.16 is useful as a template, it should not be blindly followed as being suitable for all network designs.

In terms of understanding the effects of changing individual aspects of a base station, by far the best way to learn is to use a planning tool and spend time adjusting each in turn and examining the changes. Our experience has shown that this is far more effective than simply reading about them.

## 9.7 Project Types

### *9.7.1 Introduction*

In this section, we look at some practical aspects of identifying suitable site locations in different types of project. We will look at Greenfield projects, where the infrastructure is not constrained by existing infrastructure, and the more common projects, where sites will be selected from a list or database of existing potential sites, or selected from sites already used by the organisation.

### *9.7.2 Greenfield Projects*

#### 9.7.2.1 Introduction

Sometimes, the designer has the luxury of placing infrastructure anywhere within the operational area. This is rare in practice, even for military applications, because although there may be no constraints from a social point of view (although this is seldom true either), there will be practical limitations. In this section, we look at a few typical constraints that the designer may have to consider. These are the types of constraints that may not be automatically accounted for in a radio planning tool, so the radio planning engineer may have to consider them and make manual corrections to take them into account.

#### 9.7.2.2 Definition of Green Field

A green field site is an undeveloped location that has no existing radio infrastructure. It may be that there are buildings, access to power and a suitable location to erect a radio mast or attach the antenna to an existing structure, or it may be that there is nothing, just an empty field or rough ground. A Greenfield project is thus one that allows the planner to place radio structures anywhere. Normally, however, a set of rules would need to be applied. A green field site might involve building a mast and shelter for the equipment, and even modern radio equipment has to have some protection from the local environment and the elements. Electrical power and site access for other amenities will also be an issue. This means that the designer will be constrained by a set of practical rules that will limit where the base stations can go. Such a set of rules might be as follows:

- Must be within 250 m of an existing road or track. This is to ensure that the site can be accessed easily for maintenance.

- Must be within 500 m of an existing power supply source. This is to ensure that power can be brought to the site within a reasonable cost.
- Must be within 1000 m of an existing telephone line. This may or may not be required, but if there is a requirement for this, then it must form part of the rules.
- Not located in a national park or other areas where planning permission will be difficult. This is to maximise the probability that there will be no problems associated with planning permission, which could otherwise delay the project or require a revised design to get around potential sites that have their permission refused.
- Mast height restricted to a maximum of 15 m. This is a nominal figure used for planning when in fact the real antenna heights achievable will not be known until later in the design.

The rules used must be practicable otherwise the planning will not reflect what is achievable, thus defeating the object of using any green field sites within a proposed network design. There is nothing to be gained in designing a network that cannot physically be realised within reasonable cost and effort.

#### 9.7.2.3 Site Finding Methods

Planning tools can be used to automatically identify good sites for radio infrastructure, but there are some potential pitfalls and points that the operator should use to manually improve the design. This involves a manual process of considering various aspects, such as:

- Avoid mountain tops! High hills will often be picked out as good potential radio sites because they will offer long-range coverage. However, how is the equipment to be transported there, and how will it maintained when it is placed there? Some networks do have sites on mountain tops, together with wind turbines to provide power, and helicopters are used to deliver and retrieve maintenance teams, but in general this is a high-cost solution.
- Need to have power to sites. Undeveloped areas may not have power for some considerable area around the site, and if so, power will be a problem. It may be possible to use a generator (which will need fuel) or wind turbines (which will need batteries to store the power, and wind to generate it), but these are probably undesirable unless absolutely needed. Also, generators are very popular with rural thieves! It is therefore usually necessary to be able to obtain power from an existing source without too high an associated cost. This may require obtaining availability and pricing from power distribution companies so that the viability of obtaining power can be properly assessed.
- Need access to the site. The site will require access, and maintenance engineers carrying equipment will probably not wish to hike miles across the moors to reach the site. Road access would be ideal, but for rural areas, logging trails, forestry tracks or farmers roads (with permission) will be adequate. Of course, such access may depend on having suitable off-road vehicles, so the type of vehicles available will be a factor in determining where sites can be accessed. If small roads and tracks are to be used, then suitable scale maps that show these features must be obtained for the project.
- When selecting sites, it is important to realise that it is the overall network coverage that is important, not the coverage of individual sites. Thus, the contribution of the site to the overall network is the important factor.

- If the network will be traffic-limited, long-range coverage is actively not wanted, and thus the design should reflect this. This may mean that the sites selected by automatic algorithms to determine best coverage achieved are not suitable for designing these types of network.
- In general, the aim is typically to achieve desired coverage (not exceed it) with minimum site count, but this will depend on the cost of implementation and operation. It may be that certain types of site are less expensive to develop and maintain than others, and so a higher site count with lower associated costs per site may be the preferred solution.

## 9.8 Legacy Projects

A legacy project is where the existing infrastructure of a radio network is to be used as the basis for a new network. This is often the least expensive way of developing a new network, since many of the costs associated with developing new sites have been absorbed by the previous network and will have been written off. All that is required for those sites to be retained is to remove the old equipment and install the new equipment.

Although legacy radio sites may be cheapest to develop, it is not necessarily the case that they offer the best sites for the new network. However, as long as the revised design works, it may still be desirable to accept a less than optimal design, so that the implementation cost reduction benefit can be realised.

Using legacy network infrastructure sounds like a very straightforward activity, but in our experience it can be more difficult than anticipated to realise. The reasons for this vary from project to project but usually focus on one aspect, availability of information. It is common there to be no central record of sites, or for data to be old or out of date. Few organisations appear to be capable of maintaining accurate records of sites, equipment and ancillaries available at each site. Even when it is believed that such records have been meticulously kept, in practice they are often out of date or just plain wrong. Usually, this is down to different departments being in charge of different aspects of support, re-design and maintenance, and their not informing other departments of changes made. Also, sometimes records are not kept at all, with the information being stored in the minds of employees. It is fine if 'Old Ken' knows everything – not so good when he retires and everyone suddenly realises that no one left knows anything about the network. Thus, legacy projects can sometimes be a can of worms; for consultants looking to take on such projects, it is vital to determine what information is available and how accurate it is before accepting the task.

Network coverage design projects for new networks using legacy site generally require the following design steps:

- Determine coverage obtained when using old sites but using link budget and technical parameters of new technology.
- Determine best combination of sites to provide coverage of wanted service area using minimum number of sites.
- Determine remaining coverage gaps.
- Identify suitable sites to provide coverage into coverage gaps. This will typically either be from a list of sites available (see next section), or green field sites.

## 9.9 'Existing Site' Projects (Site Databases)

Nowadays, there is a reluctance to allow new radio masts to be erected in many countries. For this reason, and often for reasons of cost, it is desirable to re-use existing radio masts for a new network. Many companies operate a portfolio of radio masts and associated infrastructure and rent out the facilities and space on the mast for a fee. These can often be the most cost effective ways of deploying radio network infrastructure; they do not require development or survey, the sites are proven as being good radio sites (but their appropriateness to the network technology needs to be assessed to make sure they are suitable) and they may well be the cheapest sites to use, since the cost of their upkeep is shared among all of their uses (plus a surcharge for the site owner). Since the sites are already up and running, the timescales required to deploy sites will also be reduced.

On the downside, these sites may have some drawbacks; the available antenna locations on the mast may be constrained by other antennas, the noise floor may well be higher than bespoke sites and the frequencies that can be used may well be constrained by the other transmissions if interference and inter-modulation products are to be avoided.

The design process when working with existing sites, especially those commercially available is on the following lines, which is very similar to the legacy situation:

- Obtain database of all sites available within the commercial site portfolio within the project area.
- Determine coverage achieved from each site.
- Determine best combination of sites to provide coverage of wanted service area using minimum number of sites.
- Determine remaining coverage gaps.
- Identify suitable sites to provide coverage into coverage gaps. This will typically either be from a list of sites available (see next section), or green field sites.

## 9.10 Coverage Design Deliverables

At the end of the coverage design process, a number of documentary deliverables should be available. The list may vary according to what has been agreed with the customer, but a typical example might include the following:

- A list of the candidate sites, including their locations.
- Antenna height, orientation and tilt for each antenna to be used at the site.
- Power to be used for each channel.
- Coverage plots per site, for each different category of service in the requirements.
- Coverage plots for the whole network, for each different category of service in the requirements.
- Best server plots for the whole network, and optionally for sub-areas of it.
- Statistics for each requirement, provided in tabular format.
- Statistics for the contribution of each site to overall coverage, based on best server coverage.
- Other information as agreed by the customer. This may include comments on difficult areas to cover; potential options for network re-design and so on.

# 10

# Traffic Demand and Capacity

## 10.1 Introduction

We have looked so far at the coverage aspects of design, in which we engineer the network to provide service over the geographic service area required. This is only one aspect of mobile radio network design; in this section we look at another, dimensioning the network to meet expected traffic demand levels.

The network will need to cater for more than one subscriber using the service at a time and this means providing a number of circuits for concurrent use. The number of circuits required is unlikely to be uniform across the whole network unless subscribers are likely to be equally distributed across the service area, which seldom happens. Usually, subscribers will tend to congregate in certain parts of the service area and these parts may vary over time. For example, subscribers of a public network may start the day at home, travel into work, be at work during the day and then return home or go out in search of entertainment in the evening. Of course, this will only be valid during the working week and their behaviour will be different at weekends and holidays. In addition, there may be less frequent and unpredictable variations to network usage caused by events that affect large numbers of subscribers; for example, bad weather affecting travel, major sports events or even social disturbances such as terrorist attacks. Networks designed for specific user groups will have subscribers that do not share the spatial and temporal distribution of the public. For example, in a network used to serve the employees of a company, subscribers will be distributed in a manner dictated by the company's business. We need to design the network to cater for subscriber call behaviour under most expected conditions. This is a major issue for those responsible for defining the network requirements, since it will have a major impact on the attractiveness of commercial networks or the ability of the network to meet specific contingencies in the case of non-commercial networks.

In this chapter, we will look at traffic dimensioning, starting by considering methods of modelling how subscribers behave. This is an essential step before we can start to consider how to design the network to meet the needs of these subscribers. As with the other design aspects we discuss in the book, the more tightly we define the requirements, the better we can engineer the network. We will use relatively simple models of subscriber behaviour and network capacity in these examples, but be assured that they can be extended to far more complex scenarios.

## 10.2 Modelling Mobile Subscribers

### *10.2.1 Determining Subscriber Types*

It may be that the mobile subscribers are of one generic type that cannot be differentiated in any way, or it may be that it is possible to specify sub-categories. If this is the case, we may be able to better represent total traffic demand throughout the service area in greater detail, and this in turn will allow us to more critically engineer the solution. To examine some of the issues, we will consider two specific types of network; a public access network where the subscribers are any member of public who signs up to the service and a public safety network, where the subscribers are individual police officers patrolling their area.

#### 10.2.1.1 Public Network Example

In this example, we state that a mobile subscriber may be any member of the public, and without express information of who each user is, where they live, where they work and where else they travel to, it is impossible to determine how any one subscriber will behave at a certain point in time. But does this mean that we cannot determine how large numbers of subscribers will behave? In fact, although it is impossible to determine the behaviour of one individual subscriber, it is possible to estimate the behaviour of large numbers of subscribers from relatively coarse factors. A simple method, which has its limitations but can give a good first indication, is simply to relate subscriber density to the ground clutter category. A simple example of a clutter map is shown in Figure 10.1, which shows a small city on a river estuary surrounded by small villages and forest to the southeast. The simple model shows four levels of clutter category from the city which may be defined as:

Code 1 = suburban, residential
Code 2 = suburban, mixed residential and commercial
Code 3 = urban, commercial
Code 4 = urban, commercial, high density

These categories will be representative of the dominant type of usage in each category type, as there will be residences in the city centre, businesses in the suburbs and so forth. It is possible to break the categories down a lot further but this depends both on the ability to capture the relevant data and also the benefit derived from refining the clutter categories down to ever smaller resolutions. The degree to which this will be necessary will depend on the particular project. In this example, we will assume that the four categories are sufficient.

Using the measurement tools typical on modern, sophisticated radio planning tools, we can determine the total area covered by each category of clutter. In this case, the areas of the categories are as follows:

Code 1 = 191 $km^2$
Code 2 = 35 $km^2$

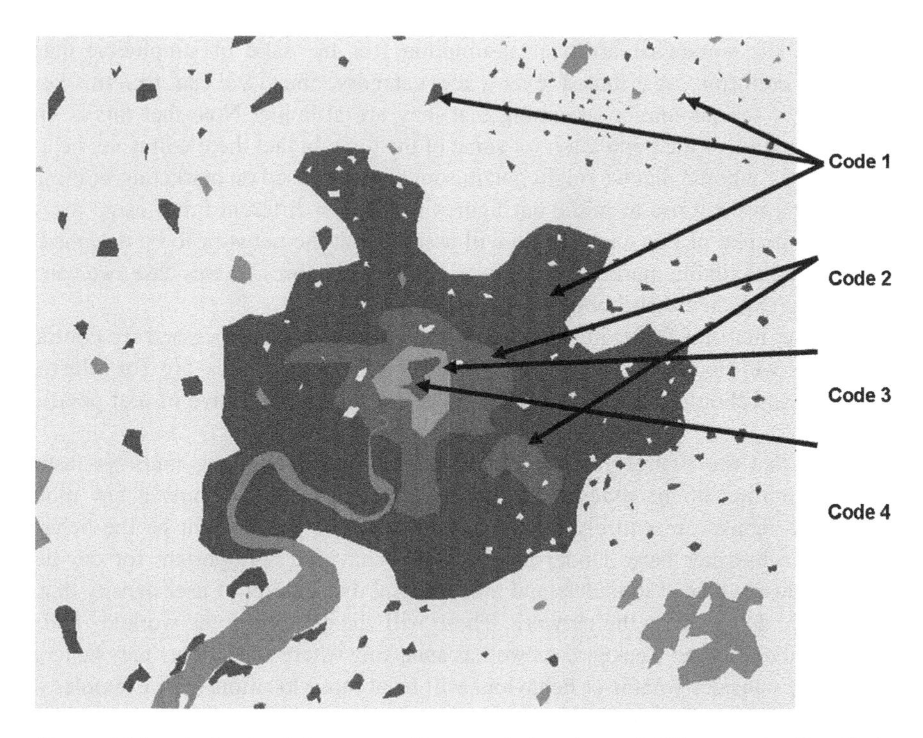


**Figure 10.1.** A simple clutter map, with example locations of different codes identified.

Code 3 = 6.7 $km^2$
Code 4 = 1.5 $km^2$

If we were to assume that the city has a population of 100 000, and that the relative split of residents per category is (determined from population statistics or some other social data source):

Code 1 = 87 %
Code 2 = 10 %
Code 3 = 2.5 %
Code 4 = 0.5 %

We can then determine the population per category and thus the relative density of subscribers when they are at home. This is illustrated in Table 10.1.

**Table 10.1.** Subscriber distribution by clutter category.

| Code | % | Area ($km^2$) | Population | Population density (per $km^2$) |
|---|---|---|---|---|
| 1 | 87 | 191 | 87,000 | 456 |
| 2 | 10 | 35 | 10,000 | 286 |
| 3 | 2.5 | 6.7 | 2,500 | 373 |
| 4 | 0.5 | 1.5 | 500 | 333 |

If we make the somewhat artificial assumption (for the sake of simplicity) that the population is uniformly distributed over each category, then we can use this as the representation of the population assuming that they are at home. Note that this is only a partial representation of the population for some of the time; in fact there will never be a case when everyone is at home, but we might obtain some figures based on marketing information collection which we can use to refine our figures further for different times using the same approach. For the rest of this analysis we will assume that the network to be designed will have 10 % of the available market (probably an optimistic figure). In this case, we can start to work out the expected distribution of subscribers.

If we assume that the following data has been collected and processed by contracted pollsters, then we can refine our figures further (all these figures are for illustrative purposes only and should not be regarded as in any way representative of real population movement).

We assume that the figures for ‘travelling’ refer to relatively long journeys lasting a significant proportion of the timescale rather than short journeys of only a few minutes. Although these figures are entirely fictitious, they illustrate what might be the behaviour typical of the subscriber base. Understanding this behaviour is important for our design processes, because we need to understand the temporal distribution of user density that will affect the traffic demands on the network. These will change during the workday and they will be different during the weekend. As well as analysing where subscribers may be located, we also need to consider what user behaviour will be at those locations. For example, will a significant proportion of people at home use their mobile phone to make calls or will they prefer to use their existing landlines in preference? Clearly, there would be no need to supply traffic capacity for a demand that is not there; this would adversely affect the financial viability of the network. This type of information can be collected by marketing survey to give results of the following kind:

Percentage of subscriber population that use network while at home[1]: 30 %
Percentage of subscriber population that use network while at work[1]: 50 %
Percentage of subscriber population that use network while travelling[1]: 70 %
[1]when they need to make a call

As a note to engineers, look at how this marketing information is essential to the engineering process. The engineering process benefits from all types of information coming into the business, and all such information should be exploited. If there is a temptation to ignore

**Table 10.2.** Subscriber distribution by time, shown as percentages at different times.

| Statistic–population distribution | Code 1 | Code 2 | Code 3 | Code 4 | Travelling |
|---|---|---|---|---|---|
| 0000–0700 | 5 | 5 | 10 | 75 | 5 |
| 0700–0900 | 10 | 10 | 15 | 30 | 35 |
| 0900–1700 | 25 | 20 | 15 | 30 | 10 |
| 1700–1900 | 15 | 15 | 10 | 40 | 20 |
| 1900–2100 | 10 | 10 | 10 | 50 | 20 |
| 2100–0000 | 10 | 5 | 10 | 70 | 5 |

information coming from other parts if the business, this should be suppressed (in accordance with the idea of the 'business-centric' approach outlined in chapter 6).

We can also look at the ratio of travellers in the service area at particular times. If we assume there are the following ways of getting around used by most people then we can examine the behaviour of mobile users on the move (with illustrative percentages):

Road – primary route (25 %)
Road – main route (15 %)
Road – secondary route (10 %)
Rail (30 %)
On foot (20 %)

In terms of the network, each individual subscriber can belong to different subscriber types at different phases throughout the day and when doing different things, for example each subscriber can be:

- At home.
- Travelling to work (by a particular method).
- At work.
- Seeking entertainment in the evening.
- Attending sports or other occasional events.

We want to determine the performance of the networks during times when mixtures of these different types of subscriber are present. This means that in our various scenarios we will effectively model a single subscriber as a different type in different circumstances (because at each instance they will effectively belong to different subscriber categories). For each of these scenarios, we will build up a composite picture of expected composite subscriber behaviour at that time and then dimension against the total traffic demand against this composite demand. It is of course important to ensure that network subscribers are not double-counted in this analysis.

Taking this approach, we can build a reasonable model of composite potential subscriber behaviour over the course of a working day. This can be collated to provide a picture of all subscribers in the network area. We do of course have to remember to modify the total population figures by the network penetration values (10 % in this case). An example is shown in Table 10.2.

As well as describing subscriber distribution, we will also need to determine a model of call behaviour during the day. Such figures can be extrapolated from measurements of traffic taken from other similar networks on the basis that the new network will probably experience roughly the same traffic levels. An example of relative traffic density for an average 24 h period is shown in Table 10.3. The traffic is dimensioned in relative terms; the 100 % value being set at the maximum expected load, which may be more or less than the network can actually cope with depending on the definition. We will need to scale these figures to the total traffic demand and also in a real project we would need to carefully validate them against actual traffic behaviour.

With this data it would now be possible to determine traffic capacity that will be necessary at each base station location. We can dimension the traffic capacity against this table – and many others like it that examine different circumstances – but we need to bear in mind that a network that just meets the 'typical' expected traffic will not be capable of meeting extraordinary load. This is effectively an issue of business risk. For example, imagine the situation where there is an extraordinary incident such as a terrorist scare (unfortunately all too imaginable) and the subscribers are unable to call their loved ones; how many would throw their phones in the nearest bin and never use that network again? This is unlikely to happen but its effect on the business would be major (the project should have a risk register to take account of this kind of effect).

### *10.2.2 Public Safety Example*

The public safety example has some common features to the previous example and some differences also. The commonalities arise from the fact that the public safety network exists to allow police and other emergency services to serve the public, and thus must be present wherever people are present. The differences arise because the probability of an event occurring that requires emergency services intervention is not necessarily directly related to population density. There may be particular 'hotspots' that require particular attention. These may include, for example:

- Specific known trouble-spots.
- Areas with nightclubs and other late night entertainment.
- Sports arenas.
- Industrial facilities with hazardous materials.
- Airports, ports and railway stations.
- Large conference venues.
- Roads.

This means that the raw population data figures need to be modified to reflect the changed emphasis in importance. This can be performed in the same way as seen in the previous example, but with the type of figures shown in Table 10.1 changed to reflect the distribution of callouts and replacing the population distributions shown. Additionally, the public service network will not only need to meet day-to-day traffic demands, but also be able to cope with large-scale emergencies. Part of this demand may be met by deployable infrastructure that can be set up to provide a temporary increase in traffic capacity in the vicinity of the incident, but there must be sufficient permanent capacity to deal with the majority of events.

**Table 10.3.** Relative traffic density by time (traffic against hour of the day) for a commercial network.

| 0001 | 0100 | 0200 | 0300 | 0400 | 0500 | 0600 | 0700 | 0800 | 0900 | 1000 | 1100 |
|---|---|---|---|---|---|---|---|---|---|---|---|
| 6.0 | 4.0 | 2.0 | 1.0 | 3.0 | 3.0 | 1.0 | 4.0 | 11.0 | 18.0 | 21.0 | 31.0 |

| 1200 | 1300 | 1400 | 1500 | 1600 | 1700 | 1800 | 1900 | 2000 | 2100 | 2200 | 2300 |
|---|---|---|---|---|---|---|---|---|---|---|---|
| 19.0 | 31.0 | 26.0 | 22.0 | 24.0 | 27.0 | 19.0 | 13.0 | 11.0 | 10.0 | 9.0 | 7.0 |

**Table 10.4.** Relative traffic by time (traffic against hour of the day) in an emergency services network.

| 0001 | 0100 | 0200 | 0300 | 0400 | 0500 | 0600 | 0700 | 0800 | 0900 | 1000 | 1100 |
|---|---|---|---|---|---|---|---|---|---|---|---|
| 20.0 | 15.0 | 10.0 | 5.0 | 3.0 | 1.0 | 1.0 | 5.0 | 15.0 | 20.0 | 18.0 | 19.0 |

| 1200 | 1300 | 1400 | 1500 | 1600 | 1700 | 1800 | 1900 | 2000 | 2100 | 2200 | 2300 |
|---|---|---|---|---|---|---|---|---|---|---|---|
| 19.0 | 25.0 | 19.0 | 18.0 | 20.0 | 25.0 | 28.0 | 15.0 | 18.0 | 25.0 | 35.0 | 37.0 |

Additionally, the variation of calls made in an emergency services network will vary in a similar form to that of the commercial network but with different values. For example, a police force with an area of bars and nightclubs might have a distribution as shown in Table 10.4, which shows an increase in calls during the evening and night.

Traffic does not necessarily just appear at one location. A call from a mobile to a base station may also cause traffic at another base station where the call is to another mobile in the network. Traffic modelling needs to account for this demand also.

A public safety network will also potentially have different types of subscribers, such as vehicle mounted systems. Also there may be different types of call, as illustrated in Table 10.5, each with its own traffic characteristics on a per call basis. Some of these calls will only involve a single base station, whereas others will travel throughout different parts of the network. These calls will be transmitted over several base stations (inter-site) and thus the traffic requirements need to incorporate the demand wherever it appears in the radio part of the network. The total traffic demand will be the sum of all types of calls, summed over the entire network.

Actual traffic metrics will be different for each type of network and need to be derived in each case. This can be a difficult task because the expected traffic demand can be hard to determine, particularly for new types of networks or commercially risky networks. In this case, it may be appropriate to try out various scenarios and determine a capacity structure able to meet the most likely versions.

**Table 10.5.** Example of TETRA traffic call types. F means the call is originated from the fixed part of the network and M indicates that a mobile starts the call. Intra-site means that the call is handled within the service area of a single base station and inter-site means that the network backhaul needs to be used for the call.

| Type of call | Network requirements | Traffic per user (E) |
|---|---|---|
| F - Individual voice call | Intra-site | 0.010 |
| M - Dispatcher voice call | Inter-site-division | 0.020 |
| M - Group #1 voice call | Intra-site | 0.015 |
| M - Group #2 voice call | Inter-site-regional | 0.010 |
| M - M intra-site voice call | Intra-site | 0.030 |
| M - M divisional call | Inter-site-division | 0.010 |
| M - M regional call | Inter-site-regional | 0.001 |

### *10.2.3 Other Methods of Representing Subscribers*

If it is not possible to represent traffic in the manner described, then there are other methods that rely on more general representations. The simplest and likely to be the least representative of the real-life situation is that of a linear distribution in which subscribers are uniformly distributed across the service area. This is unlikely to be true in the commercial environment, but may be closer to the mark for networks where users are not distributed in line with urban density, such as a military network on an open battlefield.

Another method that has been used in the absence of detailed information is the use of a Gaussian or normal distribution as a representation of an urban environment, with the greatest density of subscribers at the centre, coinciding with the centre of the town and the density reducing as distance to the centre of town increases. This allows some estimation of subscriber distribution to be made in the absence of more detailed information.

Both of these methods are quoted in system design specifications such as TETRA but in reality, more sophisticated methods should be used if at all possible.

## 10.3 Representing Traffic Demand Metrics

We will use traffic metrics in a planning tool, and represent them in spatial and temporal terms. So far, we have discussed subscriber movements, but not how often the subscriber is using the network. To do so, we use the traffic metric, the Erlang (E), which is defined as the percentage of time a circuit is busy, expressed as a fraction. Thus 1 E means that a single circuit is used 100 % of the time. No single user will use a network in this way for any length of time; so, multiple subscribers will normally be able to use a single circuit at different times.

The traffic per user is usually determined by observing subscriber behaviour. This may result in a value of, say 25 mE. If the subscriber density is 45.6 subscribers per $km^2$, then the total traffic offered per $km^2$ will be 0.025 multiplied by 45.6, which is 1.14 E. This can then be used in a planning tool by ground clutter category as shown in Table 10.6, with a calculated figure shown for each clutter code (the figures shown are purely for illustration of the technique, but would be generated in the same way as described):

If the traffic is not related to the existing ground clutter, then it may be necessary to generate a new geo-spatial model (and by multiple representations, geo-temporal) of the requirements. This should be possible to do in modern planning tools by populating a data layer with appropriate values. The exact method used will vary from tool to tool.

The time varying aspects of traffic also needs to be expressed in a planning tool. A simple example is illustrated in Figure 10.2. The values shown are those already derived from observation of subscriber behaviour displayed in Table 10.3 for a 24-h period. The figures

**Table 10.6.** Offered traffic by category.

| Category | Offered traffic per $km^2$ |
|---|---|
| Code 1 | 1.14 |
| Code 2 | 7.15 |
| Code 3 | 9.32 |
| Code 4 | 8.32 |

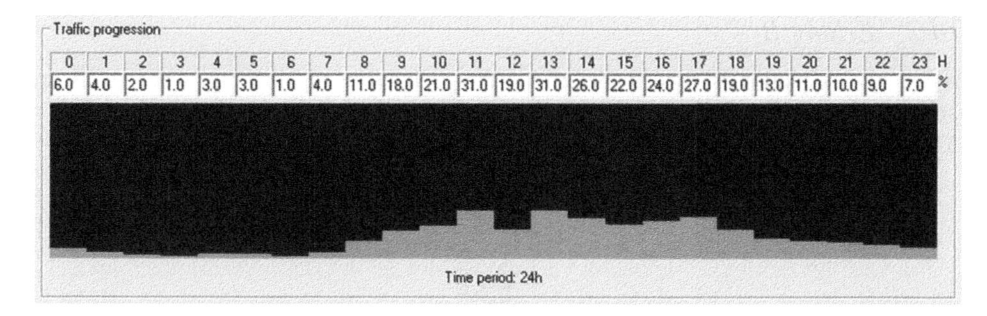


**Figure 10.2.** 24 h traffic variations, showing the percentage of traffic against a reference peak value for each of the 24 h in a day. This can also be repeated for each day in a week and for days when variations from the norm are expected. Reproduced by permission of P & D Missud.

can be changed as necessary to model different days. This model uses a coefficient to vary the amount of traffic expressed in a single time unit (typically one hour). A more sophisticated model would vary traffic mix along with call density.

Now, we have looked at various methods of describing subscriber behaviour both in geographical and temporal terms; we now need to look at relating offered traffic to the infrastructure requirements needed to serve it. After that, we will return to the planning tool to see how it all fits into the traffic design process.

## 10.4 Blocking and Queuing

### *10.4.1 Introduction*

If a network has sufficient capacity to handle all calls that might be made by all possible subscribers simultaneously, then there will be no capacity problems. In practice, however, such networks are not normally feasible except for very small user communities. Instead, the network will normally be designed to cope with an amount of traffic deemed as reasonable given certain assumptions based on the expected traffic behaviour, as already discussed. In this section we look at dimensioning networks to allow them to handle a defined amount of traffic while providing an acceptable level of service to each subscriber. In this context, two concepts are important. These are 'blocking', which is the situation in which a subscriber attempts to make a call but the network refuses the call, and 'queuing', in which the subscriber is not rejected, but the call is placed in a queue to wait for connection when capacity allows.

We will now look at two sets of mathematical formulae commonly used for capacity planning (we will not provide the derivation of these methods since this is covered in many other books, instead we will look at how they should be applied in a planning tool). These are the Erlang B and Erlang C formulae. These are not only the methods of characterising traffic but they illustrate the principles used for most traffic analysis approaches. Note that both of these formulae are based on the premise that calls arrive randomly. This is sometimes not the case in practice, so the engineer should consider the applicability of the formulae for a specific purpose.

### 10.4.2 Erlang B

The Erlang B equation is used to determine the probability of blocking, where calls are abandoned if they are not immediately successful.

The Erlang B equation is:

$$P = \frac{\frac{a^c}{c!}}{\sum_{x=0}^{x=c} \frac{a^x}{x!}} \qquad [1]$$

where

$P$ is the probability of blocking
$a$ is the offered traffic (including call set up and disconnect)
$c$ is the number of circuits available

The Erlang B formula is based on the premise that calls arrive randomly. This is typically true when looking at short durations, but is typically not true for long periods. For example, taken over a 10-min period during the day, we would expect calls to arrive randomly. Taken over an entire day, we would expect fewer calls very early in the morning compared with later in the day. Also, there can be situations where a single event causes a sudden change; for example, a goal is scored in a major football match, and many people call each other or, in a military context, a 'flash' message reporting an enemy attack causes calls to propagate through the battlefield. This means that effectively, the Erlang B equation should be applied to particular instants rather than over long periods.

A graph of the Erlang B equation is shown below in Figure 10.3. The graph shows the probability of blocking given a number of circuits ('c' in the figure; values of 1, 5, 10 and 20 are shown). The *X*-axis shows the total offered traffic in Erlangs, and the *Y*-axis shows the probability of blocking.

In terms of dimensioning networks for traffic, the key issue will be to determine the number of circuits required at a base station to achieve a given Grade of Service (GoS) to achieve a link. GoS is the percentage chance of a call failing, so it is equal to $P$ multiplied by 100. In this analysis we will restrict our analysis to the radio link and not include the other vital parts of the link such as the switches and landline components of the networks (the term 'reliability' is used to represent the end-to-end probability of call success). We therefore need a method to determine c given a value of $P = \text{GoS}$ (expressed as a probability). An example of this is graphed in Figure 10.4 below for different grades of service, with number of circuits shown against offered traffic. The graph is not a smooth curve since circuits can only be expressed as integers rather than a continuous value, so the graph increases in integer steps.

The graph in the figure has been derived using an iterative solution of the Erlang B equation for different values of $c$. This numerical method is used in place of solving a complex integral and works with sufficient precision to determine the required number of circuits, which of course can only be expressed as positive integer values. This approach makes it easy to determine the required number of channels using nothing more complex than a spreadsheet.

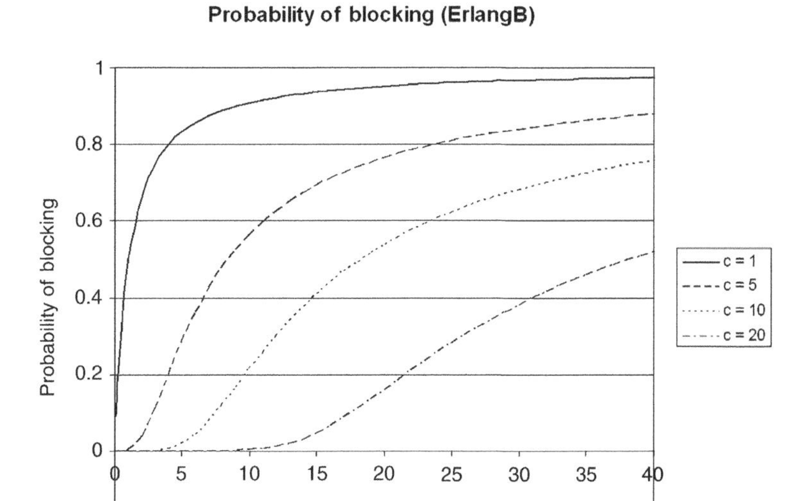


**Figure 10.3.** Probability of blocking against offered traffic using the Erlang B equation for different numbers of circuits available, c = 1, c = 5, c = 10 and c = 20.

The Erlang B equation has one major limitation: What happens if some of those who try a call that fails subsequently try again immediately? This is dealt with by the enhanced Erlang B equation, which includes a term $R$ which is the percentage of subscribers who attempt to recall the service. The process is iterative as shown below and is suitable for computational analysis.

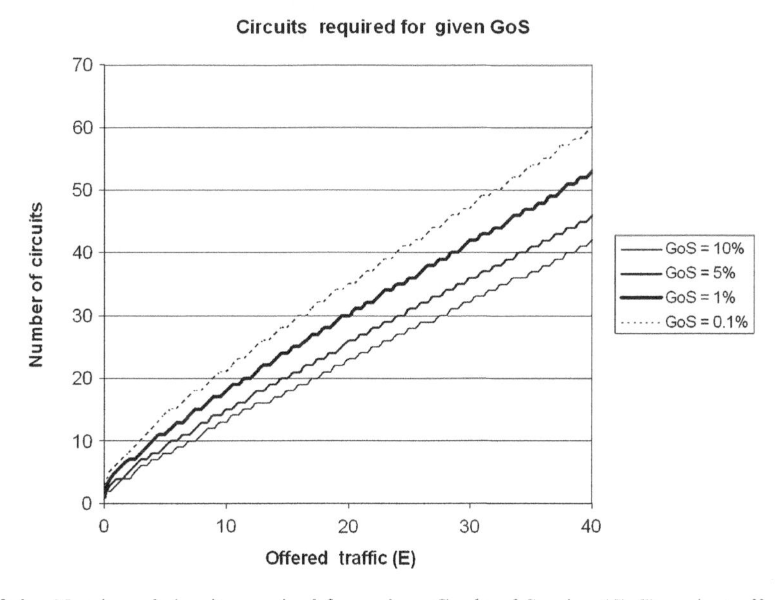


**Figure 10.4.** Number of circuits required for a given Grade of Service (GoS) against offered traffic.

### 10.4.3 Enhanced Erlang B

The enhanced Erlang B algorithm assumes that when subscribers attempt to make a call, a certain percentage of them will immediately try again. This is more representative of the true situation than the basic Erlang equation. The algorithm is a series of steps as shown below.

Enhanced Erlang B (EEB) algorithm is:

1. Use the original equation to determine $P$ (*probability of blocking*)
2. Calculate blocked and carried traffic:

$$B = P \times E \qquad [2]$$

$$C = E \times (1 - P) \qquad [3]$$

3. If

$$C + B(1 - R) = E1 \ldots \qquad [4]$$

where

$B$ is blocked traffic

$C$ is carried traffic

$E1$ is the originally offered traffic

$R$ is the recall percentage

. . .then stop as the traffic demand is not rising and the calculation is complete, otherwise go to step 4

4. Add the recalling traffic to the original traffic by

$$E = E1 + (R \times B) \ldots \qquad [5]$$

. . .and repeat the above process until the condition of step 3 is met.

If $R$ is high, then if the number of subscribers is higher than the available number of circuits, the system will soon become blocked for all practical purposes. An example of this is shown in Figure 10.5.

This illustrates that if original demand is low, then the behaviour of any subscribers suffering a blocked call is not important in terms of overall network performance. If demand is higher, so that more calls are blocked, then the behaviour of blocked subscribers becomes increasingly significant until the network effectively fails. This type of behaviour has been seen in practice, where events such as terrorist attacks have led to paralysation of network performance due to exactly this mechanism. In such a case, the effect of word of mouth will also tend to increase the load on the network as more and more people attempt to make call, and keep trying in the event of not being successful. Designing a network to cope with such an event is probably financially unviable.

### 10.4.4 Erlang C

Calls may not necessarily be lost if the network is busy. Instead, there may be a queuing system that allows calls to be held and then processed sequentially as capacity becomes available. This situation is modelled by the Erlang C equation, which gives the probability that a call is put into a queue rather than lost.

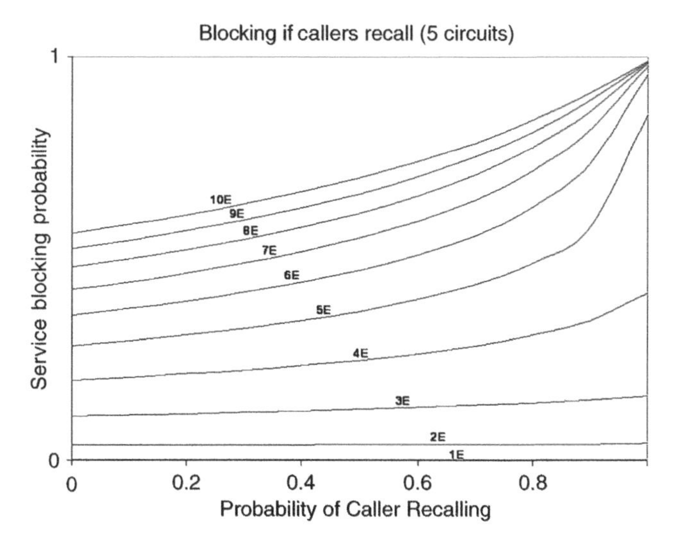


**Figure 10.5.** Enhanced Erlang-B blocking probability against probability of callers recalling on failed initial call.

The Erlang C equation is:

$$P = \frac{\left(\frac{a^c}{c!}\right)\left[\frac{c}{c-a}\right]}{\left\{\sum_{x=0}^{c-1}\frac{a^x}{x!} + \frac{a^c}{c!}\left[\frac{c}{c-a}\right]\right\}} \qquad [6]$$

where $P$, $a$ and $c$ are the same as for equation [1].

Or, expressed in terms of the Erlang B equation

$$P = \frac{c \times ErlangB(c,a)}{c - a(1 - ErlangB(c,a))} \qquad [7]$$

where *ErlangB(c,a)* is equation [1].

The Erlang C equation is illustrated in Figure 10.6.

For dimensioning of queuing systems, the desired metric is the number of circuits required to provide a given probability that a call waiting time is not exceeded. This is given by:

$$P(W > W_0) = P \times ErlangC(a,c) \times \exp\left[-(C-A)\frac{W_0}{H}\right] \qquad [8]$$

$P(W > W_0)$ is the probability that queuing time exceeds $W_0$

$W_0$ is the acceptable waiting time

$H$ is the average call duration

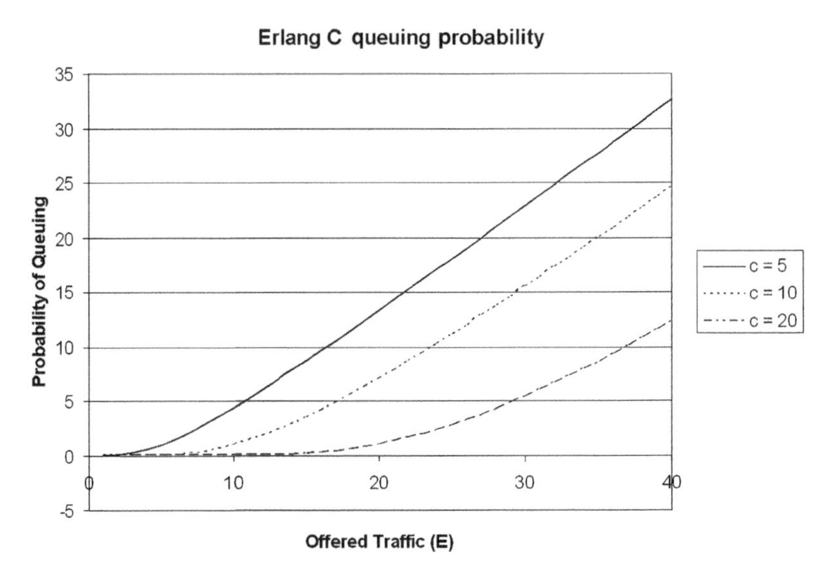


**Figure 10.6.** Erlang C queuing probability; this is the probability of being placed in a queue rather than the call getting through immediately.

So given a requirement that the probability for a given waiting time does not exceed a certain probability, we have to solve this equation for $C$, as illustrated in Figure 10.7.

The results shown in the graph can be obtained by a table of probabilities, or calculated iteratively in a spreadsheet. Note that the variables are the required GoS, the offered traffic, the target waiting time in seconds and the average call duration in seconds.

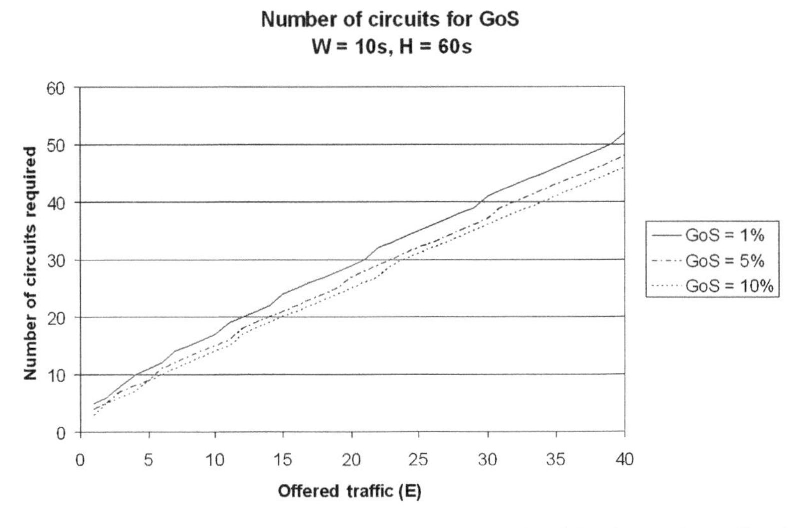


**Figure 10.7.** Erlang C; number of circuits for a given Grade of Service against offered traffic.

## 10.5 Determining Traffic Demand on a Site-By-Site Basis

We now have methods of expressing subscriber behaviour and also relating traffic demand to the number of circuits that have to be provided to meet the performance objectives for both blocking and queuing systems, given offered traffic. The next step is to determine the total traffic expected to be offered to each base station throughout the network, and this can be estimated using the best server approach (on the assumption that the base station with the strongest signal will carry the traffic of that subscriber – if not, we have to vary the approach to ensure the coverage represents true network behaviour).

Our approach to this aspect of network design is to sum the total expected traffic for the given service area of each base station. This is not the same as the coverage, because there will be overlap between adjacent base stations, which is why we use the best server approach as the simplest expression of effective service area per base station. If necessary, we can generate a more complex model, with for example transition regions where coverage is split between base stations, or discrete models where subscriber behaviour is modelled on a discrete basis as samples of subscriber behaviour, including voting to particular base stations based on defined rules. Presently, however, we shall illustrate the process using the best server method.

Recall from Chapter 9 that the best server coverage is where one station provides the strongest signal to mobile subscribers. To contrast this with coverage area, Figure 10.8(a) shows the coverage area of a base station without reference to other stations, and (b) shows the actual contribution to the network where the station provides the strongest signal (note that this is not a very well designed site, to illustrate the principle). We need to sum the traffic within the service area shown in (b), not (a).

As we illustrated in Section 10.3, we can determine the approximate traffic distribution by using a linear distribution over different clutter categories. Thus, the total area of a given category covered by a best server can be used to sum the total traffic presented. Table 10.7 shows the total area covered by the base station under study, split into different clutter categories. The traffic density has been estimated on a per-km basis for each clutter category (on the basis of a similar

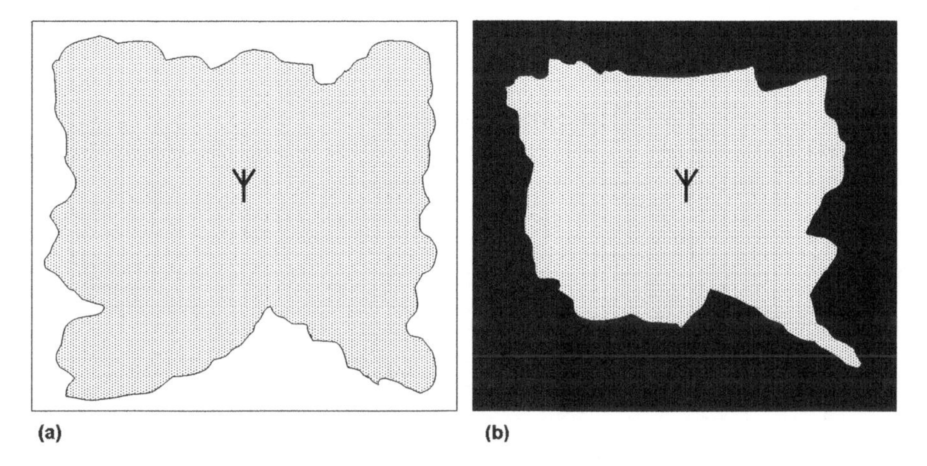


**Figure 10.8.** (a) Total extent of coverage of a site, without considering other base stations. (b) Best server coverage, limited by other base stations (their coverage shown black).

**Table 10.7.** Offered traffic.

| Clutter category | Area served (km$^2$) | Traffic density (E/km$^2$) | Offered traffic (E) |
|---|---|---|---|
| Rural | 156 | 0.0048 | 0.75 |
| Suburban | 10 | 0.02 | 0.2 |
| Urban low density | 18 | 0.0271 | 0.4875 |
| Light forestry | 10 | 0.005 | 0.05 |
| Water | 4 | 0.003125 | 0.0125 |
| **Total** | **198** | | **1.5** |

study to that carried out in Section 10.3). It is a simple task to multiply the served area in each category by the traffic density to determine the total offered traffic served, as shown in the table.

If we assume that the system is a blocking one, and that calls arrive truly randomly, then we can use the Erlang B equation to determine that for a desired GoS of 5 %, we need to have 4 circuits available at the base station.

Note that the more detailed we make our subscriber modelling, the more detailed we can make our traffic analysis. But also remember that traffic will vary from hour to hour, day to day and over longer periods, so it does not do to critically engineer for one specific set of circumstances since this may make the network vulnerable to poor performance under a different set of circumstances.

## 10.6 Capacity Planning for Traffic

We have looked at determining the number of circuits required for a given amount of offered traffic, using a model of subscriber behaviour and the Erlang B equation. It is important to recognise that there will be variations in offered traffic. For example, consider the following three variations of offered traffic over a period of a few weeks, shown in Figure 10.9.

Sequence one shows a relatively static picture with variations over the day and between workdays and weekends. The traffic does not show any major unexpected variations. Sequence two shows typically lower traffic than sequence one, except on five occasions over the period when traffic suddenly ramps up to a high level. What could cause such a

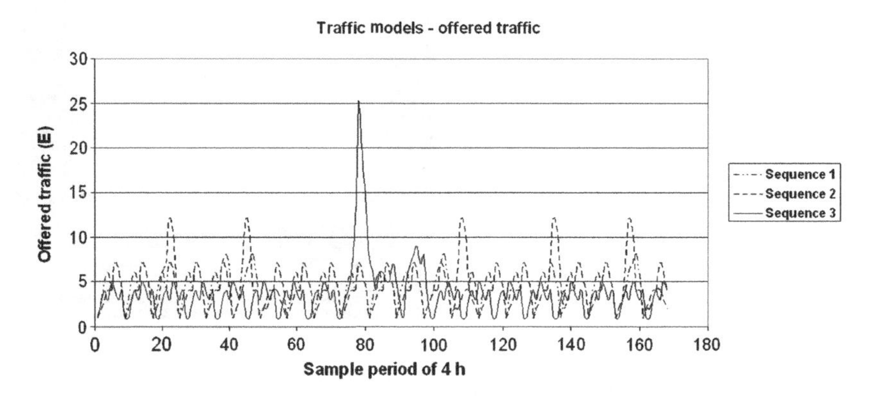


**Figure 10.9.** Variations in offered traffic over time; fairly stable traffic amounts plus some peaks where demand is considerably higher than the average.

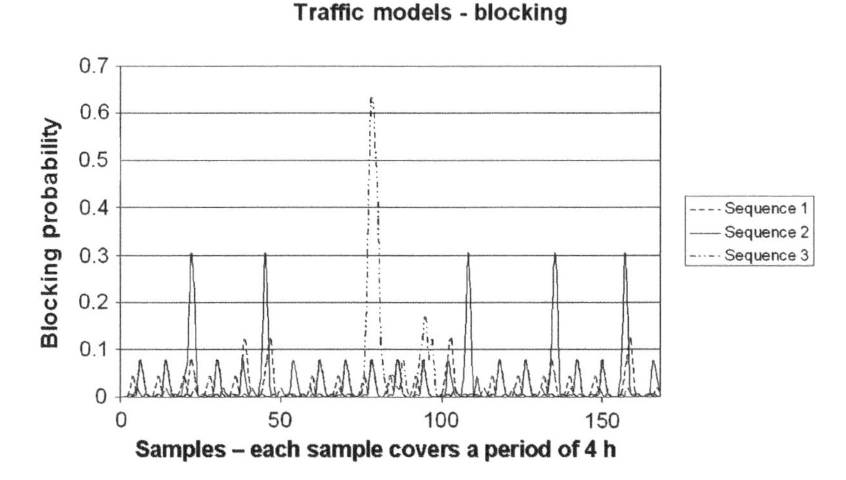


**Figure 10.10.** Variations in network blocking over time; this is directly related to the amount of offered traffic, given that the network capacity is fixed.

thing? An example would be a football stadium located in the suburb of a town. Normally, traffic levels are dictated by the local people who live and work there, but when a major match is on, demand suddenly increases dramatically. Sequence three shows relatively stable demands except on one occasion when the demand suddenly increases to levels not seen in the other models. This could be the effect of disruption to travel due to very bad weather or strike, or some other major event that simultaneously affects a large number of subscribers. We can use the traffic metrics in the model to determine blocking probability for these different conditions as shown in Figure 10.10 (we have assumed 10 circuits are available).

Analysis of the blocking probability exceeding 0.1 (chosen as an arbitrary acceptable limit) shows the following:

- Sequence one exceeds the required blocking probability for 2 % of the time.
- Sequence two exceeds the required blocking probability for 6 % of the time.
- Sequence three exceeds the required blocking probability for 5 % of the time.

What can we deduce from these figures? On their own, not a great deal, but by also comparing the figures with the graphs and knowledge of the triggering factors we can make certain deductions.

- Sequence one was fairly stable, so this means we can expect the figures produced to be indicative of future traffic. We therefore need to ask the question of whether the 2 % blocking probability that the network offers a poorer performance than wanted is acceptable.
- Sequence two was also stable, considering the fact that the coverage area includes a football stadium. We probably need to adjust the network to provide an acceptable level of coverage during match days and accept the over-capacity at other times. If we do not do this, we risk alienating many customers on a regular basis by not offering them a good service.

- Sequence three was unstable and was influenced by an unforeseen event. We need to determine how likely this is to happen again and what the impact on our service will be – in other words, carry out a standard risk assessment. Ultimately, the decision of how much capacity is added to the network infrastructure to deal with abnormal conditions is a management decision, on the basis of the financial model and business risks. The network designers must then work to design the network to meet the stated requirements. Of course, if network engineers or designers identify specific risks or problems that they do not think have been considered previously, then professionalism dictates that they must raise the issue with management and let them decide (even if the decision is to do nothing). In these days of increased litigation, it is more important than ever that engineers do this (and record the fact that it has been done).

Traffic analysis need not be particularly difficult to perform, but as is often the case in radio network engineering, the key aspect is to correctly identify the requirements in sufficient depth to be able to design against them. For traffic engineering, this will involve considering traffic demand under a variety of expected conditions. Modern planning tools normally provide tools for analysing traffic under different conditions. The expected traffic density can be expressed on a per data point basis, linked to a data level within the planning project. This does imply the possibility of quantisation error and if the actual traffic varies more than expressed by the data, then errors may be made.

An alternative approach based on modelling specific scenarios is discussed in the next chapter.

## 10.7 Modelling Mixed Services

So far we have restricted our analysis to single services (for example voice), but what happens when there are mixed services with different characteristics on the same network (for example voice and data at different rates)? Again, modern planning tools provide the ability to model the network using mixes of different traffic. The behaviour of the network in the mixed traffic scenario will depend on how these services are provided. For example, Table 10.8 shows an example for a network where mixed services are provided at different bandwidths.

In this case, the traffic demand for each service area will be the summation of the traffic mix across it. Although this is not technically much more difficult to do than the single service example, it is more fraught simply because there are more degrees of freedom and it is easier to incorrectly characterise the traffic. This is more a business issue than a technical one, and it is important that sufficient effort is expended in producing the metrics for traffic demand and that there is a bit of 'wriggle room' in the event that the original figures prove to be in error. An

**Table 10.8.** Traffic by bandwidth.

| Bandwidth (kHz) | Percentage of total traffic |
|---|---|
| 12.5 | 30 |
| 25 | 35 |
| 38.4 | 25 |
| 200 | 10 |

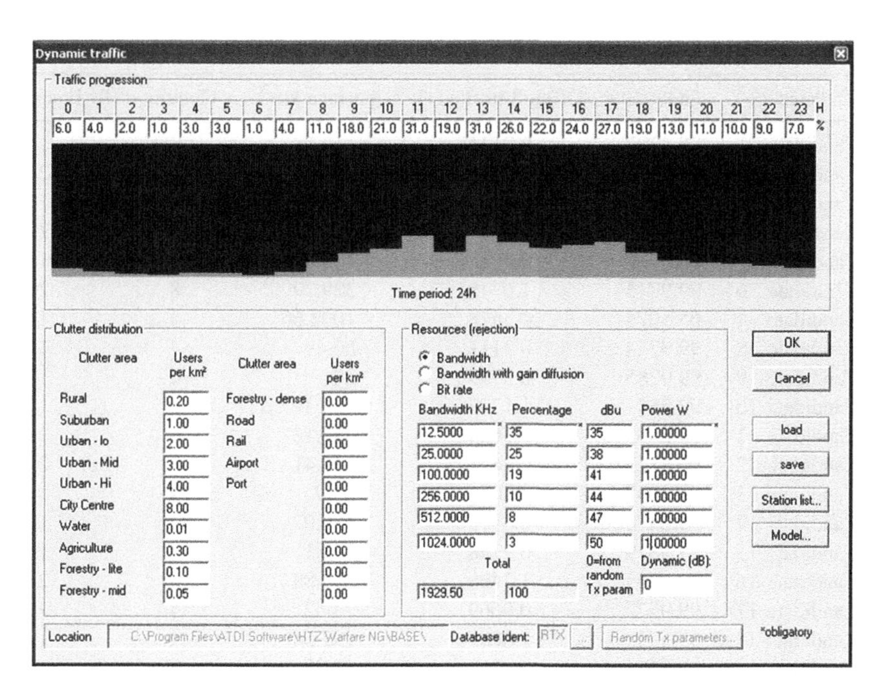


**Figure 10.11.** Configuration of user traffic by user density and call type. This is essential for analysing performance when the network offers several types of service concurrently. Reproduced by permission of P & D Missud.

example of representation of traffic over a period and including mixed services and subscriber distribution based on the clutter environment is shown in Figure 10.11.

Such analysis tools allow detailed analysis of traffic demand, but always remember that the design ultimately depends on the quality of the data for predicted subscriber behaviour.

## 10.8 Designing for Traffic

A typical result of a traffic analysis is shown in Table 10.9. This shows the grade of service for each candidate base station used in the simulation. The designer must then consider whether the wanted GoS is met for each service and, if not, then what actions should be taken to modify the design to ensure that the requirements are met. As we have not looked at interference or frequency assignment, we will ignore these for the moment.

Table 10.9 shows that most of the candidate base stations provide a grade of service above 99 %, but stations 1, 4, 7, 10 and 11 do not provide service to this level. Where the level of performance is only marginally missed and other adjacent sites have some spare capacity, then it may be possible to adjust the tilt of the antenna to reduce the coverage area somewhat. This can be achieved either mechanically or electrically depending on the antenna type. This is likely to be true for stations 1, 4, 10 and 11, but the situation for site 7, which only achieves a grade of service of 65.5 %, is more severe and this approach is unlikely to work. Possible options could include adding more candidate sites in the vicinity of site 7, but an alternative would be to increase the traffic capacity at the single site, and a

**Table 10.9.** Grade of service analysis.

| BS | Callsign | GOS(%) | Offered traffic (E) | Surface km$^2$ | Circuits | Tx Frequencies |
|---|---|---|---|---|---|---|
| 1 | Candidate 1 | 98.8297 | 0.9144 | 176.68 | 4 | 1 |
| 2 | Candidate 2 | 99.9942 | 1.3104 | 265.89 | 8 | 2 |
| 3 | Candidate 3 | 99.6982 | 0.6032 | 119.23 | 4 | 1 |
| 4 | Candidate 4 | 98.5626 | 3.3736 | 737.12 | 8 | 2 |
| 5 | Candidate 5 | 99.6919 | 0.6068 | 115.65 | 4 | 1 |
| 6 | Candidate 6 | 99.9713 | 1.6745 | 329.79 | 8 | 2 |
| 7 | Candidate 7 | 65.5024 | 4.3678 | 1002.66 | 4 | 1 |
| 8 | Candidate 8 | 99.9714 | 0.3111 | 60.49 | 4 | 1 |
| 9 | Candidate 9 | 99.9285 | 0.3999 | 28 | 4 | 1 |
| 10 | Candidate 10 | 97.555 | 1.1708 | 239.34 | 4 | 1 |
| 11 | Candidate 11 | 94.6965 | 1.5604 | 337.12 | 4 | 1 |
| 12 | Candidate 12 | 99.4827 | 2.7475 | 573.41 | 8 | 2 |
| 13 | Candidate 13 | 99.9937 | 0.208 | 33.9 | 4 | 1 |
| 14 | Candidate 14 | 99.6472 | 0.6316 | 150.07 | 4 | 1 |
| 15 | Candidate 15 | 99.9819 | 0.2748 | 52.81 | 4 | 1 |
| 16 | Candidate 16 | 99.9633 | 1.7408 | 359.88 | 8 | 2 |
| 17 | Candidate 17 | 99.9477 | 0.3669 | 74.62 | 4 | 1 |
| 18 | Candidate 18 | 99.5733 | 0.6685 | 137.56 | 4 | 1 |
| 19 | Candidate 19 | 99.8527 | 0.4901 | 98.92 | 4 | 1 |
| 20 | Candidate 20 | 99.197 | 0.8112 | 175.87 | 4 | 1 |
| 21 | Candidate 21 | 99.9932 | 0.2118 | 37.53 | 4 | 1 |
| 22 | Candidate 22 | 99.996 | 0.1849 | 37.13 | 4 | 1 |
| 23 | Candidate 23 | 99.9978 | 0.1585 | 29.3 | 4 | 1 |
| 24 | Candidate 24 | 99.9996 | 0.0997 | 19.37 | 4 | 1 |
| 25 | Candidate 25 | 99.999 | 0.1273 | 18.2 | 4 | 1 |

good way of doing this would be to replace the omni-directional antenna with a number of sectored antennas that provide coverage over a specified sector. This is illustrated in Figure 10.12, which shows a close up of a site with three antennas oriented to provide best server coverage in different directions around the site. Both the angle the main beam is directed towards and the antenna horizontal pattern are shown in the figure.

In this case, the original traffic is split into three, on the basis of the best server coverage of each antenna. This will then allow three times as must traffic to be handled in the same part of the network.

For any network with high traffic demand, sectored sites are the norm rather than the exception. Although this means there needs to be more equipment at each site, the overall site count is reduced. This is usually more cost-effective. As well as producing sites with sectored antennas, the design of base stations should be optimised to provide the best server coverage that best meets the traffic demands. This often means designing sites that provide sub-optimal coverage for wide areas and thus the mechanisms for achieving the objectives are different from those used to maximise coverage. For example, it may mean:

- Reducing antenna height (for increased coverage it is often useful to increase antenna height).
- Selecting an antenna that only covers a sector around a base station (therefore reducing the effective coverage area of each sector compared with an omni-directional system).

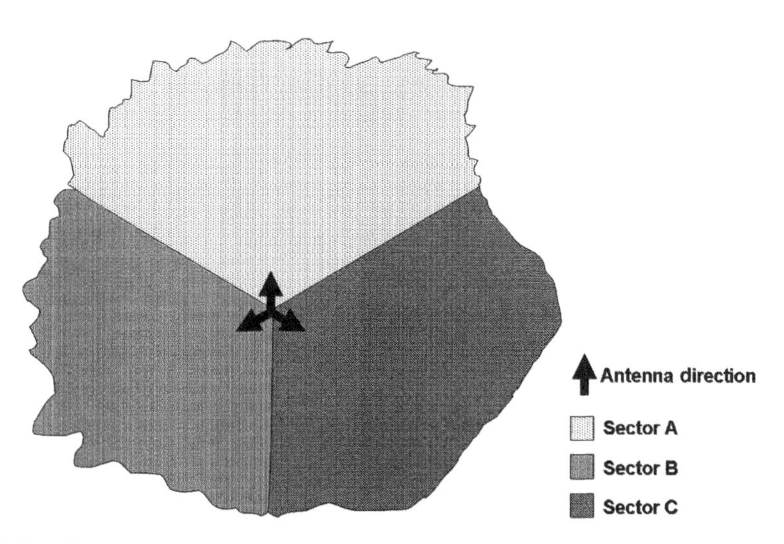


**Figure 10.12.** Sectored antenna; three sectors with the antenna pointing North, 120 and 240 degrees. This simple best server plot shows the transitions between the service area of each transmitter. Typically, the coverage at the edge of each service area would be limited by other stations, but this is not shown in this example.

- Electrically or mechanically tilting the antenna down so the main lobe in the vertical plane is orientated towards the ground rather than towards the radio horizon as would be desirable to achieve maximum range.
- Reducing transmitted power, as opposed to increasing it to achieve longer links.

It should be clear that network design for coverage and capacity is a trade-off and that an iterative design process may be needed to achieve all the objectives. This is what typically occupies design engineers on a day-to-day basis. Planning tools with an ability to work in an iterative manner can ease the design process considerably. For technologies based on CDMA, they are effectively essential.

## 10.9 Traffic Analysis Deliverables

At the end of the traffic planning process, it will be necessary to provide following information for each candidate site in the proposed network design:

- A list of the candidate sites, including their locations.
- The number of sectors to be used.
- Antenna height, orientation and tilt for each antenna to be used at the site.
- The grade of service estimated for each sector.
- The number of frequencies required for each site (to be used in the frequency assignment process).
- Power to be used for each channel.
- Other, technology-specific information as required.

Such information is typically provided as a spreadsheet such as the segment shown in Table 10.9.

# 11

# Network Design Methods

## 11.1 Network Limiting Factor

When starting the design, it is important to identify the limiting factor for the network. This may be related to coverage, traffic, interference or some other nontechnical business-related issue. This limiting factor will determine how the entire project will progress in the sense that it will determine 'cost' in monetary or other terms. In this section, we identify some common 'rollout' strategies (rollout is the term used to identify the point at which services and service areas are made available to subscribers). The strategy will typically be chosen to suit the business need rather than engineering considerations and this is entirely the right way round; engineering should support the operational requirements, not the other way. In this chapter, we describe some simple methods of rolling out a network to illustrate some suitable approaches.

## 11.2 Rollout Strategy

### *11.2.1 Introduction*

Rollout strategy will normally depend on business constraints. Such constraints may depend on a variety of factors, such as the following example considerations:

- The technology is largely unproven and is too risky to rollout the whole network in one move. To militate against this risk, a small system will be created so that the technology can be tested and problems ironed out before the full network is implemented.
- The finances available do not allow for a full rollout of the network in one step. Instead, part of the network can be implemented, and the revenue generated from this can be used to finance other parts of the network.
- Customer demand is expected to be low initially but to build up over a number of years. Thus, the traffic on the network will be low initially and the system needs not to be capable of handling high volumes of traffic. Eventually, however, it must be possible to increase traffic handling capability as demand increases.

Mobile Radio Network Design in the VHF and UHF Bands: A Practical Approach
*Adrian W. Graham, Nicholas C. Kirkman and Peter M. Paul* 

- The hardware for planned services may not be available yet. This may be a new mobile units or the fixed infrastructure required to provide the service. A typical example might be the ability to send and receive video from a mobile phone; it may be that initially the network has to be designed purely to provide voice calls, but with the intention to be capable of handling video in the future.

Each of these strategies will have an impact on the designer, and we will look at some possible options next.

### *11.2.2 Pilot System*

The purpose of a pilot system is to test performance before rolling out the whole network. This is particularly beneficial for high-risk networks using new technologies or when the network is particularly vital. A pilot system will typically be implemented across a small portion of the wanted service area, ideally in an area that is representative of the whole service area for the full system. The pilot system can also be used to refine the design process for the whole network and also to gain a better understanding of the link budgets used.

A pilot system will ideally be planned to provide exactly the same service as the full system, so the planning methods used will be the same. Sometimes, only a subset will be implemented. In this case, it is expected that it will be possible to extrapolate parameters for the services not implemented from those that are.

The pilot system approach has a number of advantages and disadvantages, such as

- It helps to de-risk the main project by proving the technology and allowing technical issues to be identified and resolved. This is the main advantage of this approach.
- It allows calibration surveys to be carried to link performance with received field strength and other metrics such as mean opinion score (MOS) and BER. This can be used to refine the link budget values, propagation model and also help to refine clutter category attenuation values (if the model uses them). This helps to improve the design of the main project.
- Traffic metrics can be obtained to help the design of the main project. This is far better than relaying on generic data.
- A pilot system approach is more expensive than other approaches. This is particularly the case if the pilot system will be abandoned rather than incorporated into the main system, which often happens.
- The pilot system approach tends to delay full network rollout, so the overall project duration is longer than would otherwise be the case.
- In the ideal case, the target subscribers will use the pilot system, rather than engineers or other nontarget users. In this case, the subscribers will still be doing their normal job, and it is important that this is not disrupted. This will typically mean that for the duration of the pilot study, these subscribers will have to carry two radios; the one for the pilot system under test, and the other for the legacy system to be replaced. This can be inconvenient for the subscriber, and can lead to resentment about the entire project.

In general, pilot systems are appropriate for high-risk networks, but for less-risky ventures they add unnecessary cost and inconvenience.

### 11.2.3 Regional Rollout

Sometimes the financial constraints on the network will mean that it is necessary to rollout on a regional basis. This means that the network will be designed according to the same standard set of rules, but individual areas will be implemented at a time. This leads to an 'island' design approach, as illustrated in Figure 11.1. Each area is designed individually according to the overall rollout plan.

As the network develops and it is necessary to link the 'islands', then network coverage, traffic and interference design are applied to the gaps in between, until the network coverage is contiguous.

The regional rollout strategy can be desirable from a cash-flow point of view, but it will typically be more expensive than a single design exercise. Also, the problems due to interference and frequency assignment are likely to be more severe since often these have not been considered in the design of each island. The regional rollout strategy has features including the following:

- Full network capacity can be provided over a restricted region from the start.
- Marketing and sales activity can be focussed on the served area, providing a more focussed approach.
- Customers who stray from the service area will lose service, which is undesirable.

If this approach is to be taken, it is beneficial to design the main parts of the entire network and work backwards for the design of each island. An illustration of this is shown in Figure 11.2. This shows a completed design for a network over an entire projected service area.

If necessary, it is now possible to reduce the design down to an individual area, as shown in Figure 11.3. This shows part of the full network design to be implemented on a regional rollout basis.

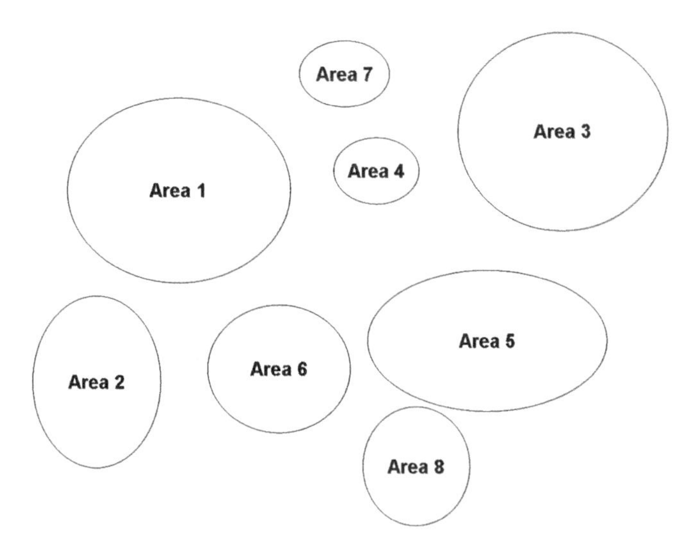


**Figure 11.1.** The 'island' approach to network design; each small area is considered on its own before considering coverage between them.

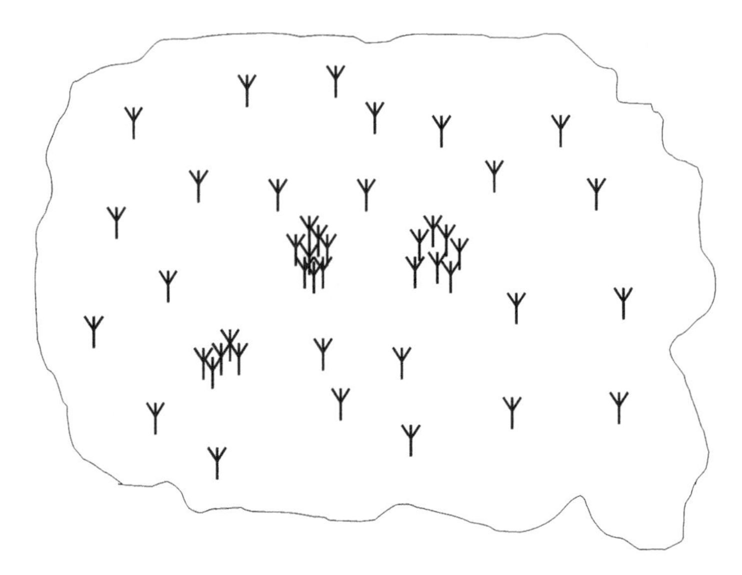

**Figure 11.2.** A full network design, showing base station locations and service area boundary.

The benefit of working this way round is that when the network design advances, the composite network will work effectively without problems of coverage, capacity or interference. Considering that the cost of network design is likely to be considerably less than that of the network infrastructure, performing the full network design at an early stage is not likely to be very expensive, but major savings will be made over the life of the project.

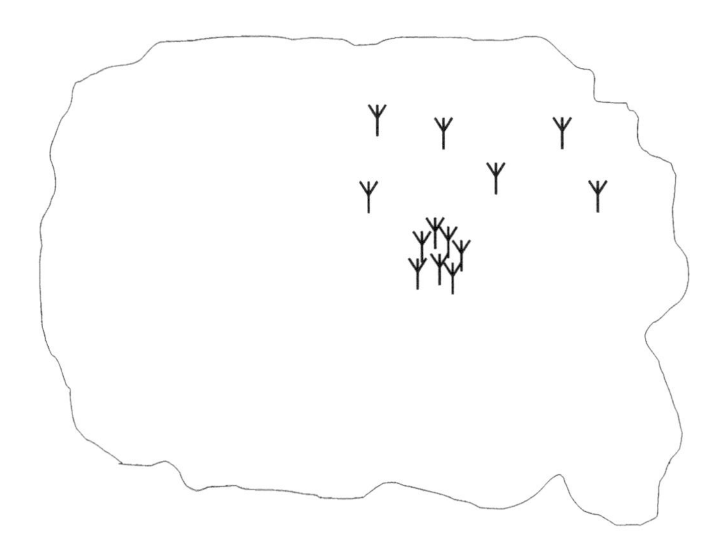

**Figure 11.3.** A partial network design, drawn as a subset from the full network design, covering only one part of the full service area.

Unfortunately, in many cases this does not occur and the overall costs for network design and implementation are higher than could have been the case.

## *11.2.4 Sparse Network*

If network demand is expected to increase during its life but may well start relatively low, then it may be possible to start with a sparse network and then progressively add capacity as required. The aim in this case is to provide a full service to those subscribers that have joined the network but not to provide extra capacity when it is not required. This situation is illustrated in Figure 11.4, where traffic demand is shown to increase over time.

The question is, should the network be designed to cope with the traffic expected in several years, or should it be designed to cope with traffic in the early years, and then add capacity over time? If the extra capacity is added at the beginning, then this will add to the cost of development early on and might endanger the financial viability of the entire network. If it is designed to meet the early traffic only, then it might be difficult or even impossible to add capacity later. This is a difficult question to answer and should be examined for each network. Ideally, it should be possible to add capacity as required without reversing early design decisions.

Ideally, it is not desirable to change the network design to meet extra capacity later by adding new sites, but it might be acceptable to provide more capacity at existing sites. One way of achieving this is illustrated in Figure 11.5. In Figure 11.5(a), each element has an omni-directional antenna. In Figure 11.5(b), each site has a three-sectored antenna, with the direction of the main lobe shown by an arrow. This effectively triples traffic capacity.

In this case, each site starts off with an omni-directional antenna. Later, as traffic demands increase, then each site can have a sectored antenna system installed to cope with demand. Additionally, more channels can be added (on different frequencies) to specific sites to provide additional capacity in high-demand areas.

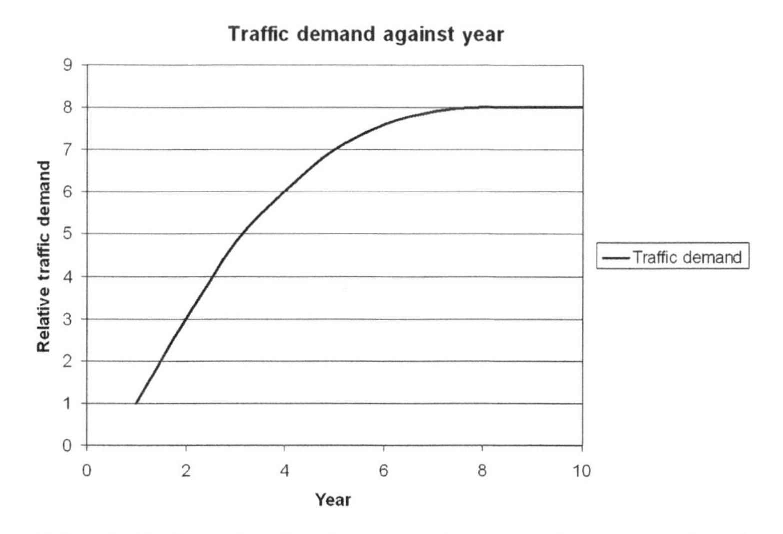


**Figure 11.4.** Traffic demand against time; as service take up increases, so does the traffic.

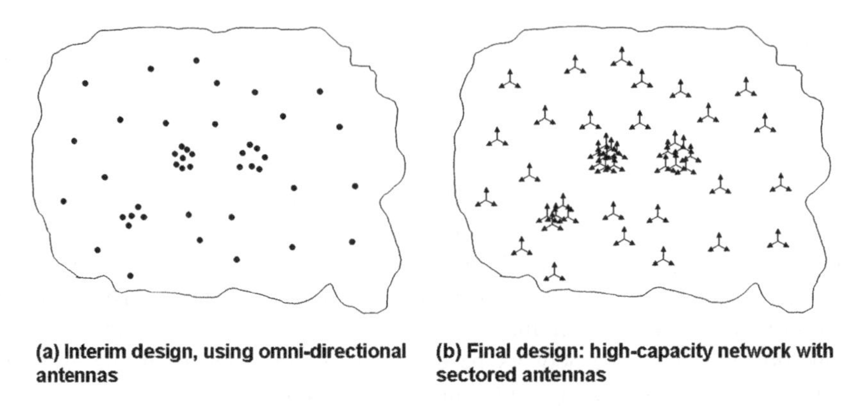


**Figure 11.5.** A partial network design derived from a full network design.

The same approach can be used to the sites themselves to determine when they will be rolled out. Figure 11.6 shows a network where the design is performed for year three, but some sites are selected for rollout in year one and year two. These sites are selected to provide coverage over the service area but limited traffic capacity.

The extra sites in year two and three provide some additional coverage and more traffic capacity. Designing the network for year three and then selecting a subset for earlier rollout

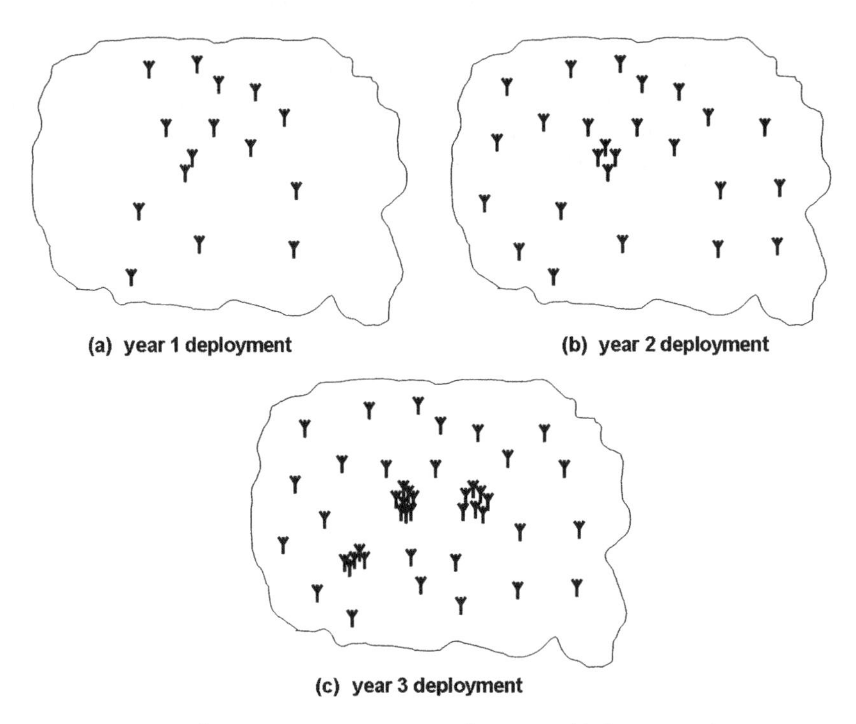


**Figure 11.6.** Sparse rollout over 3 years; extra sites are added to extend coverage and add capacity.

means that the additional parts of the network can be rolled out without unforeseen problems. This is the ideal situation; in practice it may not be possible, but if it can be done, it can reduce the overall cost of rollout.

Sparse network design can help spread cash flow in network design, but there is the risk that if the network is too limited, then customers may become disillusioned with its performance. Once this occurs, the situation is very difficult to recover.

### *11.2.5 Limited Service*

In some cases, when there is an intention to provide multiple services via a network, it will be appropriate to rollout a partial service before implementing all of the planned services. Some of the services may require higher field strength at the receiver and thus will require shorter links between base station and mobile. In this case, the sparse network design methodology may be the best approach to take.

## 11.3 Re-Broadcast/Relay Links

A particular common design problem is to implement re-broadcast (also referred to as 're-bro') links between areas of coverage. This is illustrated in Figure 11.7, which shows two separated coverage areas linked by a re-bro station. This allows subscribers in coverage area one to talk to coverage area two and vice versa.

The re-bro link may be implemented by equipment operating in the mobile frequency band or at microwave frequencies. In either case, mobile design methods can help to identify good locations for a re-bro site. In this case, the planning tool is configured with the characteristics of the link (not the link budget for the base station to mobile) and then the re-coverage plot feature is used (see Chapter 9).

This is illustrated in Figure 11.8. This shows a re-coverage plot calculated for two sites. Each of the two sites will be the base stations that need to be linked by a re-bro station. The re-coverage area shown in the light colour is where both sites provide a working signal. This

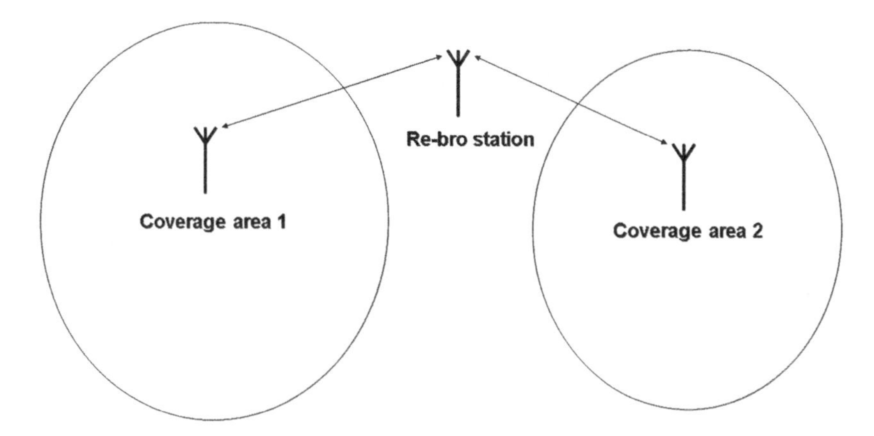


**Figure 11.7.** Re-broadcast ('re-bro') station to link two area of mobile coverage served by the base stations shown. The re-bro link is designed to link both base stations together.

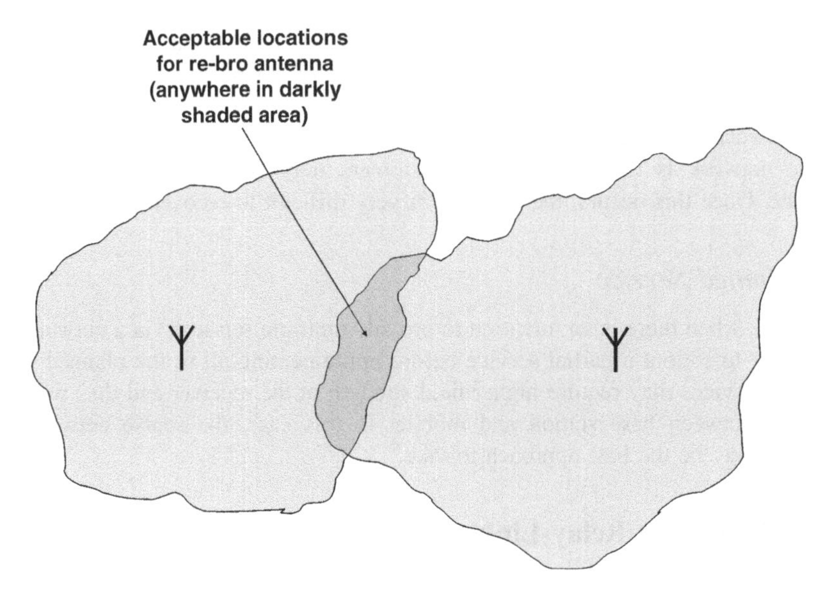


**Figure 11.8.** Overlapping coverage for re-bro link positioning; anywhere where both base stations will be able to communicate with the re-bro will be acceptable from a radio planning point of view. The designer will then have to choose a suitable candidate site from within the acceptable area.

means that, subject to an acceptable noise floor, the re-bro station can be positioned anywhere in the light coloured area, and it will be able to communicate with both stations.

This simulation provides an area-based solution, meaning that engineers can look at maps for the best area and select a candidate re-bro site based on road access, power availability and so on.

## 11.4 Future Proofing

Most radio networks will evolve during their lives, and successful networks will often stay in service longer than initially planned. When the network is initially designed it will not be possible to accurately predict what form mid-life updates will be, nor how demand may change over time. It is, however, useful for the network designers to consider some of the more likely scenarios and to test the proposed network design to see how it would be able to cope. Ideally, the designers can indeed prepare plans for likely changes so that if the situation arises then the time to re-configure the network to meet the new need is short. In general, this is probably more relevant to emergency service networks and the likes, since commercial operators may be reluctant to spend money and effort on plans that may not be needed.

# 12

# Backhaul

## 12.1 Introduction

In the rest of the book, we focus on radio planning for the network access link between the mobile subscriber and one or more base stations. What we have not considered is how the call is treated after this point. Unless the called subscriber is within the service area of the same base station as the calling subscriber, the call must be routed through the network and sent to the correct destination. At any one moment, there will be many calls in progress, each with a unique caller and a unique destination, and these must be concurrently handled. This is a complex situation and it is dealt with by the line structure of the network, which is often referred to as the 'backhaul'. We do not intend to cover line-related backhaul planning in this book, but we will look at the radio aspects of backhaul to illustrate the differences between fixed planning and mobile networks. The engineering required for fixed links, which operate at far higher frequencies than the mobile traffic, is very different than that of mobile radio engineering, and this chapter will serve as an introduction to the subject. Microwave planning can take up many books on its own. In this book, we merely wish to introduce the subject and identify the key design processes and factors that must be considered. This chapter also serves as a useful illustration of the contrast between mobile radio propagation and propagation at microwave frequencies for fixed links. Most of the techniques described in this section have been derived from ITU-R P.530 (version 10), unless a different recommendation has been identified for a specific process or expression.

## 12.2 Background to Backhaul

The key differences between backhaul and the mobile systems so far covered include the following:

- Backhaul links are always between fixed locations.
- Backhaul can be provided by different methods:
  - 'Line' or 'fibre' plant, which is based on cables, either old-fashioned telephone lines or fibre optic cable. This can either be owned by the network operator or leased from a fixed line provider.

Mobile Radio Network Design in the VHF and UHF Bands: A Practical Approach
*Adrian W. Graham, Nicholas C. Kirkman and Peter M. Paul* 

  - Microwave point-to-point radio links (normally in a 'star' or 'ring [circular diversity]' configuration, as shown in Figure 12.1).
  - Microwave point-to-multipoint systems. In this case, some nodes in the network link to a variety of other nodes, usually either via 'star' type architecture or a 'mesh' approach (shown in Figure 12.1).
- Backhaul will in general need to deal with high levels of traffic, far higher than that experienced in the mobile element of the network.
- The performance of the network is usually expressed in terms of 'outage'. Outage is made up of reliability, which includes the effect of equipment failure, and availability made up of propagation effects (as for the mobile case).

The choice of which system is selected depends very much upon cost and the traffic requirements. As a general rule, traffic requirements of greater than STM1 (155 Mbps) would generally use a fibre line plant solution, be it leased or owned, while traffic requirements of less than STM1 could be served by either line plant or microwave systems. For traffic requirements of less than about 2 Mbps, point-to-multipoint systems become attractive. The network designer will select which solution to use firstly on financial grounds and secondly on operational grounds. The financial calculations will include the cost to implement new links and lines compared to the cost of renting services provided by telecommunications companies. The operational considerations will include the existing infrastructure already available and the technical feasibility of using microwave links.

In many cases, particularly for larger systems, it is likely that a mix of approaches will be taken. This is illustrated in Figure 12.1. This shows a hypothetical network with a

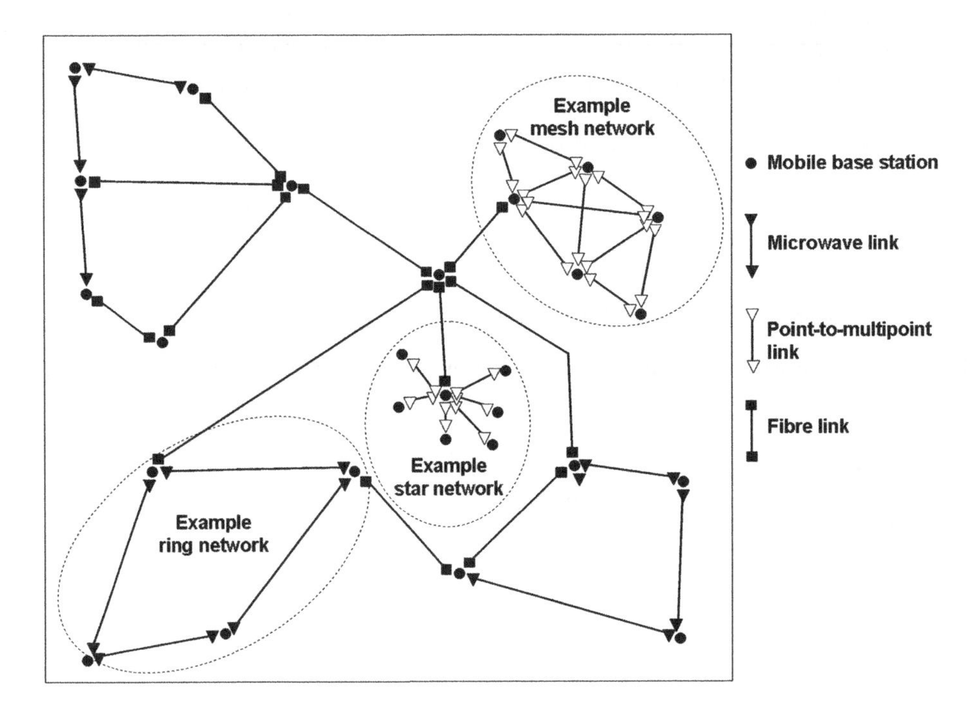


**Figure 12.1.** Schematic of mixed backhaul for a mobile network.

mixture of point-to-point, point-to-multipoint and fibre links used for mobile network backhaul.

This type of diagram is intended to be a schematic and may not be representative of the physical locations of each node in the network. Figure 12.1 also shows examples of the ring network architecture; in which microwave links are connected in a loop (this provides redundancy should any one link fail, because the traffic can still be routed via the other direction), the star architecture; where a central hub element radiates links out to a number of nodes and the mesh architecture; where many elements are linked together to provide multiple paths to any destination.

Next, we will consider the design of individual microwave links. These can be used for either point-to-point or point-to-multipoint technologies. We will focus on point-to-point networks since the design of point-to-multipoint is typically a subset of this, and is often merely a 'try-it-and-see' type of approach due to the lower costs and higher numbers of such links used in a network.

## 12.3 Specifying Link Performance

In an ideal world, every time a subscriber makes a call it should provide a working link to the called subscriber in an error-free manner 100 % of the time. In practice, any link may fail for a variety of reasons. Apart from the possibility that the link may fail due to mobile availability, any link in the backhaul system may also fail causing loss of service for the user. Typically, for any individual link, these failure modes can be categorised into three areas:

- Radio propagation failures, due to the path loss temporarily exceeding the maximum that can be sustained.
- Equipment failure, when electrical equipment fails.
- Other causes, such as power cuts or planned outages for maintenance (in which case back up capability should ideally be provided, or the downtime should be scheduled for a quiet period when the network usage is low).

Network designers can militate against equipment failure by the selection of reliable equipment and by incorporating redundancy into the design. They can also assess the probability of link failure by using standard statistics, as we shall show towards the end of the chapter. Power outages tend to be out of the control of the designer except in certain circumstances (for example military microwave systems will probably run on diesel generators provided for the purpose). Historical and statistical data can be generated to determine the likelihood of this for the situations where this is not considered force majeure, and this can be taken into account for reliability calculations. The results of a reliability calculation will normally be expressed as a percentage, or in terms of time in seconds per month or minutes per year. We will see how these calculations are arrived at later in this chapter.

Before any link can be designed, the required performance of the link needs to be specified. This performance is based on ITU-T Recommendations ITU-R G.826, G.828 and G.829 for PDH and SDH paths. The availability objectives can be calculated using ITU-R F.1668 and ITU-R F.1703 but in practice the regulator usually defines the availability that the link can be designed too. We will assume that the availability required for the example link, designed later in this chapter, is 99.99 % available.

For most of this chapter, we will as usual focus on the radio engineering aspects of the design.

## 12.4 Radio Link Design Aspects

### *12.4.1 Line of Sight Condition*

For the link between mobile and base station, line of sight clearance is very seldom achieved. Indeed, in some technologies, it is disadvantageous. For fixed links, however, achieving radio line of sight between both terminals is absolutely essential; in fact, in most cases it is also necessary to achieve Fresnel zone clearance of at least 0.6 of the first Fresnel zone (FFZ). This is illustrated in Figure 12.2.

Some microwave links might be very long, of the order of several ten's of kilometres, and in these conditions, atmospheric conditions, will have a major impact on the effective Earth radius used to determine the $k$-factor and hence the radio line of sight (see Chapter 4). Since we must achieve 0.6 FFZ in almost any atmospheric conditions that arise, in microwave engineering it is common to determine the clearance not using the typical $k$-factor of 4/3, but rather a far more pessimistic value that will only be experienced on a very small number of occasions throughout the life of the network. The smaller the $k$-factor is, the smaller the apparent radius of the Earth becomes and the more pronounced the Earth bulge is; thus, to account for the atmospheric conditions that might be the least desirable from a path clearance point of view, a small value is used; a value of $k = 0.6$ is often selected by microwave engineers. This value is based on experience, but the actual $k$-factor prevalent at a certain percentage of the time of the year can be deduced from the ITU tables of atmospheric refractivity or, where available, from historical measured data.

Most microwave links will fail to work when the path is within 0.6 FFZ and so, for any microwave link engineering, the first task is to ensure that this clearance can be achieved for the path of interest. It should be noted that in the millimetre (microwave) bands, the 0.6 FFZ can be measured in centimetres. This means that in practice at these frequencies physical clearance rather than Fresnel clearance is more important.

Note also that under these conditions, the free space loss (FSL) model is the most applicable, and thus there is no complex propagation required to account for terrain or clutter. We will, however, see that other forms of complexity do present themselves in microwave link budget calculations.

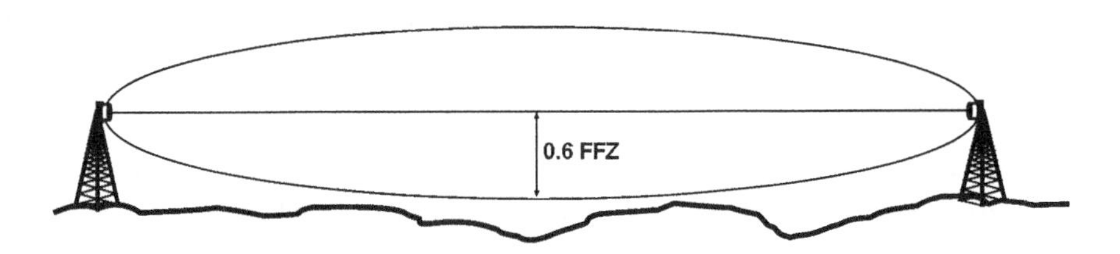


**Figure 12.2.** Path profile showing the condition of 0.6 First Fresnel Zone (FFZ) clearance.

## 12.4.2 Availability Calculations for Fixed Links

### 12.4.2.1 Introduction to Microwave Availability

Having determined that microwave links require different link characteristics for their path profiles compared to mobile links, we can now state a commonality; the radio part of the path performance is based on availability, just as in the mobile case. However, there are a number of differences that must be considered. First, the fading statistics are unlikely to be of the Rayleigh form commonly found in mobile systems; they are far more likely to follow a Ricean or lognormal distribution. Secondly, at microwave frequencies, it is necessary to take into account a number of other factors. These are:

- The attenuation effects of the gases in the atmosphere. These are negligible for VHF and UHF frequencies, but have a significant impact in the microwave bands particularly above 10 GHz.
- The effect of rainfall. Rainfall is not continuously present, but when it is, it will cause reduction in signal strength at the receiver due to attenuation and scattering, and this effect is more significant above 10 GHz.
- Selective fading. This mechanism is caused by destructive interference caused by more than one propagation path arriving at the receiver. This can be due to reflection from solid objects or refraction through permeable ones.

Since these mechanisms are of great importance to microwave network design, and since they have not been covered elsewhere in the book, we will now look at each of these in turn to determine how their effects are accounted for. As they required complex calculations, we will introduce each in turn.

The path profile shown in Figure 12.3 has been generated for a real link and will be used as an example for the calculations shown through this chapter.

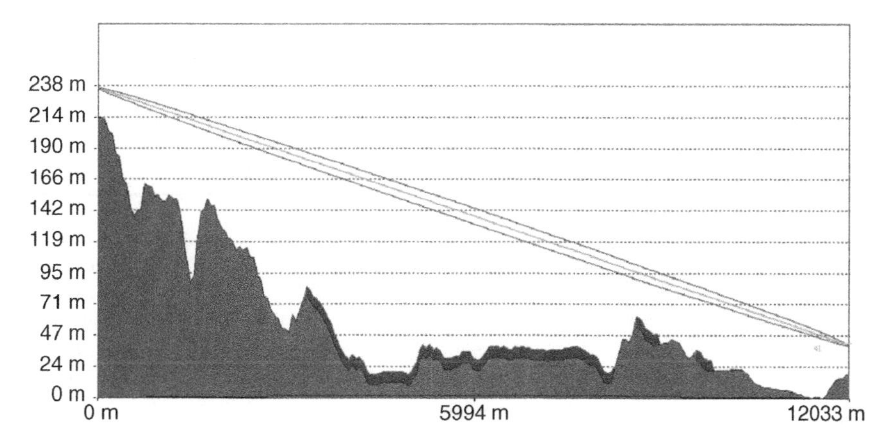


**Figure 12.3.** Example of a real path profile. The raised, shaded area above the terrain shows urbanization present along the path. The direct radio line of site and Fresnel ellipse is also shown, and the total link length is just over 12 km. Reproduced by permission of P & D Missud.

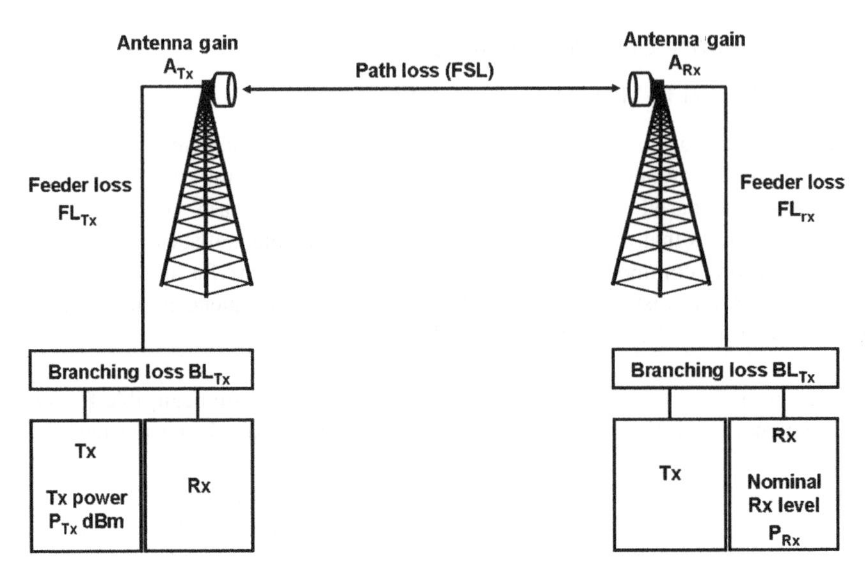


**Figure 12.4.** Link elements.

### 12.4.2.2 The Basic Link Budget

Just as for the mobile case, the core of microwave link planning is the link budget. This is illustrated in Figure 12.4 and expanded in Table 12.1.

The link margin can be calculated in nonfading conditions by the following equation, solving for the nominal receive level:

$$P_{Rx} = P_{Tx} - BL_{Tx} - FL_{Tx} + A_{Tx} - FSL + A_{Rx} - FL_{Rx} - BL_{Rx} \qquad [1]$$

$P_{Rx}$ is the unfaded signal power received at the receiver. Just as for the mobile case, each receiver will have a given sensitivity but this is termed the threshold values for microwave systems $R_{Th}$, and the availability can be calculated from the difference between the unfaded

**Table 12.1.** Representation of elements in microwave link budget.

| Abbreviation | Description | Units |
|---|---|---|
| $P_{Tx}$ | Transmitter power | dBm |
| $BL_{Tx}$ | Branching losses | dB |
| $FL_{Tx}$ | Feeder loss | dB |
| $A_{Tx}$ | Antenna gain | dBi |
| FSL | Path loss (free space loss) | dB |
| $A_{Rx}$ | Antenna gain | dBi |
| $FL_{Rx}$ | Feeder loss | dB |
| $BL_{Rx}$ | Branching losses | dB |
| $P_{Rx}$ | Unfaded receive level | dBm |

receive level and the threshold value. The fade margin (A) is simply:

$$A = P_{Rx} - R_{Th} \quad [2]$$

The fade margin is used to determine the availability of the link by using the relationship between the margin and the fade statistics. Before that, however, we have other terms to compute.

In Chapter 4, we introduced the concept of FSL and showed the equation in terms of the most useful units (distance in km and frequency in MHz). For microwave engineering, we keep the distance units in km but change the frequency units to GHz. This gives us the following expression for FSL:

$$\text{FSL} = 92.45 + 20\log d + 20\log f \quad [3]$$

The only change is to the constant value of the equation, due to the change in units.

As an example, if we assume that the total path length is 12.033 km and the operating frequency is 28 GHz, then Equation 3 becomes:

$$\text{FSL} = 92.45 + 20\log 12.033 + 20\log 28 = 143\,\text{dB}$$

To illustrate the working of the link budget, which we will use over the next few sections, Table 12.2 shows example figures against each parameter, using the FSL calculated.

Thus, for the link illustrated, the fade margin is 59 dB. The fade margin is used in the calculation of the link availability as we shall see, but before we apply this to the availability calculation, we have to consider another set of factors. The first of these is attenuation due to absorption by the gases in the atmosphere, as discussed in the next section. However, before we do so, we note now that the path loss (PL) can be described as follows:

$$\text{PL} = 92.45 + 20\log d + 20\log f + L_g \quad [4]$$

where $L_g$ is the additional loss due to gaseous absorption.

**Table 12.2.** Example link budget figures.

| Abbreviation | Value | Units | Calculation |
|---|---|---|---|
| $P_{Tx}$ | 30 | dBm | |
| $BL_{Tx}$ | 0.2 | dB | |
| $FL_{Tx}$ | 0.8 | dB | |
| $A_{Tx}$ | 42 | dBi | |
| **EIRP** | **71** | **dBm** | $P_{Tx} - BL_{Tx} - FL_{Tx} + A_{Tx}$ |
| FSL | 143 | dB | |
| $\mathbf{P_{Eq}}$ | **−72** | **dBm** | **EIRP - FSL** |
| $A_{Rx}$ | 42 | dBi | |
| $FL_{Rx}$ | 0.8 | dB | |
| $BL_{Rx}$ | 0.2 | dB | |
| $\mathbf{P_{Rx}}$ | **−31** | **dBm** | $P_{Eq} + A_{Rx} - FL_{Rx} - BL_{Rx}$ |
| $R_{Th}$ | −90 | dBm | |
| **A** | **59** | **dB** | $P_{Rx}$ − RTH |

### 12.4.2.3 Gaseous Absorption

At typical mobile frequencies, the absorption of radio energy due to the gases that make up the atmosphere and the water content present within it is negligible and is not normally accounted for in link budget calculations. This is not the case for microwave frequencies. Figure 12.5 shows approximate excess loss per kilometre for a given set of circumstances. Notice that there are separate sets of figures for dry air and suspended water vapour, which are summed to give a total attenuation per kilometre of a link. This graph has been derived from ITU-R P.676, which also includes formulae for the accurate calculation of gaseous attenuation.

To determine attenuation at a specific frequency, a microwave engineer can either use the calculations or read off the values from an accurate graph. For 28 GHz, the specific attenuation per kilometre is approximately 0.095 dB per km, and for our link of 12.033 km, this works out at an additional 1.14 dB. The FSL for our example was calculated to be 143 dB. With the

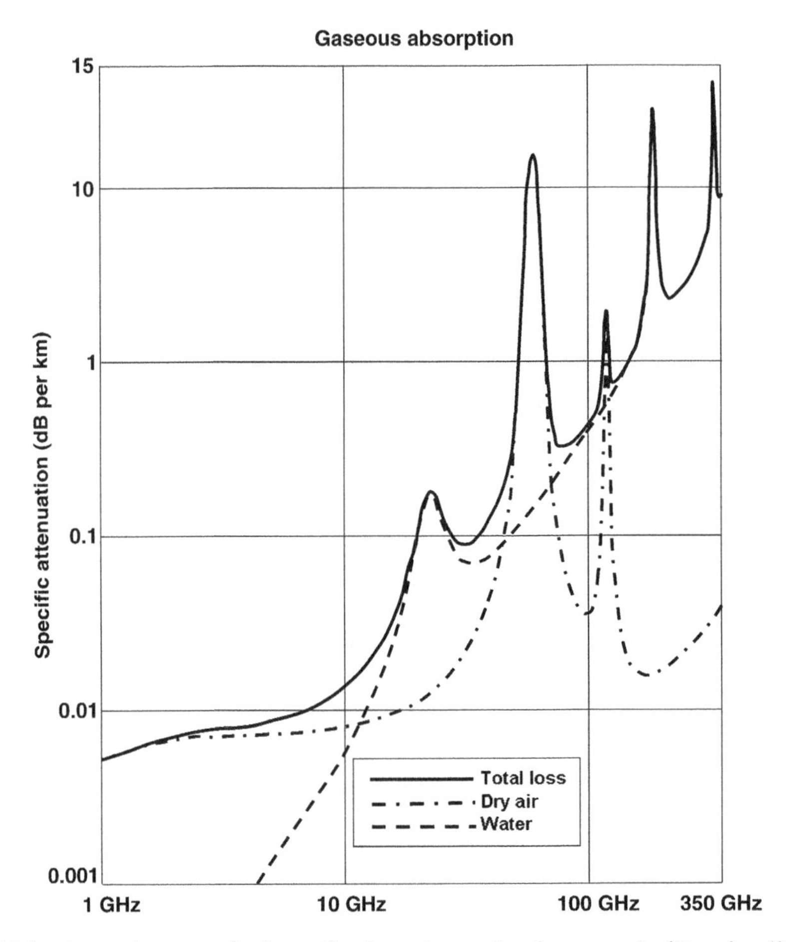


**Figure 12.5.** Approximate graph of specific absorption against frequency In dB per km (15 degrees C; 1 013 hPa; 7.5 g/m$^3$ water content).

addition of gaseous absorption, this becomes 143 + 1.14 = 144.14 dB, and the fade margin is reduced from 59 to 57.86 dB.

Atmospheric attenuation will always be present in a link and can be considered a continuous phenomenon. In addition to this, to determine overall link availability figures, we must also include transient effects that will affect the link. Of these mechanisms, rainfall is the most important.

#### 12.4.2.4 Rainfall Attenuation

Mobile transmissions are unaffected by rainfall in the link. However, microwave transmissions at 10 GHz or above are seriously affected by rain precipitation. When a radio path's Fresnel zone has water droplets from a rain shower within it, each raindrop contributes to an additional attenuation of the radio transmission. When rain becomes particularly intense and raindrops increase in size, they depart from their spherical shape and adopt a more flattened shape. This has a greater effect on horizontally polarised radio waves more than vertically polarised waves, and this explains why vertical polarisation is favoured when designing for higher frequencies. At lower frequencies, either vertical or horizontal polarisation is used (unlike most mobile applications, which typically only use vertical polarisation).

The amount of attenuation experienced in a link during a downpour is dependant on the frequency, size and density of raindrops. The principle mechanisms of attenuation are scattering, in which energy is re-directed from its path, and absorption, in which the rain transfers some radio energy into a small amount of heat. In the situation where the wavelength is fairly large relative to the raindrop, scattering is the predominant mechanism. In the converse situation, absorption dominates.

It is possible to determine rainfall attenuation by following a sequence of steps, as outlined in ITU-R P.838. The steps are:

- Determine the rainfall rate exceeded for a given reference percentage of time, such as 0.01 % of the time during an average year. Worldwide data for rainfall rates are published by the ITU in ITU-R P.837, although this data suffers from two shortcomings; firstly, it is rather coarse in nature and does not account for local variations. Secondly, the data is rather historical and may not represent the most recent weather trends. Other sources will include meteorological organisations and sometimes spectrum regulators; for example in the United Kingdom, Ofcom provides more accurate figures for parts of the United Kingdom. For the purpose of our example, we will assume a rainfall rate ($R$) of 30 mm/h as being typical of the rainfall rate experienced for 0.01 % of the time.
- Next, this information can be used to determine the specific attenuation $\gamma_r$ in dB/km by using the following mathematical relationship:

$$\gamma_r = kR^{\alpha} \qquad [5]$$

where coefficients $k$ and $\alpha$ are given in Table 12.3 for horizontal ($H$) or vertical ($V$) polarisation.

The above calculation for rain attenuation applies to either vertical or horizontal linear polarisation along horizontal paths in which the altitude of both ends is roughly similar. If the path has a significant elevation angle, or has a polarisation angle other than horizontal or

**Table 12.3.** Coefficients for the calculation of rainfall attenuation.

| Frequency (GHz) | $k_H$ | $k_V$ | $\alpha_H$ | $\alpha_V$ |
|---|---|---|---|---|
| 1 | 0.0000387 | 0.0000352 | 0.9122 | 0.8801 |
| 1.5 | 0.0000868 | 0.0000784 | 0.9341 | 0.8905 |
| 2 | 0.0001543 | 0.0001388 | 0.9629 | 0.9230 |
| 2.5 | 0.0002416 | 0.0002169 | 0.9873 | 0.9594 |
| 3 | 0.0003504 | 0.0003145 | 1.0185 | 0.9927 |
| 4 | 0.0006479 | 0.0005807 | 1.1212 | 1.0749 |
| 5 | 0.001103 | 0.0009829 | 1.2338 | 1.1805 |
| 6 | 0.001813 | 0.001603 | 1.3068 | 1.2662 |
| 7 | 0.002915 | 0.002560 | 1.3334 | 1.3086 |
| 8 | 0.004567 | 0.003996 | 1.3275 | 1.3129 |
| 9 | 0.006916 | 0.006056 | 1.3044 | 1.2937 |
| 10 | 0.01006 | 0.008853 | 1.2747 | 1.2636 |
| 12 | 0.01882 | 0.01680 | 1.2168 | 1.1994 |
| 15 | 0.03689 | 0.03362 | 1.1549 | 1.1275 |
| 20 | 0.07504 | 0.06898 | 1.0995 | 1.0663 |
| 25 | 0.1237 | 0.1125 | 1.0604 | 1.0308 |
| 30 | 0.1864 | 0.1673 | 1.0202 | 0.9974 |
| 35 | 0.2632 | 0.2341 | 0.9789 | 0.9630 |
| 40 | 0.3504 | 0.3104 | 0.9394 | 0.9293 |

vertical, then $k$ and $\alpha$ from the table above need to be modified. The process for doing so will not be covered here, but is detailed in ITU-R P.838.

Table 12.3 does not directly have a value for 28 GHz, but the value can easily be calculated using linear regression. If this is done, then for a vertically polarised link, values of $k_v = 0.1454$ and $\alpha_v = 1.0108$ can be derived for a link 28 GHz using the rain rate ($R_{0.01}$) of 30 mm/h.

Applying these figures gives $\gamma_r = 0.1454 \times 30^{1.0108} = 4.5252\,\text{dB/km}$.

When this type of information has to be used frequently, it is often useful to compute graphs from which the attenuation value can be read directly. An example of this is illustrated in Figure 12.6.

- Rainfall, particularly heavy rain, is often not present along the entire length of a link, but rather in one or more 'cells'. It would therefore be wrong to apply this attenuation along the entire length of a path, but instead an effective distance ($d_r$) affected by a rain cell must be estimated. Measurements have allowed a simple estimation to be made via the expression:

$$d_r = \frac{1}{1 + \dfrac{d}{d_o}} \qquad [6]$$

where

$d$ is the total path length in km

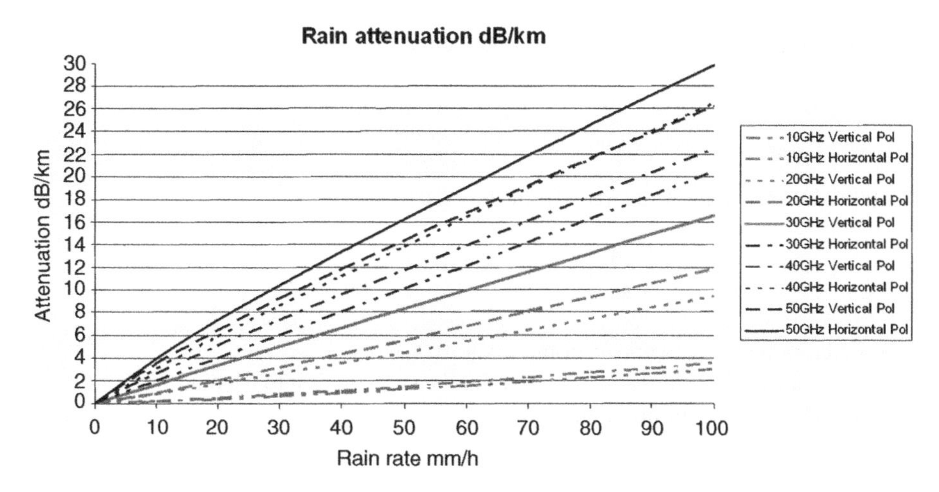


**Figure 12.6.** Rainfall attenuation rates.

$d_o$ is calculated based on the empirical expression:

$$d_o = 35e^{-0.015R_{0.01}} \qquad [7]$$

where $R_{0.01}$ is the rainfall rate for 0.01 % of the time.

For our example, where $d = 12.033$ km and $R_{0.01} = 30$ mm/h,

$$d_o = 35e^{-0.015\times 30} = 22.32$$

and thus

$$d_r = \frac{1}{1 + \frac{12.033}{22.32}} = 0.65.$$

For 0.01 % of the time, which is used as a reference, the following expression can be used to compute overall link attenuation due to rainfall:

$$A_{0.01} = \gamma_r d d_r \qquad [8]$$

For our example,

$$A_{0.01} = 4.5252 \times 12.034 \times 0.65 - 35.4\,\text{dB}$$

- If it is necessary to predict attenuation for different percentages of the time other than the reference, then it is possible to apply corrections to the 0.01 % figures. This is influenced by the latitude of the link location, and depending on location, the expressions to do this are:

For latitudes greater than 30 degrees from the equator:

$$\frac{A_p}{A_{0.01}} = 0.12p^{-(0.546+0.043Log_{10}p)} \qquad [8]$$

Or, for latitudes less than 30 degrees from the equator:

$$\frac{A_p}{A_{0.01}} = 0.07p^{-(0.85+0.139Log_{10}p)} \qquad [9]$$

where $p$ is the probability ordinate in the range 0.001 – 1 %.

Having calculated the attenuation due to rain, we can take it into account for our path loss equation by adding the rainfall attenuation:

$$\mathrm{PL} = 92.45 + 20\log d + 20\log f + L_g + A_p \qquad [10]$$

For our example, the new figure for path loss for 0.01 % of the time is:

$$\mathrm{PL} = 143 + 1.14 + 35.4 = 179.5\,\mathrm{dB}$$

Thus, during rainfall of 30 mm/h, the path loss exceeds 179 dB.

From our partially completed example, we can examine the effect on the fade margin during rainfall of this level by adding the additional attenuation due to rain. This is 57.86 – 35.4 = 22.46 dB. This is a considerable reduction in the fade margin compared to the clear air conditions.

Having now derived an appropriate fade margin, it is now possible to feed this into the availability calculation.

### 12.4.2.5 Availability Calculations

Calculation of availability for microwave links is far from simple and involves a variety of considerations and computations. The availability calculation shown below is the 'quick' method described in ITU-R P.530-10 – a more detailed calculation is also available within this recommendation.

The availability for the link can be computed according to the following method, which includes the concept of the average 'worst month' outlined in ITU-R P.452. The aim of this analysis is to calculate the percentage of time ($P_w$) that a fade depth ($A$) is exceeded in the average worst month. We will also calculate another factor, known as the multi-path occurrence factor ($P_o$). Both of these are used to determine availability based on all the known factors. The process is to determine both of these as described below.

First, an estimate of the Geo-climatic factor $K$ for the average worst month from fading data for the geographic area of interest is made, by applying a value for $dN_1$ (the refractivity gradient in the lowest 65 m of atmosphere not exceeded for 1 % of an average year) to the following expression:

$$K = 10^{-4.2-0.0029dN_1} \qquad [11]$$

The value for $dN_1$ can be obtained from ITU-R P.453. For the UK a typical value of $dN_1$ is $-210$.

Using this value gives a value of $K = 10^{-4.2-0.0029\times-210} = 2.56 \times 10^{-4}$

The next step is to determine the path inclination in milli-radians, using the following expression:

$$|\varepsilon_p| = \frac{|h_r - h_e|}{d} \qquad [12]$$

where

$h_e$ is the antenna height of the transmitter above sea level in metres.

$h_r$ is the antenna height in meters above sea level

$d$ is the path length in kilometres.

For our example,

$$h_e = 40\,\text{m}$$
$$h_r = 233\,\text{m}$$

and thus

$$\varepsilon_p = (233 - 40)/12.034 = 16.038\,\text{mrads}$$

Having calculated this figure, the percentage of time ($P_w$) that the fade depth ($A$) is exceeded in the average worst month is given by the following expression:

$$P_w = Kd^{3.0} \times \left(1 + |\varepsilon_p|\right)^{-1.2} \times 10^{0.033f-0.001h_L-\frac{A}{10}} \qquad [13]$$

Taking the values we have identified thus far for our worked example:

$$K = 2.56 \times 10^{-4}$$
$$d = 12.033\,\text{km}$$
$$\varepsilon_p = 16.038\,\text{mrads}$$
$$f = 28\,\text{GHz}$$
$$h_L = 40\,\text{m}$$
$$A(\text{no rain}) = 57.86\,\text{dB}$$
$$A(\text{rain}) = 22.46\,\text{dB}.$$

We can now calculate $P_w$ for both the dry and rain conditions,

*$P_w$ – No Rain*

Assuming flat fading across the bandwidth of the radio link.

$$P_{w_{Flat}} = 2.56 \times 10^{-4} \times 12.033^{3.0} \times (1 + 16.038)^{-1.2} \times 10^{0.0033\times28-0.001\times40-\frac{57.86}{10}}$$
$$P_{w_{Flat}} = 2.7 \times 10^{-8}$$

This means that the percentage of time that the fade depth is exceeded in dry conditions is $2.7 \times 10^{-8}$ (which works out a 0.016 min/year, or about 1 s). The percentage of time that the link will be available in dry conditions is the remainder of the time, or

$\text{Availability}_{\text{Dry}} = (1 - 2.7 \times 10^{-8}) \times 100 = 99.999997\,\%$ of the time.

$P_{\text{w}}$ – *Rain*

We can also perform the same calculation for the condition when rain is present by using the fade margin, taking into account rain attenuation.

$$P_{\text{W}_{\text{Rain}}} = 2.56 \times 10^{-4} \times 12.033^{3.0} \times (1 + 16.038)^{-1.2} \times 10^{0.0033 \times 28 - 0.001 \times 40 - \frac{22.46}{10}}$$

$$P_{\text{W}_{\text{Rain}}} = 9.51 \times 10^{-5}$$

Thus

$\text{Availability}_{\text{Wet}} = (1 - 9.51 \times 10^{-5}) \times 100 = 99.9905\,\%$ (about 50 min/year).

The availability in wet conditions is therefore lower than for the dry condition, which is as expected.

#### 12.4.2.6 Selective Fading

As if the computations of the last section were not sufficient, we now need to consider another mechanism that has to be accounted for in availability calculations for microwave links. This is the mechanism of multi-path fading. This is caused by energy arriving at the receiving antenna from the transmitter by different paths and causing destructive interference and signal distortion. Typically, the different paths are caused by atmospheric effects that lead to the signal components being diffracted and reflected by layers in the atmosphere having different refractive indices. This mechanism can cause flat fading across the whole bandwidth as well as selective fading in parts of the bandwidth.

The multi-path occurrence factor ($P_{\text{o}}$) is derived from the following expression:

$$P_{\text{o}} = Kd^{3.0}\left(1 + |\varepsilon_{\text{p}}|\right)^{-1.2} \times 10^{0.033f - 0.001h_L} \qquad [14]$$

This is similar to the availability calculation [13] above except the fade margin ($A$) is not used.

$P_{\text{Sel}}$ which is the outage probability due to selective fading can then be calculated according to the following formula:

$$P_{\text{Sel}} = 2.15\eta\left(W_{\text{M}} \times 10^{-B_{\text{M}}/20}\frac{\tau_{\text{m}}^2}{|\tau_{\text{r, M}}|} + W_{\text{NM}} \times 10^{-B_{NM}/20}\frac{\tau_{\text{m}}^2}{|\tau_{\text{r, NM}}|}\right) \qquad [15]$$

where

$$\tau_{\text{m}} = 0.7\left(\frac{d}{50}\right)^{1.3} \qquad [16]$$

$$\eta = 1 - \exp^{-0.2P_{\text{o}}^{0.75}} \qquad [17]$$

$d$ is path length (km)

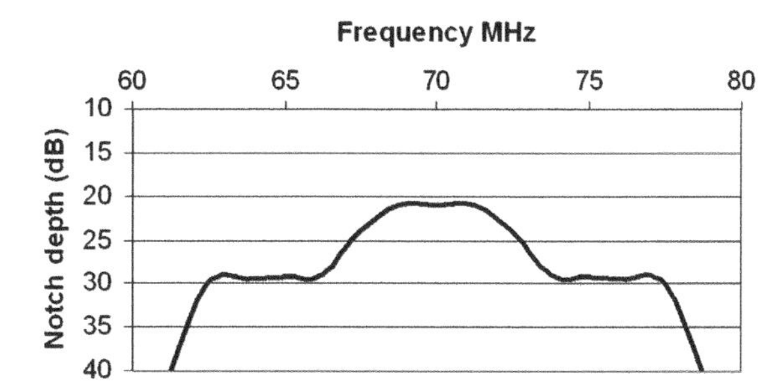


**Figure 12.7.** Minimum phase signature.

The signature information (*W*, *B* and $\tau$) is equipment specific information that needs to be obtained from the equipment manufacturer.

$W_M$ is the signature width (GHz) for the minimum phase average value across the signature bandwidth

$W_{NM}$ is the signature width (GHz) for nonminimum phase average value across the signature bandwidth

$B_M$ signature depth (dB) for the minimum phase

$B_{NM}$ signature depth (dB) for the nonminimum phase

$\tau_{r,M}$ is the reference delay (ns) used to obtain the signature for the minimum phase

$\tau_{r,NM}$ is the reference delay (ns) used to obtain the signature for the nonminimum phase

Figures 12.7 and 12.8 show typical minimum and nonminimum phase signatures for a 155 Mb/s radio system, with BER of $1 \times 10^{-3}$ and delay of 6.3 ns. These signatures are generally provided for high bit rate/bandwidth systems by the manufacturers, along with the characteristics required for the selective fading calculations.

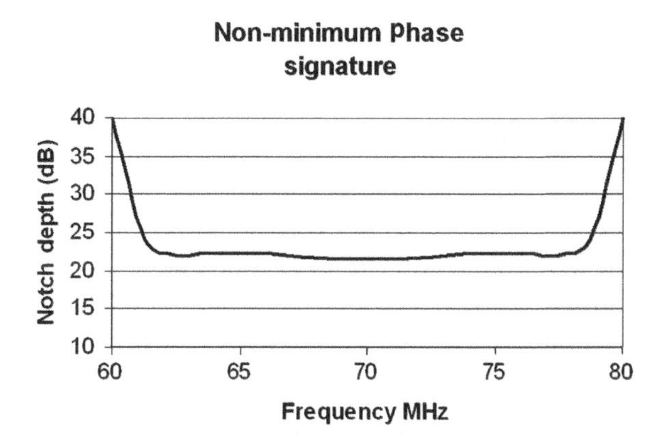


**Figure 12.8.** Non-minimum phase-signature.

From the signatures above, the following characteristics are applicable:

$$
\begin{aligned}
W_{\mathrm{M}} &= 0.026\,\mathrm{GHz} \\
W_{\mathrm{NM}} &= 0.028\,\mathrm{GHz} \\
B_{\mathrm{M}} &= 26\,\mathrm{dB} \\
B_{\mathrm{NM}} &= 22\,\mathrm{dB} \\
\tau_{\mathrm{r,M}} &= \tau_{\mathrm{r,NM}} = 6.3\,\mathrm{ns}.
\end{aligned}
$$

Therefore based on the link above, we can calculate $P_o$, $\tau_m$ and $\eta$ from the expressions given in [14], [16] and [17].

$$
\begin{aligned}
P_o &= 2.56 \times 10^{-4} \times 12.033^{3.0} \times (1 + 16.038)^{-1.2} \times 10^{0.0033\times 28-0.001\times 40} = 0.0168 \\
\tau_m &= 0.7\left(\frac{12.033}{50}\right)^{1.3} = 0.1099 \\
\eta &= 1 - \exp^{-0.2\times 0.0168^{0.75}} = 9.289 \times 10^{-3}
\end{aligned}
$$

We can then evaluate $P_{\mathrm{Sel}}$:

$$
P_{\mathrm{sel}} = 2.15 \times 9.289 \times 10^{-3}\left(0.026 \times 10^{-26/20}\frac{0.1099^2}{|6.3|} + 0.028 \times 10^{-22/20}\frac{0.1099^2}{|6.3|}\right)
$$

$$
\mathrm{P}_{\mathrm{sel}} = 1.35 \times 10^{-7}.
$$

The selective availability is $(1 - P_{\mathrm{Sel}}) \times 100$, which is greater than 99.999987 %.

Finally, after calculating the nonavailability of each component, we can compute the overall nonavailability:

$$
P_{\mathrm{Total}} = P_{\mathrm{W_{Flat}}} + P_{\mathrm{W_{Rain}}} + P_{\mathrm{Sel}} \qquad [18]
$$

Thus, the total nonavailability is

$$
P_{\mathrm{total}} = 2.7 \times 10^{-8} + 9.51 \times 10^{-5} + 1.35 \times 10^{-7} \approx 9.53 \times 10^{-5}.
$$

And the total availability is $(1 - P_{\mathrm{total}}) \times 100 = 99.9905\,\%$, which equates to about 50 min/year.

In this example, the rainfall nonavailability is the dominant mechanism, but this has not become apparent until the end of the calculation. In other cases, some of the other mechanisms will have an important contribution.

### *12.4.3 Diversity Techniques to Improve Link Performance*

Diversity techniques are used to combat all three factors that cause failure in a link.

- Propagation related nonavailability.
- Loss of function due to equipment failure.
- Failure due to other factors (such as power outage and maintenance).

We will look at space diversity, frequency diversity and some other methods used to improve link availability.

#### 12.4.3.1 Space Diversity

Space diversity relies on the use of two receive antennas that are physically separated in the vertical direction. It can improve the nonavailability due to propagation effects such as fading and changes in the atmospheric refractive index. It does not improve availability due to other factors. This is illustrated in Figure 12.9.

The space diversity improvement factor on overland paths can be estimated from this expression, which once more comes from ITU-R P.530-10:

$$I = \left[1 - \exp\left(-0.04 \times S^{0.87} f^{-0.12} d^{0.48} p_0^{-1.04}\right)\right] 10^{(A-V)/10} \qquad [19]$$

where:

$$V = |G_1 - G_2| \qquad [20]$$

and:

$A$ is the fade depth (dB) for the unprotected path
$P_0$ is the multi-path occurrence factor (%)
$S$ is vertical separation (centre-to-centre) of receiving antennas (m)
$f$ is the frequency (GHz)
$d$ is the path length (km)
$G_1$and $G_2$ are the gains of the two receiver antennas (dBi).

It should be noted that this estimation is valid for only for the following conditions:

- Path length ($d$) of between 43 and 250 km.
- Carrier frequencies ($f$) between 2 and 11 Ghz.
- Space diversity antenna spacing ($S$) between 3 and 23 m.

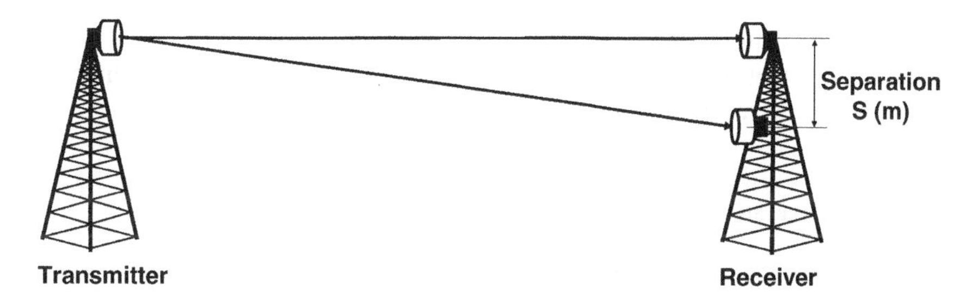


**Figure 12.9.** In space diversity, two antennas are used on the receive side of the link. The antennas are vertically separated in distance to provide an improvement in link performance.

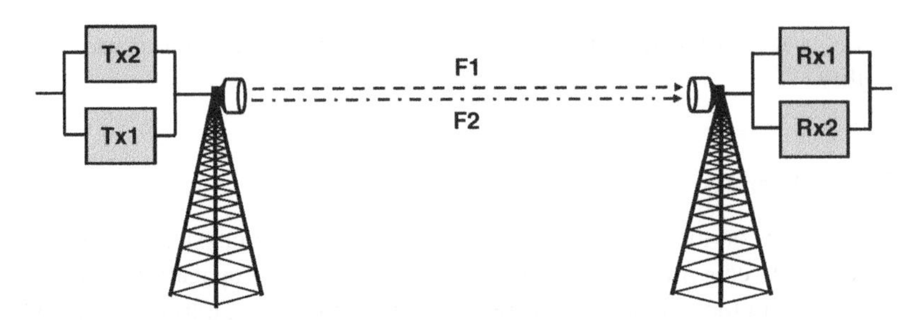


**Figure 12.10.** In a frequency diversity system, traffic is sent down two transmissions to the receiver between single antennas. Although this results in an improvement in link performance, it is spectrally wasteful and may not be allowed by national regulators in most cases.

#### 12.4.3.2 Frequency Diversity

Frequency diversity, as shown in Figure 12.10, can be used on a single link to provide redundancy. Frequency diversity can be used to combat all three of the causes of unavailability, since the failure of a receiver or transmitter does not cause an outage and during routine maintenance, the traffic can be forced onto one of the Tx/Rx paths. The drawbacks of frequency diversity are that it requires additional equipment and is thus more expensive than normal microwave links, and the requirement for additional spectrum is wasteful and thus regulators will need a very good reason before they will authorise the use of frequency diversity.

Like space diversity systems, frequency diversity offers an improvement factor to the propagation part of the link nonavailability. This is given by:

$$I_{\text{ns}} = \frac{80}{fd} \times \frac{\Delta f}{f} \times 10^{\frac{F}{10}} \qquad [21]$$

where

$f$ is the carrier frequency (GHz)

$d$ is link length in km

$\Delta f =$ frequency in separation (if the difference is greater than 0.5 GHz, then a value of 0.5 GHz is used)

$F$ is the flat fade margin (dB).

It should be note that frequency diversity will also improve the equipment reliability and hence improve the overall availability of the link.

#### 12.4.3.3 Other Diversity Techniques

A variety of other diversity techniques are available, including:

- Route diversity for microwave links. In this case, there is more than one link available to carry traffic from one location to another. Should one link be suffering from nonavailability due to rain or selective fading, then the traffic can be sent via

another link not suffering from this problem at the same time. This method also provides equipment diversity, so that failure of one set of equipment should not disrupt the link.
- Hot standby. This is where additional equipment is held in reserve against the failure of the primary system. It would normally only be done for vital parts of the network equipment infrastructure.
- Hybrid systems, which are a mix of half-space and half-frequency diversity. This is used where full space diversity of two-spaced dishes at each end of the link cannot be achieved.

The addition of diversity will increase the cost of a network due to the extra equipment and reconfiguration systems, and there are also issues about the time taken to switch between equipment in the event of failure or a planned switch; delay times of only a few milliseconds can take mobile systems off the air.

## 12.5 Calculating Microwave Reliability

We have so far looked at microwave link availability calculations, and identified the need sometimes to include some element of equipment diversity into a network. In this section, we will look at the question of overall link reliability. For this, we not only need to consider the effects of link outage due to propagation effects but also the effect of equipment failure. In our analysis, we will assume that the equipment failure rates are constant throughout their lives; in practice, they will be higher for new equipment and as the equipment approaches the end of its life.

The equipment manufacturer will normally quote a value for mean time between failure (MTBF) for each module in a radio system. Let us assume that a radio link has the modules shown in Figure 12.11. Each module has the quoted MTBF value.

Given the above simplified block diagram of a uni-directional '1 + 1' system with the individual module MTBF figures, the failure rate ($\lambda$) for each unit can be calculated using the simple formula:

$$\lambda = \frac{1}{\text{MTBF}} \qquad [22]$$

We can also introduce the concept of mean time to repair (MTTR) which, as its name suggests, is the average time taken to repair the equipment in the event of failure and which

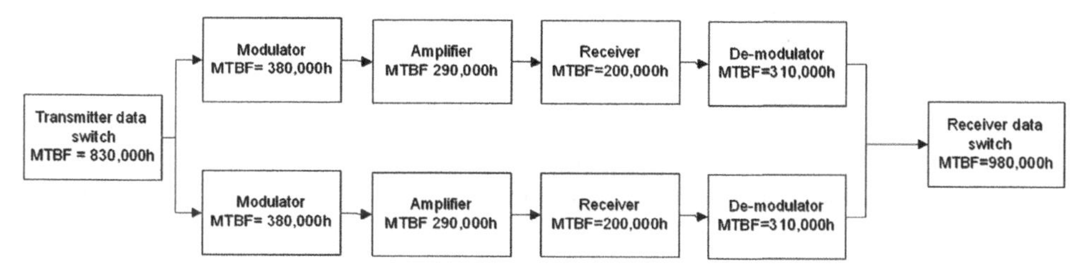


**Figure 12.11.** Modules in a microwave link and their nominal MTBF figures.

**Table 12.4.** Non-availability rates for the modules in the link.

| Module | MTBF | Failure rate | $N$ assuming MTTR = 5h |
|---|---|---|---|
| Transmitter data switch | 830000 | $1.20 \times 10^{-6}$ | $6.02 \times 10^{-6}$ |
| Modulator | 380000 | $2.63 \times 10^{-6}$ | $1.32 \times 10^{-5}$ |
| Amplifier | 290000 | $3.45 \times 10^{-6}$ | $1.72 \times 10^{-5}$ |
| Receiver | 200000 | $5.00 \times 10^{-6}$ | $2.50 \times 10^{-5}$ |
| Demodulator | 310000 | $3.23 \times 10^{-6}$ | $1.61 \times 10^{-5}$ |
| Receiver data switch | 980000 | $1.02 \times 10^{-6}$ | $5.10 \times 10^{-6}$ |

is normally quoted in hours. With this, we can express the nonavailability ($N$) of each module by the expression:

$$N = \frac{\text{MTTR}}{\text{MTBF} + \text{MTTR}} \qquad [23]$$

MTTR must include travel time as well as the time taken to repair the system; in remote areas this can be 24 h or more and can be far longer than the repair time itself.

Using the calculation for $N$, we can tabulate nonavailability for each component, as shown in Table 12.4. Note that in practise, MTTR is measured in hours and MTBF is measured in thousands of hours. Thus Equation [23] can usually be simplified to just MTTR/MTBF.

The entire link system will fail if any of the common units fail (transmitter data switch and receiver data switch), or one unit in each of the 1 + 1 paths fail. We can now look at the probability of this using standard statistical theory.

Given that the unreliability $N_s$ for a series system with $n$ components, such as either section from the modulator to the de-modulator, is given by:

$$N_s = 1 - \Pi_{i=1}^{n}(1 - N) \qquad [24]$$

And that the unreliability for a parallel system, such as the two parallel systems between the transmitter data switch and the receiver data switch, can be described by:

$$N_p = \Pi_{i=1}^{n}(N) \qquad [25]$$

For the figures quoted in our example, the unreliability for each of the serial branches is:

$$1 - ((1 - 1.32 \times 10^{-5}) \times (1 - 1.72 \times 10^{-5}) \times (1 - 2.50 \times 10^{-5}) \\ \times (1 - 1.61 \times 10^{-5})) = 7.15 \times 10^{-5}$$

This is the probability that either of the serial branches fail. If we wished, we could use Equation [25] to calculate the probability that both fail.

The unreliability for the common units (transmitter data switch and receiver data switch) is:

$$1((1 - 6.02E - 06 \times 10^{-6}) \times (1 - 5.10 \times 10^{-6})) = 1.11 \times 10^{-5}$$

As can be seen, the composite of the serial modules dominates the equipment unreliability. We can calculate the unreliability for the link, given that both branches have to fail, by calculating the parallel unreliability and putting it in series with the composite unreliability figure already calculated;

$$1 - ((1 - 1.11 \times 10^{-5}) \times (1 - 7.15 \times 10^{-5} \times 7.15 \times 10^{-5}))$$
$$\approx 1.11 \times 10^{-5} \text{ (about 6 min/year)}$$

Thus the reliability when considering equipment failure is:

$$\text{Reliability}_{\text{Equip}} = 1 - N = 99.9989\,\%$$

Since the probability of equipment failure is not correlated with nonavailability due to propagation effects, the total unreliability is simply the composite of these effects. Thus, the total link outage is $50 + 6 \approx 56\,\text{min/year}$.

## 12.6 Microwave Noise and Interference

### *12.6.1 Noise in Microwave Systems*

For radio systems, noise comes from several types of sources, some of which we will be discussing next in Chapter 13. These include;

- Equipment noise.
- Interference.
- Man-made noise.
- Atmospheric noise.
- Galactic noise.

For microwave link systems we can generally ignore the last three noise sources in this list, since they are swamped by the equipment noise. Thus in this chapter we will only consider equipment noise and interference from other systems.

### *12.6.2 Equipment Thermal Noise*

The fundamental lower limit on the noise power generated in the front end of a receiver is given by the expression:

$$P_n = kTB \qquad [26]$$

where $k$ = Boltzmann's constant ($k = 1.38 \times 10^{-23}\,\text{J/K}$)
$T$ = temperature in Kelvin (K)
$B$ = bandwidth of noise spectrum (Hz)

This can also be expressed in logarithmic terms, and in general, units of dBm are quoted. This is determined by:

$$P_{n_{dBm}} = 10\log_{10}P_n + 30 \qquad [27]$$

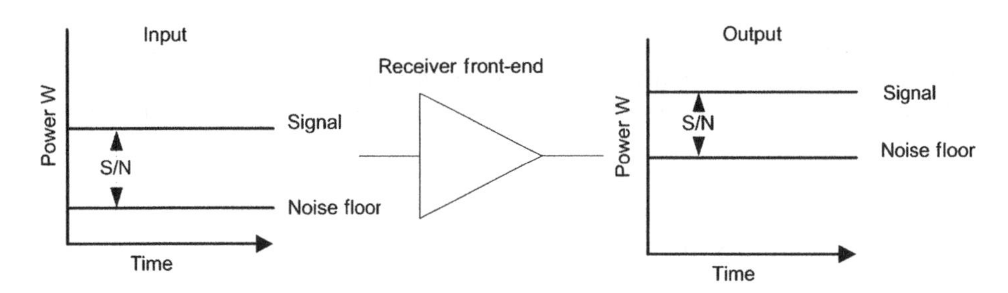


**Figure 12.12.** Noise in the front end of a receiver; amplifiers not only increase the level of the wanted signal, but also any noise present at the input to the amplifier.

Although this is the fundamental lower limit, it is never actually achieved in practice. Practical radio systems will also generate noise above this, and where amplifiers are used they will not only amplify the input signal but also any noise present at the input of the amplifier. These are characterised by the noise factor (*F*), the principle of which is shown in Figure 12.12.

The noise factor is calculated by the following expression:

$$F = \frac{\frac{S}{N}(\text{input})}{\frac{S}{N}(\text{output})} \quad [28]$$

This can also be expressed in dB terms by the normal power conversion process:

$$f = 10\,Log_{10}\,F \quad [29]$$

With the addition of the noise factor, a more accurate definition of the noise in the receiver is:

$$P_t = kTBF \quad [30]$$

Which can be expressed in dBm as:

$$P_{t_{dBm}} = 10\log_{10} P_t + 30 \quad [31]$$

The receiver system will normally be made up of a number of stages, each of which will contribute to the total noise. The procedure described in Section 13.3 can be followed for microwave systems as well as for mobile systems.

In analogue systems, the noise degrades the intelligence of the speech or information being transmitted, while in digital systems it causes errors in the data stream. The errors are measured over time and a bit-error-rate (BER) is derived. If we take the example of a standard 64 kb/s codec, a BER of 1 error in $10^6$ bits is not audible but an error rate of 1 error in $10^3$ bits causes a noise degradation on the audio signal, but the audible signal is still just understandable.

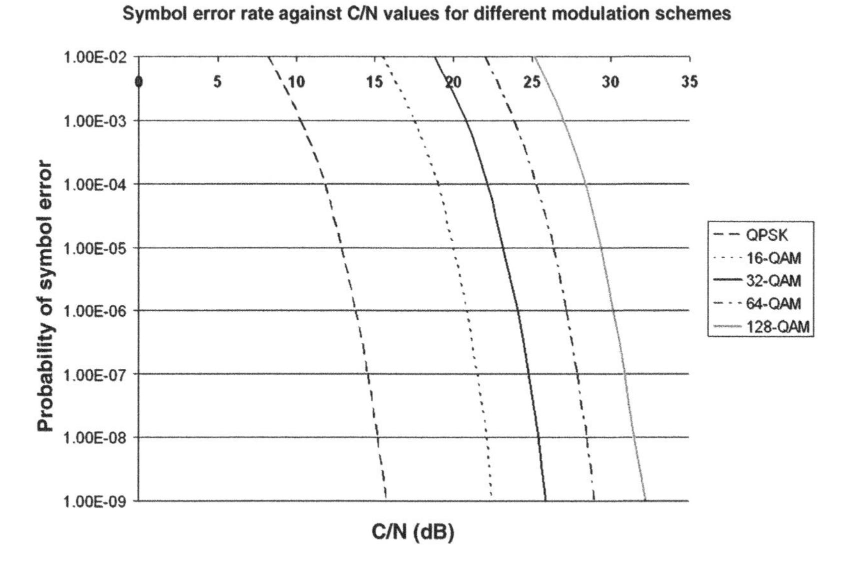


**Figure 12.13.** Theoretical symbol error rates values for some common modulation schemes.

The excess signal power over noise required at the front end of a receiver depends on the modulation and coding schemes being used. The Figure 12.13 shows the theoretical *S/N* required for various common modulation schemes.

## *12.6.3 Interference*

### 12.6.3.1 Threshold Degradation

Combining the *C/N* against probability of error above with the noise of the receiver gives the practical receive level against BER as shown in Figure 12.14.

The continuous graph shown in Figure 12.14 is a signal to BER graph for a noise-limited receiver only, since the only noise being taken account of is the noise generated from the receiver electronics themselves. The dashed line shows what can happen to the minimum necessary signal level for a given BER when interference is added; in order to achieve a target BER, the signal level must increase to compensate for the extra noise. This additional requirement is due to receiver threshold degradation caused by noise. When we express this in terms of the radio signal necessary at the input to the antenna system, we need to increase the signal to protect the amount of fade margin required to achieve system performance objectives. This is illustrated in Figure 12.15.

If we assume that the example shown in Figure 12.15 shows a receiver with a 26 MHz bandwidth and a receiver noise figure of 5 dB, then the thermal noise floor without interference is −95 dBm (calculated by Equations [30] and [31]).

If an interfering signal at −92 dBm were now to occur, the effective noise floor would be raised to:

$$\text{New noise floor} = 10 \times \log_{10}(10^{-(95/10)} + 10^{(-92/10))} = 90\,\text{dBm}$$

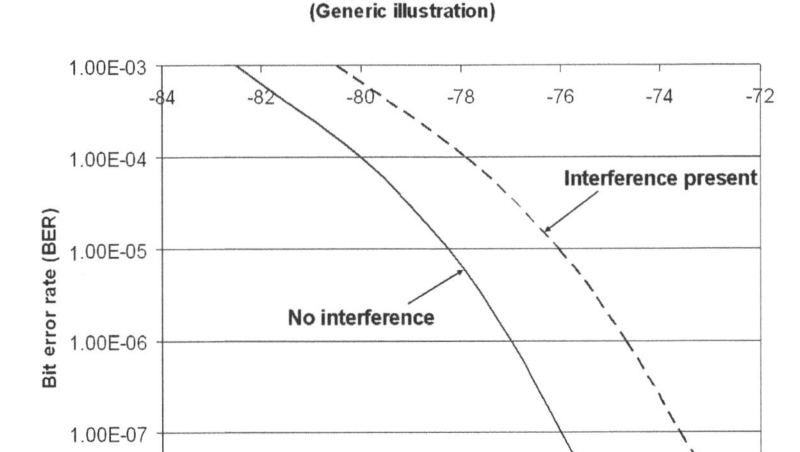


**Figure 12.14.** Received signal requirements for a given BER.

If the receiver requires a signal to interference ratio (*S/I*) of 20 dB for a given BER, this gives a required threshold level of −75 dBm. The additional interference has caused the noise floor to increase by 5 dB, and hence the threshold level must also increase by 5 dB or expressed conversely the threshold has been degraded by 5 dB. This will also feed through to affect the fade margin of the link by the same value so, for example if the fade margin without interference is 30 dB, with interference it would be 25 dB. To calculate link availability, it would be necessary to perform the calculations listed previously in this chapter to determine the availability when the interferer is present.

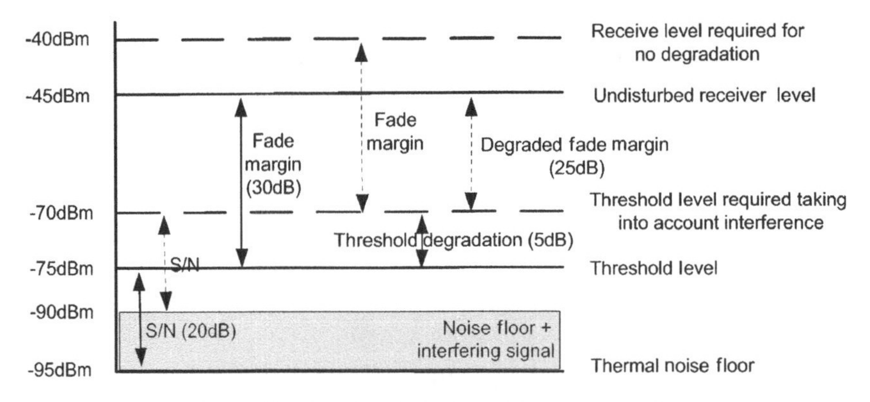


**Figure 12.15.** Threshold degradation due to noise.

#### 12.6.3.2 Interference Calculations

When a link is designed, it needs to be coordinated with all other links in the same frequency band. Figure 12.16 shows an example where this has not happened. Both links 1 and 2 share the same two frequencies. Energy from Site A at frequency F1 arrives at Site D on the same channel as the wanted signal from Site C, and energy from Site B at frequency F2 arrives at Site C along with the wanted signal. In both cases, interference exists and the link will be degraded.

It is important when considering microwave interference under specific fading conditions in order to account for the real interference effects. This is because the fading experienced in the path from the interferers to the victims will not be correlated with the fading of the wanted path. In general, two conditions are normally considered:

- The minimum median unwanted signal, in relation to the faded wanted signal. In this case the calculation assumes that the wanted link has faded to its threshold level, and that the path from the unwanted link is in average conditions.

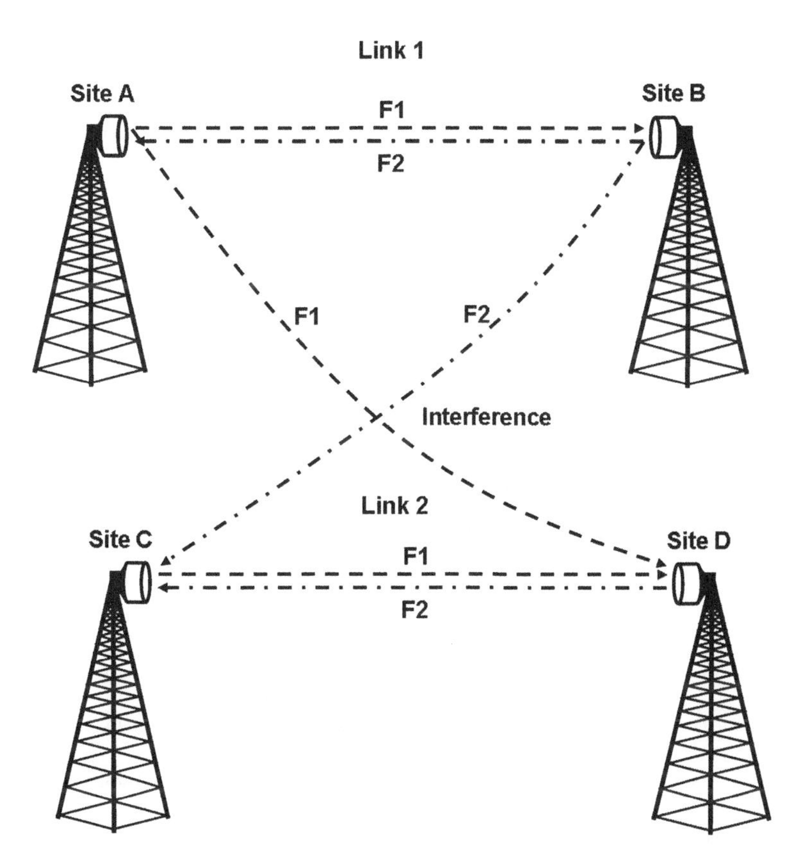


**Figure 12.16.** Microwave network, showing interference from link 1 to link 2.

- The minimum enhanced unwanted signal, in relation to the median wanted signal. In this case the wanted link is operating in average condition and the path from the unwanted system is enhanced (caused by ducting) for small percentages of time.

These calculations are usually undertaken with ITU-R recommendation P.452, which details how to calculate microwave interference between stations on the earth's surface at frequencies above 0.7 GHz and allows users to calculate the signal levels for very small percentages of time.

Based on the interference illustration, we showed in the previous section, the initial interference calculation based on the first bullet point above would be the minimum median unwanted signal would be −92 dBm, and the faded wanted threshold would be −75 dBm, which gives a wanted to unwanted (*W/U*) ratio of 17 dB.

For the second bullet point the enhanced unwanted signal calculation, we would need to calculate the nonavailability time required by the wanted link (i.e. if the wanted link requires to be 99.99 % available, then the unwanted interfering signal must be calculated for 100 % − 99.99 % = 0.01 % of time). If we assume that this is −62 dBm, then the wanted links average operating condition is −45 dBm and therefore the *W/U* is again 17 dB.

The discussion thus far has assumed that the interferers share the same channel as the wanted system. In practice, interference can occur even when there is channel separation. Most of the bands in which backhaul systems operate have regulator-approved frequency plans for different capacity and bandwidth systems, ranging from 2 Mb/s systems with 1.75 MHz of bandwidth to 140 to 155 Mb/s with 28 MHz bandwidth. Where systems do not have the same carrier frequency and bandwidths, corrections need to be applied. For instance a 2 Mb/s 1.75 MHz system can operate in the same band as a 155 Mb/s 28 MHz system; we need to be able to account for the effects of this in the interference calculation. To enable these two systems to operate in the same band, minimum wanted to unwanted levels are published by ETSI (for example) and usually by the national regulator. An example of this is shown in Table 12.5.

The use of Table 12.5 can be illustrated by continuing our example. If we assume that both the wanted and unwanted links are operating at 2 Mbps in 1.75 MHz and are co-channel, then the wanted/unwanted ratio can be read off as 27 dB from the first line of the table for

**Table 12.5.** Typical wanted to unwanted matrix.

| Wanted system (Mbits/s in MHz) | Wanted / Unwanted ratio (dB) versus channel spacing | | | | | | | | | | | | | | |
|---|---|---|---|---|---|---|---|---|---|---|---|---|---|---|---|
| | 0 | 1 | 2 | 3 | 4 | 5 | 6 | 7 | 8 | 9 | 10 | 11 | 12 | 13 | 14 |
| 2 Mbps in 1.75 Mhz | 27 | 6 | −7 | −40 | −40 | −40 | −40 | −40 | −40 | −40 | −40 | −40 | −40 | −40 | −40 |
| 2 × 2 Mbps/ s in 3.5 Mhz | - | 25 | 20 | −1 | −11 | −40 | −40 | −40 | −40 | −40 | −40 | −40 | −40 | −40 | −40 |
| 8 Mbps/s in 7 Mhz | - | 23 | 22 | 14 | 3 | −5 | −10 | −16 | −40 | −40 | −40 | −40 | −40 | −40 | −40 |
| 2 × 8 Mbps/ s in 14 Mhz | - | 20 | 20 | 20 | 18 | 11 | 5 | −2 | −4 | −6 | −9 | −12 | −15 | −18 | −40 |

zero channel spacing. This would mean that to prevent interference, the unwanted signal must be 27 dB or less than the wanted signal in the two propagation conditions described above. For example taking one of these conditions the wanted receiver threshold value of −75 dBm, this would mean that if the interferer was co-channel, the maximum interference level allowed would be −75 − 27 = −102 dBm minimum median unwanted signal level. However, we have stated above that the interferer signal arrives with a median value of −92 dBm, which is above this value by 10 dB hence the wanted system would be interfered.

One solution would be to move the wanted or unwanted systems frequency. Moving them one channel apart (1.75 MHz) means that the *W*/*U* required is now 6 dB, hence the maximum interference level allowed would be −75 − 6 = −81 dBm. In this case the two systems would not interfere. However, if the unwanted system was a 2 × 8 Mbps system in 14 MHz bandwidth, the unwanted system would have to be at least five channels away so as to not cause interference; −75 − 11 = −86 dBm. To calculate how far this is in MHz the narrowest bandwidth (in this case 1.75MHz) is used, hence the unwanted systems carrier frequency must be at least 5 × 1.75 Mhz = 8.75 MHz away from the wanted systems carrier frequency.

The above calculations can be applied from single interferers though in the case above allowance has been applied for multiple interferers in the *W*/*U* table.

## 12.7 Summary

The planning and design of microwave links for backhaul of mobile networks is very different from that of the mobile case. This is due in part to the propagation conditions prevalent at microwave frequencies. We have also shown some equipment-related aspects for unreliability calculations that do also apply to the mobile case, but are covered in more depth elsewhere.

Although the calculations shown in this chapter appear to be reasonably complex, in practice, the microwave engineer will use planning tools and spreadsheets to automate much of the calculation process. This is not only easier for the engineer but also helps to reduce errors.

It is essential that the planning engineer has a good understanding of the principles and calculation methods used in order to recognise when the results given by an inaccurately programmed planning tool or spreadsheet are erroneous.

## References and Further Reading

[1] ITU-R P.530-10: Propagation data and prediction methods required for the design of terrestrial line-of-sight systems.
[2] ITU-R P.838-2: Specific attenuation model for rain for use in prediction methods.
[3] ITU-R P.676-5: Attenuation by atmospheric gases.
[4] ITU-R P.525: Calculation of free space attenuation.
[5] ITU-R P.452: Prediction procedures for the evaluation of microwave interference between stations on the surface of the Earth at frequencies above about 0.7 GHz.
[6] ITU-R F.1703: Availability objectives for a real digital fixed wireless links used in 27 500 km hypothetical reference paths and connections.

# 13

# Network Interference

## 13.1 Introduction

Radio links cannot be created over infinite distances and will be limited in range by background noise. In addition, noise from other radio systems or intentional jamming may further restrict range. In this chapter, we will be looking at the behaviour of radio receivers in the presence of noise and interference. We will also look at the design tools that can assist the militation of interference in the radio network design. As ever, the approach will involve examining the fundamental factors and then seeing how they are applied in modern radio planning tools by radio designers and engineers. We start by introducing the fundamental concept of the noise floor.

## 13.2 Thermal Noise Floor, Receiver Noise Floor and Receiver Sensitivity

In the absence of intentional or unintentional energy, the fundamental limit on radio communications is caused by the thermal noise floor. This is caused by random fluctuations of electrons in the system, and is described by the expression:

$$N = kTB \qquad [1]$$

where

$N$ is the noise floor
$k$ is the Boltzmann constant $= 1.38 \times 10^{-23}(J/K)$
$T$ is the noise temperature of the receiver in Kelvin
$B$ is the bandwidth in Hz

Often, the concept of reference noise power is used, with $T = 290$ K. For a 1 Hz bandwidth, this is equal to $4 \times 10^{-21}$W, which is $-204$ dBW (in a 1 Hz bandwidth). It should be noted that technically, the bandwidth is not the same as the 3 dB limit normally quoted on equipment data sheets, but rather is the area under the power transfer curve (see reference [1]). However in practice, the two values are normally typically fairly similar to it is normally reasonable to use the normal bandwidth figure.

From Equation 1, it is clear that the thermal noise floor is predicated on the temperature of the receiving system and also on the bandwidth of the system. This means that highly sensitive systems should be cooled, and the system should be limited in bandwidth if possible.

In practice, physically realisable systems suffer from additional noise above this fundamental limit. To account for this, the concept of noise figure ($F_r$) is used. This is described as an additional noise component that adds to the thermal noise floor to produce the same response actually measured from a receiver. The noise figure is expressed in decibels and it varies according to the design of the radio. To obtain the total noise in the system, the following expression is then useful. For a more detailed examination of this equation and additional factors that may need to be considered, see reference [2].

$$P_n = F_r - 204 + B \qquad [2]$$

where

$P_n$ is the receiver noise floor in dBW

$B$ is 10 log (bandwidth in Hz)

Any practical signal must exhibit desired characteristics with respect to the receiver noise floor. In non-CDMA systems, the signal level must be at a higher level than the noise floor (known as the Signal to Noise Ratio or SNR) in order for the receiver to achieve a given level of performance such as SINAD or BER. is illustrated in Figure 13.1. The noise floor varies due to small fluctuations around a constant value.

The thermal noise floor of a receiver is shown at the bottom, with the receiver noise floor shown $F_r$ dB above it, and the minimum level of signal required to achieve a given degree of performance is shown SNR dB above that. In practice the noise floor value used may need to take account of the probabilistic distribution it obeys, and a correction based on a wanted ordinate value may be needed in order to perform this, but for the moment we will assume white Gaussian noise and ignore greater complexities. This is because in practice, the mobile radio network engineer will use appropriate SNR figures provided by equipment manufacturers or technology specifications that will already have accounted for any necessary complexities.

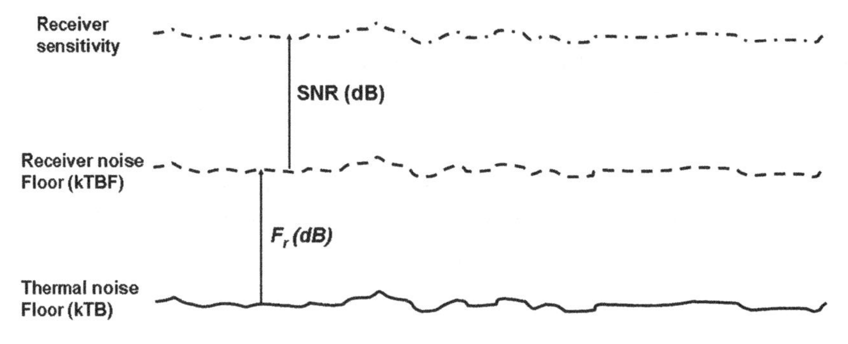


**Figure 13.1.** Noise floor and receiver sensitivity. Noise floor will constantly vary around the mean due to small fluctuations. These variations occur over time.

To illustrate this basic approach with a practical example, assume that a receiver has a bandwidth of 25 kHz, a noise figure of 4 dB and an SNR value required of 10 dB, thus

$$B = 10 \text{ Log } (25{,}000) = 44 \text{ dB (approximately)}$$
$$N = -204 + B = -160 \text{ dBW}$$
$$P_n = N + F_r = -160 + 4 = -156 \text{ dBW}$$

Receiver sensitivity $= P_n + 10 = -146$ dBW $= -116$ dBm
The radio receiver sensitivity is the minimum signal level at which the desired performance will be achieved. This value can then be used in the link budget along with all the aspects of radio prediction and fading required, as normal.

## 13.3 Noise in the VHF and UHF Bands

The discussion in Section 13.2 does not consider that in addition to the thermal noise, there may be additional noise provided by natural and manmade phenomena that is present at the antenna. In this context, we do not include energy radiated by other radio systems (which is considered later). This additional noise may arise from such sources as electrical and electronic equipment, power generation and transmission lines, vehicles, atmospheric noise due to lightning discharge, energy from celestial sources and noise from obstructions in the antenna beam. In general, atmospheric noise can be disregarded for most mobile radio networks in the VHF and UHF range.

To account for this, and the other noise sources that may be present between the antenna and the receiver itself, the diagram shown in Figure 13.2 is recommended to represent the situation.

This is a slightly simplified diagram from ITU-R P.372 (reference [2]). With this simplified diagram, the system noise can be assessed from:

$$f = f_a + (f_c - 1) + l_c(f_l - 1) + l_c l_l(f_r - 1) \qquad [3]$$

where

$f_a$ is the antenna noise factor, which is described next
$f_c$ is the noise factor associated with the antenna circuit
$l_c$ is the loss figure associated with the antenna circuit

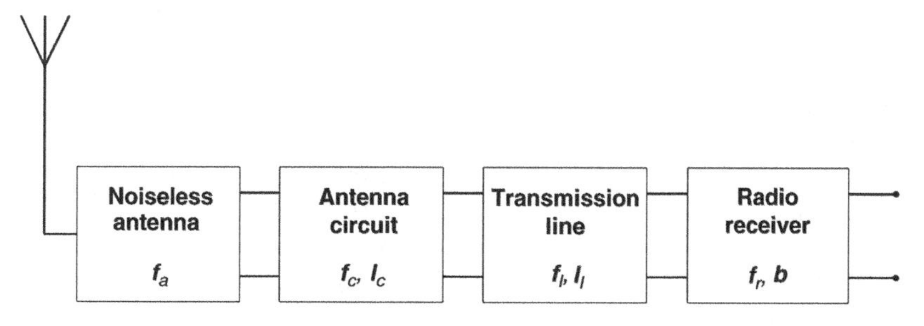


**Figure 13.2.** The elements of the front end of a receiver.

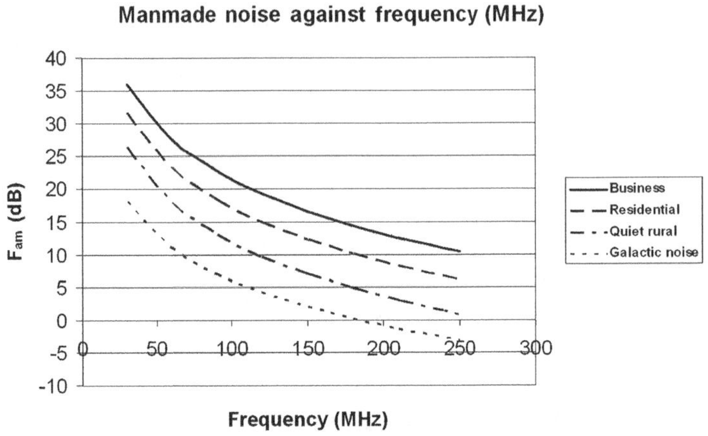

**Figure 13.3.** Manmade noise against frequency.

$f_l$ is the noise factor associated with the transmission line

$l_l$ is the loss figure associated with the transmission line

$f_r$ is the noise factor of the receiver, which is the linear version of $F_r$

Apart from the antenna noise factor, all of the other factors are dependent on the radio design itself. The antenna noise factor is dependent on the environment that the receiver is present in. This will depend on the dominant mechanism, which is often caused by manmade noise for mobile radio systems in the VHF and UHF bands.

A graph of the median value of radio noise estimated from measurements taken in the 1970s is shown in Figure 13.3. This shows the median value, around which there will be variations, and also these figures might not be representative for modern environments since noise sources will change over the years.

For frequencies between 200 and 900 MHz, ITU-R P.372 recommends the following expression:

$$F_{\text{am}} = 44.3 - 12.3 \log f \qquad [4]$$

Where $f$ is frequency in MHz

This expression is based on limited data and again more recent measurements would be far more useful. An example of a more recent study carried out in the UK can be found on the UK Ofcom website, as illustrated in Figure 13.4, which has been produced from the results included in reference [3].

Whatever the source of the noise figures used, their use can be used as illustrated in the following example.

Assume that the design of the radio has the following characteristics:

Man-made noise at the receiver = 16 dB, so $f_a = \exp(\text{noise}/10) = 39.81$

Omni-directional antenna, with loss factor = 0 dB, so $f = \exp(0/10) = 1$

0 dB antenna circuit loss factor = 0 dB, so $l_c = \exp(0/10) = 1$

0 dB noise or loss in the transmission cable, so $f_l = l_l = 1$

$F_r = 10$ dB, so $f_r = \exp(10/10) = 10$

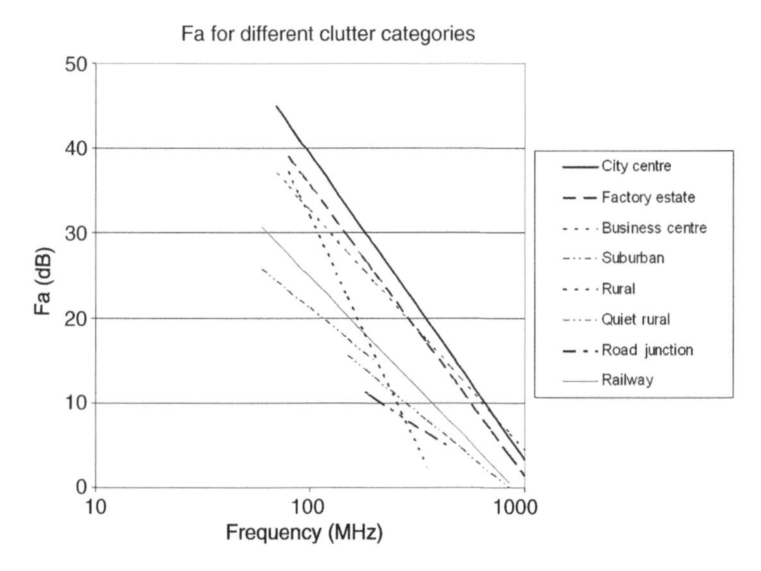


**Figure 13.4.** Manmade noise against frequency – over a wider frequency range.

Thus, from equation [3],

$$f = f_a + (f_c - 1) + l_c(f_l - 1) + l_c l_l(f_r - 1)$$
$$f = 39.81 + (1 - 1) + 1 \times (1 - 1) + 1 \times 1 \times (10 - 1) = 48.81$$

System noise F = 10 log(48.81) = 16.9 dB

From this the noise power for a system with a bandwidth of 25 kHz can be calculated from:

$P_n = F - 204 + 10\log(25000) = 16.9 - 204 + 44 = -143.1\,\text{dBW} = -113.1\,\text{dBm}$

If no sources of interfering radio energy is present and the system is limited by the noise power as shown, then the system is known as noise-limited. In planning tools, such systems are modelled using the coverage predictions we discussed in Chapter 9. Now, we will look at the situation where the noise produced by other radio systems is dominant, and the system is said to be interference-limited.

## 13.4 Interfering Radio Noise at the Receiving Antenna

We have identified that a system will require a given SNR value above the noise floor present at the receiving system, and that if environmental noise is also present, then the SNR must be above the total noise floor. If other systems are present, then the noise floor may be raised to include interference, as shown in Figure 13.5. In this hypothetical situation, a mobile subscriber suddenly moves into a region where the noise floor is increased by manmade or some other additional noise, and then another radio system is activated, and the noise floor increases to include that interference.

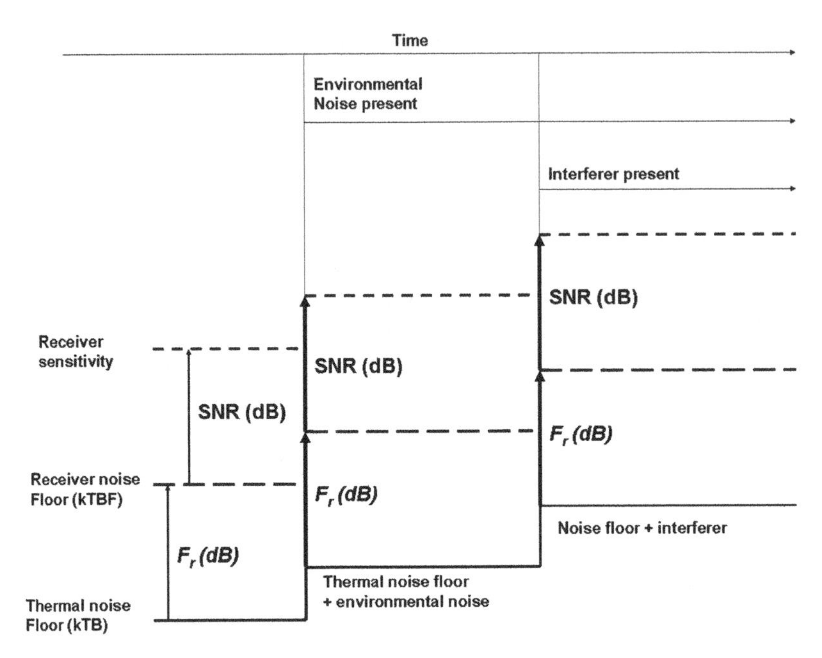


**Figure 13.5.** Noise plus interference; examples with environmental noise and with an interferer, Showing how the receiver sensitivity is decreased in the presence of noise and interference.

This means that the receiver sensitivity is effectively de-sensitised by this increase in the presence of these additional factors. This is what happens at the edges of coverage for radio links, where the received energy from the transmitter is at its weakest to still allow a workable link at the receiver, and the effect is that the maximum workable coverage area of a base station is decreased. We can examine the practical effect of a generalised increase in noise floor over the whole of the coverage by looking at the effect on a coverage plot if the generalised noise floor is increased by 10 dB. This is shown in Figure 13.6, where the darker coverage area is the effective coverage with the increased noise. The lighter grey areas show where coverage would no longer be present.

In many books, the reduction is shown as a reduced-radius circle, but as can be seen from the figure, the effect also increases coverage holes throughout the service area.

For the other condition in Figure 13.5, where an interferer is also present, the interfering signal will not be constant but will vary in accordance with the radio propagation conditions. If we assume that the interferer is at some distance from the victim radio link, then an idealised picture of the relative median signal levels can be drawn as shown in Figure 13.7.

The scale of the distance on the $x$-axis is from a base station in the direction of an interferer. The signal level of the wanted link falls off with distance, in this case shown as an idealised inverse-exponent. At some point, the signal will reach the noise-limited region (shown by the right hand vertical dashed line at about 25 km from the base station). The signal level of the interferer is also shown as the inverse exponent curve from the right hand side of the figure. If we assume that the wanted signal can tolerate an interferer that is a certain number of decibels down on the wanted signal (often referred to as a Carrier-to-Interference (C/I) ratio), then the

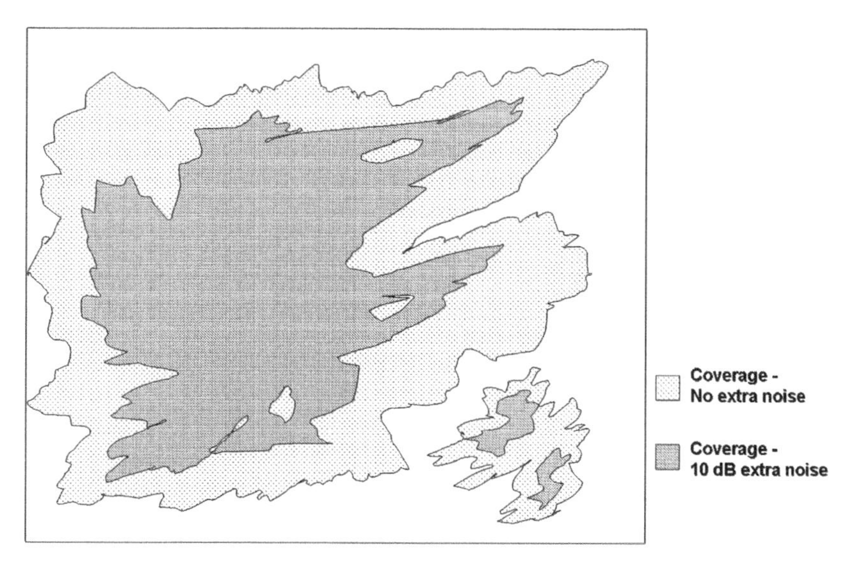


**Figure 13.6.** Effective of increased noise on a receiver; decreased service area.

maximum tolerated interference level can be plotted as shown by the dashed curve tracking the wanted signal level. Interference will occur when the interference level is higher than this maximum tolerated interference curve. The transition occurs at the vertical dashed line shown as the interference limited range. To the left of this point on the graph, the link will be effective, and to the right, it will suffer from interference. The region between the two vertical lines is the reduction in range caused by the interference.

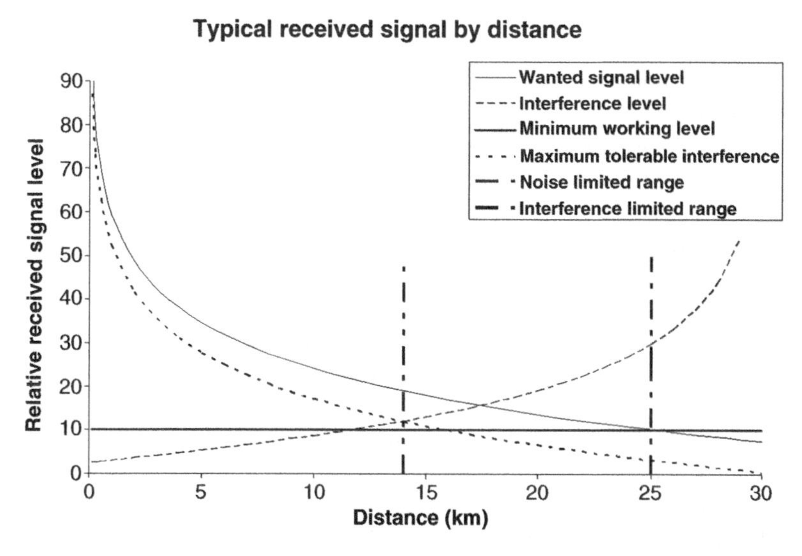


**Figure 13.7.** Typical received signal by distance, using a simple inverse-exponent model.

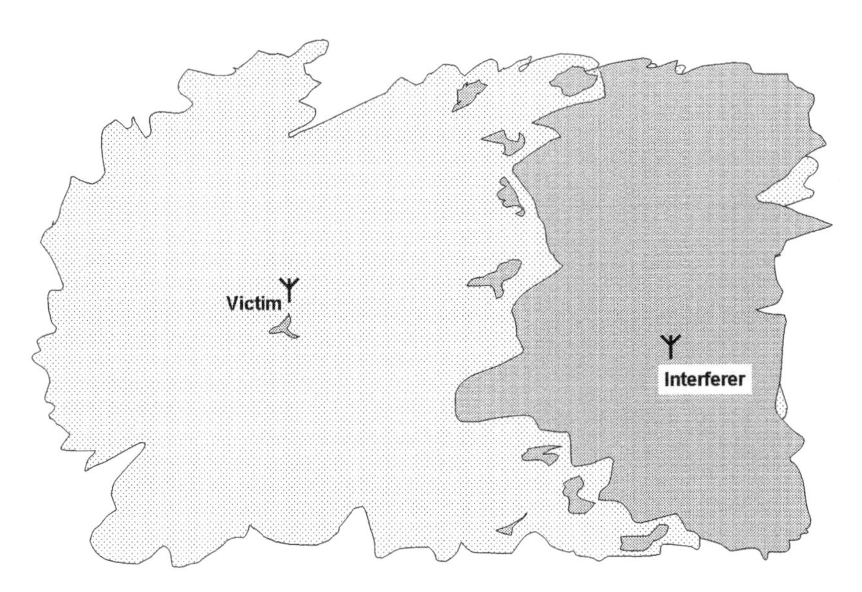


**Figure 13.8.** Typical interference plot; the interfering signal reduces the service area of the wanted server (the victim), most frequently closest to the interferer but, depending on terrain, it is possible for an interferer to cause problems even close to the victim (as illustrated in this diagram).

Of course this is an idealised version of what will typically happen in practice. Figure 13.8 shows an example of a radio interference plot. In this figure, the light grey area shows the coverage area of the victim base station that is not interfered. The dark grey area shows where usable coverage would exist if the interferer was not present, but is in fact interfered with.

Notice that in this example, which is heavily terrain limited, interference even exists close to the victim base station, where it may be expected that the victim signal strength would be high. However, because of adverse terrain effects, in practice the interferer, which may have a better path to these locations, does interfere.

This example only accounts for single interferers. Later we will look at interference from multiple interferers, but we also have to consider the situation the possible interactions that radio systems may have with each other in the frequency domain, which we shall look at next.

## 13.5 Interference Engineering

### *13.5.1 Introduction*

One of the primary tasks of the network-planning engineer will be to manage interference issues in the design. This does not necessarily mean that at the end of the design process, there will be no interference anywhere in the network, but rather that the prevalence of interference is minimised and, if unavoidable, is placed where it will do the least harm (for those technologies that allow this to be done). There are a number of ways to do this, of which the most effective is frequency assignment, which is covered in the next chapter. Others will be discussed towards the end of this chapter, but first we will examine the issues of intra-net interference – in other words interference caused by other transmitters in the same network. We will look at the modelling of co-channel interference, where both the wanted system and the interferer are tuned to the same channel, and then extend this to look at interference from interferers tuned to other channels, and also at the composite effects of multiple interferers. For most mobile networks, we will be examining uplink and downlink interference separately.

## *13.5.2 Co-Channel Interference*

### 13.5.2.1 Description of Interference Tolerance for Co-channel Interference

The simplest case is when both the wanted and the interferer are tuned to the same frequency and occupy the same frequency band with identical spectral characteristics. In this case, a simple C/I figure is sufficient to describe the wanted system's tolerance to interference caused by another member of the same network.

### 13.5.2.2 One-to-One Interference

This is the consideration of one interferer against one victim for the downlink, although the actual victims are the mobile subscribers rather than the base station itself. Interference to the mobile subscribers is likely to come from another base station on the same downlink frequency. Interference is less likely to come from mobile subscribers of the interferer base station for two reasons; firstly, the uplink is likely to be on a different channel; and secondly, the mobile units are likely to be close to ground level and their signals are unlikely to be a problem to other mobile subscribers in different areas. A simple display of one-to-one interference is illustrated in Figure 13.9.

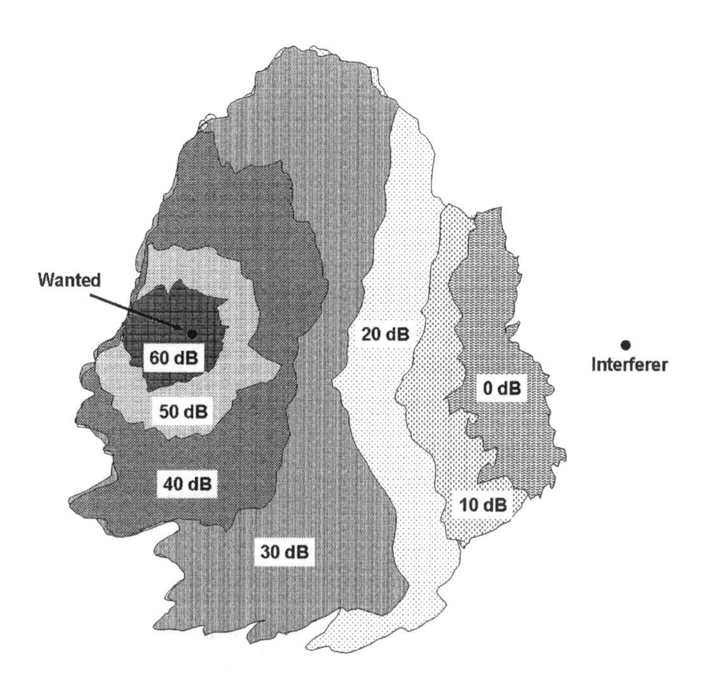


**Figure 13.9.** C/I plot; the figures shown are the number of dB that the wanted signal is stronger than the interfering signal. This is only shown in the service area of the wanted base station.

This figure shows an example of one-to-one interference. This is a composite illustration, where the wanted C/I figure is varied between 0 and 60 dB in 10 dB steps. The higher the wanted C/I figure is (which means that the signal must be a larger number of dB above the interfering signal), the smaller the effective service area. As is typical, the interference is most pronounced in the direction of the interferer, but as the wanted C/I figure becomes larger, eventually the service area becomes surrounded by interference in all directions with only a small area near the base station unaffected. This effect is seen for unintentional interferers and also for jamming systems where the wanted system range reduction is the intended effect; the aim is to reduce the service area to such a small area that it is effectively useless for the enemy.

An alternative interpretation of Figure 13.9 is that it shows the protection of the wanted signal above the interferer.

The purpose of the one-to-one interference tool is to allow detailed analysis of individual interference issues. In most cases, networks will consist of far more than base stations and thus interference issues may be more complex, but the ability to resolve the issues between individual sites remains important.

### *13.5.3 Adjacent and Other Channel Offset Interference*

All of the above calculations have assumed that the interferer has the same carrier frequency and bandwidth as the wanted system. This is not always the case; sometimes where spectrum is scarce, a frequency plan with less than full channel spacing will be used, and also (and increasingly) there is the possibility that different services will share the same spectrum. In this case, we need to consider the degree of spectral overlap in order to determine potential interference effects. This is illustrated in Figure 13.10.

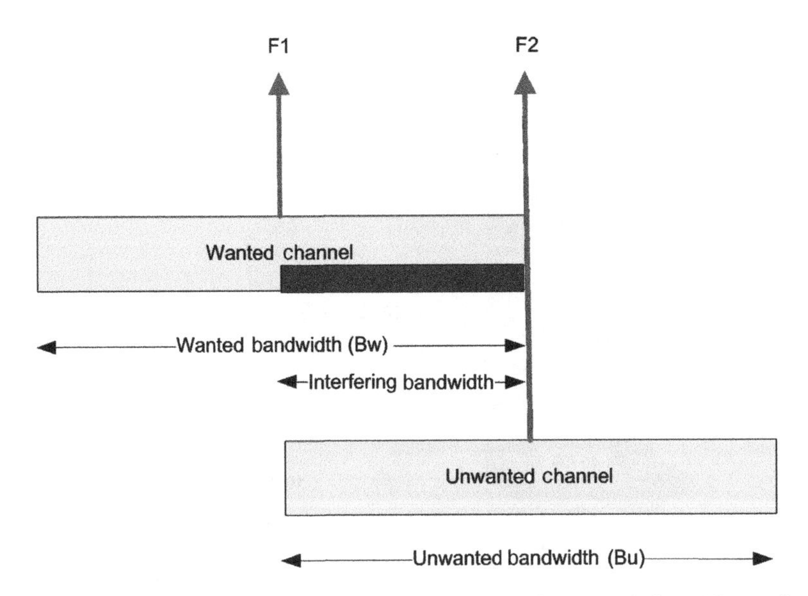


**Figure 13.10.** A spectrum display showing two overlapping radio transmissions. The vertical arrows indicate the centre frequency of each system. The wanted and unwanted bandwidths are shown, with the interfering bandwidth highlighted.

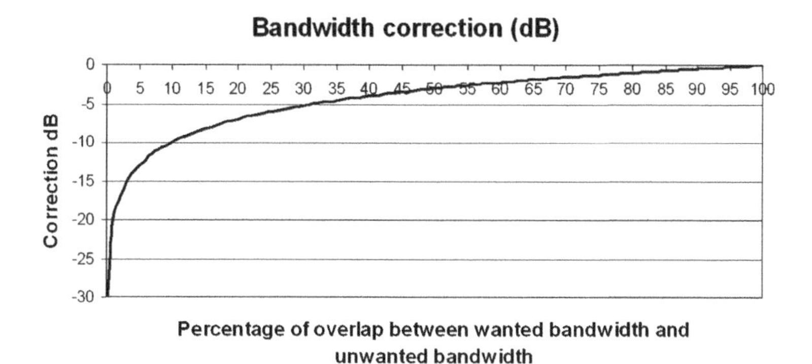


**Figure 13.11.** Bandwidth correction between overlapping signals, based on percentage overlap.

Figure 13.10 shows two carriers with similar bandwidth, which are tuned $\frac{1}{2}$ of their bandwidth apart. In this case half the unwanted channel's bandwidth is overlapped with the wanted channel and hence half the power (3dB down on full power) may cause interference. This is a slight simplification of the real situation, since neither signal will have the bandwidth response shown but in practice it is a good approximation. The following formula can be used to calculate the unwanted interference contribution in to the wanted signal in this situation.

$$P_{ei} =; P_I - 10\,\mathrm{LOG}_{10}\,BW_{ol} \qquad [5]$$

where

$P_{ei}$ is the effective level of the interfering signal

$P_I$ is the power of the interferer

$$BW_{ol} = ((B_w + B_u)/2 - |F_2 - F_1|)/B_w \qquad [6]$$

where $B_u$ overlaps $B_w, BW_{ol} \leq B_u$ and $BW_{ol} \leq B_w$

If $B_u$ is contained within $B_w$ no correction is needed. Figure 13.11 shows a graph of correction against the percentage of overlap.

Interference will not just occur when the bandwidths of two signals are overlapping in the way described above. This is because the transmitted energy will cover a wider band than the nominal bandwidth, and also a receiver will pick up energy over a wider frequency range that the nominal bandwidth. This is the difference between reality and the normal representation shown in Figure 13.12.

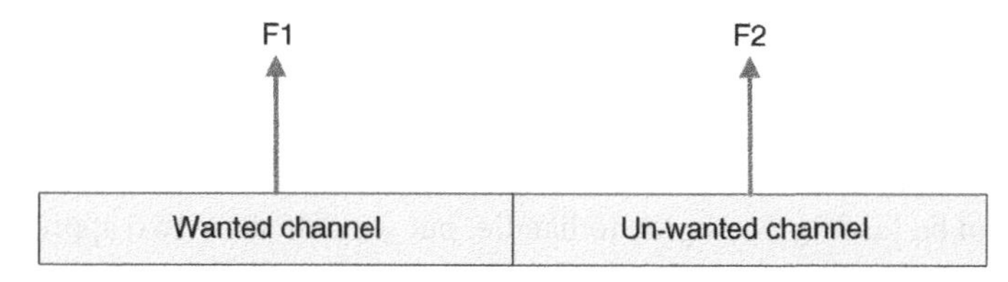


**Figure 13.12.** Normal representation of signals tuned to adjacent channels. In practice, energy will be transmitted outside of the unwanted channel, and the receiver will pick up energy transmitted outside the channel it is tuned to.

Protection ratio

Multiple C/I (dB) mask - Priority 3

Compare Tx/Rx bandwidths

| N | Value | Used | N | Value | Used |
|---|---|---|---|---|---|
| N=0 | 0 | ☑ used | N=8 | -30 | ☐ used |
| N=1 | -6 | ☑ used | N=9 | -30 | ☐ used |
| N=2 | -25 | ☑ used | N=10 | -30 | ☐ used |
| N=3 | -40 | ☑ used | N=11 | -30 | ☐ used |
| N=4 | -40 | ☑ used | N=12 | -30 | ☐ used |
| N=5 | -40 | ☑ used | N=13 | -30 | ☐ used |
| N=6 | -30 | ☐ used | N=14 | -30 | ☐ used |
| N=7 | -30 | ☐ used | N=15 | -30 | ☐ used |

**Figure 13.13.** Example of a protection ratio table. Reproduced by permission of P & D Missud.

Usually the adjacent channel response is published with the equipment parameters. For example a typical adjacent response for a PMR 12.5 kHz system is −6 dB. This means that to cause interference, energy in the adjacent channel must be 6 dB higher at the receiver than the wanted signal level. This can be repeated for adjacent channels more than one channel away to produce a protection ratio table as shown in Figure 13.13.

In the case, above a carrier two channels away could be 25dB greater than the wanted receive signal and carrier 3 channels away, 40 dB greater. This system works when the bandwidths of the wanted and unwanted carriers are the same. When the carriers are not the same, a similar principal can be applied but the situation is more complex. In this case, a matrix of Net Filter Discrimination (NFD) values needs to be generated. This is based on measured or calculated interaction figures based on a transmitter's power spectral density in W/Hz and a receiver's rejection filter characteristics. An example is shown in Figure 13.14.

The net filter discrimination response is used in place of the C/I table. It allows the interactions of dissimilar systems to be modelled and thus is a far more flexible approach. Its drawback is that it requires significantly more information than the C/I approach and it can be difficult to find that information.

### *13.5.4 Multiple Interferers*

Thus far we have considered the effects of a single interferer. In practice there can be several interferers and in this case we need to consider the composite interference effects. This can be far more complex to handle, but we will show two approaches. These are:

- Power sum method.
- Simplified multiplication method.

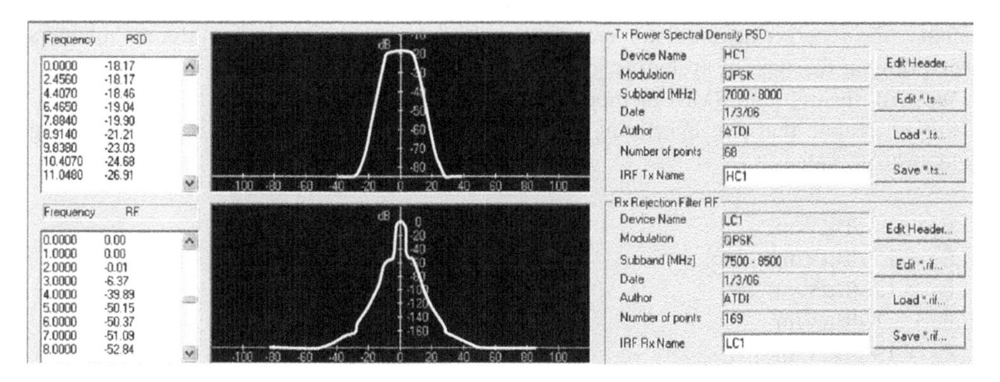


**Figure 13.14.** Net filter discrimination can be calculated by convolving the transmitter power spectral density and the receiver interference rejection characteristics. This approach allows the interference rejection characteristics to be determined for any combination of radio networks, and for any frequency offset. Reproduced by permission of P & D Missud.

#### 13.5.4.1 Power Sum Method

The power sum method is the simplest and is a nonstatistical method for summing the interference signal levels. It is only valid for the median case (50% locations) and it takes no account of the phase relationship between the received signals. It also assumes full correlation between wanted and unwanted signals. Additionally, it does not take account of noise degradations, rather just the contribution of individual noise sources.

The power sum method is:

$$P_i = 10\log_{10}\left(\sum 10^{(I/10)}\right) \qquad [7]$$

where

$P_i$ is the total interfering power (dBm or dBμV/m)

$I$ is the interfering power level (dBm or dBμV/m)

Obviously, the units used must be consistent for both power terms. The power sum method has the advantage of being simple to compute, but it does not characterise the real situation as well as the simplified multiplication method.

#### 13.5.4.2 Simplified Multiplication Method

The Simplified Multiplication Method (SMM) is a statistical computation procedure which gives the interference level assuming log normal distributed known mean values and standard deviations for the arriving signals. It also assumes there is no correlation between arriving signals (this tends to be true if the signals are coming from different locations. If some of the interferers are coming from the same location they should be summed using the power sum method described above first, and then the resultant interference level used in the SMM method), and that there is a dominant interferer. As with power sum method, noise degradations are not taken into account. The SMM method is only useful for computing the effect of unwanted interfering signals because although wanted signals may add in a similar manner, the receiver/demodulator reacts to multiple wanted signals in different manners depending on the type of system (e.g. OFDM).

We will illustrate the SMM method by working through an example.

Let us assume that five interfering signals arrive at a given location with the following field strength values:

- $E_1 = 10$ dBμV/m.
- $E_2 = 15$ dBμV/m.
- $E_3 = 12$ dBμV/m.
- $E_4 = 21$ dBμV/m.
- $E_5 = 19$ dBμV/m.

We will further assume that the standard deviation to account for fading for the signals ($S_d$) is 8.3 dB (a typical value for land mobile fading).

In order to compute the SMM, the first step is to generate a seed value ($E_s$) to start the calculation. This is the largest signal level for all the interferers plus 6 dB, and in the case of the interferers listed, the value is $E_s = 21 + 6 = 27 \text{dB}\mu\text{V/m}$. Once this value has been calculated, the first approximation can be made as shown in Table 13.1.

**Table 13.1.** Approximations to the SMM interference method.

**Approximation 1**

| *Interferer* | $E_i$ | $Z_i = E_s - E_i$ | $X_i = Z_i \frac{Z_i}{S_d\sqrt{2}}$ | *Probability Integral* $P(x_i)$ *– see below* | $L(X_i) = Px_i/2 + 0.5$ |
|---|---|---|---|---|---|
| $E_1$ | 10 | 17.0000 | 1.4483 | 0.8525 | 0.9263 |
| $E_2$ | 15 | 12.0000 | 1.0223 | 0.6929 | 0.8465 |
| $E_3$ | 12 | 15.0000 | 1.2779 | 0.7985 | 0.8992 |
| $E_4$ | 21 | 6.0000 | 0.5112 | 0.3912 | 0.6956 |
| $E_5$ | 19 | 8.0000 | 0.6815 | 0.5047 | 0.7523 |
| | | | | Product of $L(X_i)$ | 0.3690 |
| | | | | $\Delta = (0.5 - L(X_i))/0.05$ | **2.6207** |

**Approximation 2**

New Seed $E_s$ = original seed plus $\Delta = 27 + 2.6207 = 29.6207$; then same process repeated

| *Interferer* | $E_i$ | $Z_i = E_s - E_i$ | $X_i = Z_i \frac{Z_i}{S_d\sqrt{2}}$ | *Probability integral* $P(x_i)$ *– see below* | $L(X_i) = Px_i/2 + 0.5$ |
|---|---|---|---|---|---|
| 1 | 10 | 19.6207 | 1.6716 | 0.9058 | 0.9529 |
| 2 | 15 | 14.6207 | 1.2456 | 0.7868 | 0.8934 |
| 3 | 12 | 17.6207 | 1.5012 | 0.8668 | 0.9334 |
| 4 | 21 | 8.6207 | 0.7344 | 0.5374 | 0.7687 |
| 5 | 19 | 10.6207 | 0.9048 | 0.6341 | 0.8171 |
| | | | | Product of $L(X_i)$ | 0.4991 |
| | | | | Delta = (0.5 – L(xi))/0.05 | **0.0186** |

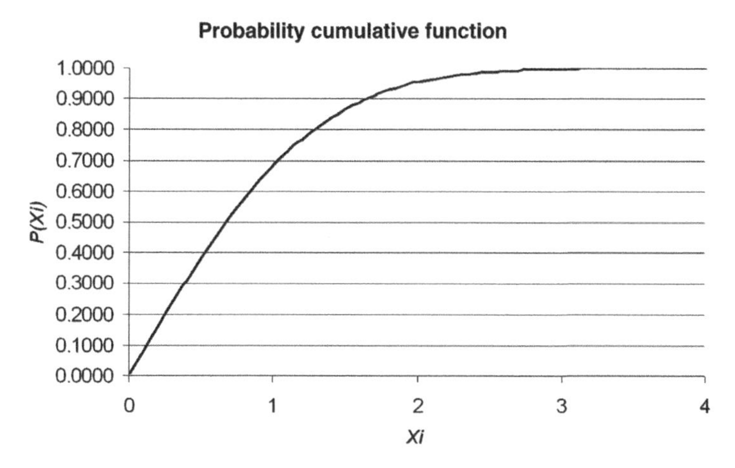


**Figure 13.15.** Probability cumulative function used in the SMM method.

Where the probability integral $P(X_i)$ is given by the cumulative probability function shown in Figure 13.15, which is obtained from the expression:

$$P(X_i) = \frac{-2}{\sqrt{2\pi}} \int_0^{x_i} \exp\left(\frac{-x_i^2}{2}\right) dx \qquad [8]$$

The second approximation of the SMM method gives a cumulative interference level of 29.639 dBμV/m. Further iterations could be applied, but since the second iteration only changed the result by 0.0186 dB and further iterations will be less than this, it is not worth doing so. The value calculated at the second iteration can now be used as the cumulative interference field strength from the interfering stations.

## 13.6 Interference Propagation Models

For interference over short distances such as within the elements of the same network, the behaviour of interfering signals can be modelled using the same propagation models as used for the planning process, because the mechanisms are identical. However, in this case it will be necessary to consider how the fading characteristics of the interfering signal should be handled. Recall that in the radio propagation discussions in Chapter 5, additional margins were added to allow us to model to an availability level of greater than 50% of locations. We may for example add a margin of the required number of decibels to ensure that the median value of the predicted signal at the cell edge ensured that 90% of locations within the short sector receive a working signal. For interference comparisons, two things should be taken into account; firstly, it is unlikely that the interfering signal fading correlates with the fading of the wanted signal, because the path to the receiving location will be different. Secondly, we are unlikely to want to consider interference only when it affects 90% of locations; it is more likely that we will want to consider the situation where 10, 5 or 1% of locations are interfered with so that we can examine intermittent interference effects that will only be present part of the time. Thus, rather than adding a margin to the median signal value to increase the availability, we subtract an amount to correct to the required value below the 50% value. Thus to represent the transition region where interference starts to occur, we need to consider the situation where the wanted system signal level including fade margin minus the interfering signal level plus the fade margin (positive since it is less than zero) is equal to the threshold C/I value for interference.

For the situation where the interferer is geographically separated from the wanted system, we will need to use specialist interference prediction models. For networks in the VHF and UHF bands, there are two propagation models that are often used to model interference. These are:

- ITU-R P.1546.
- ITU-R P.452.
- ITU-R P.528.

The 1546 model covers the frequency range 30 MHz to 3 GHz and can be used for mobile and broadcast systems. The 452 model works in the range 700 MHz and above. Although it was designed for microwave systems, experience has shown that the methods of the 452 model are applicable to mobile networks as well. The 528 model is applicable to aeronautical applications. Further information of the applicability of these (and other ITU models) can be found in ITU-R P.1144, and the original recommendations contain additional information both on their applicability and to their modes of operation.

## 13.7 Interference Mitigation Approaches

There are several ways of militating against interference. These include:

- Power management of the interferer.
- Selection of antenna with suitable polar response.
- Antenna height of both interferer and victim.
- Antenna tilt of both interferer and victim.
- Sectored antennas or null steering for both interferer and victim.
- Frequency re-assignment for either interferer or victim.

The selection of which mitigation method is applied is usually based on the amount of interference and the implications of the potential changes. Some typical considerations include:

- Power management: one approach is to reduce the power of the interferer, to benefit from the change in relative signal strength. However, reducing the power of the interferer will also reduce its service area and this will have an effect on network coverage and capacity in the region around the interfering base station.
- Antenna height: the antenna heights of either the interferer or victim can be reduced (or both). This will have the effect of reducing the range of the interferer if its antenna height is reduced (again this may have a knock-on effect to the overall network coverage and local capacity). The victim antenna height can also be reduced. This can help in

the situation where the reduction in height will result in benefits from terrain or clutter shielding. Again, of course, the coverage area of the victim will be reduced in this case.

- Antenna tilting: for antennas that have vertical directivity, it is possible to orient antennas used for paths within the horizon such that the energy radiated towards the horizon is reduced. This will have the effect of maintaining the coverage in the wanted area but reducing interference outside of it.
- Sectored antennas and null steering: antennas with directional patterns in the horizontal plane can be used to minimise interference in the direction of the victim or interferer (as required). Antennas that have a null (very low response in a particular direction) can be used to spatially filter out particular interferers. This approach can only be used when the direction of the interfering energy is known, and thus it is applicable to the condition of base station to base station interference.
- Frequency assignment: either the victim or interfering system can have their frequencies changed to prevent interference between them. This of course may cause interference with other spectrum users and thus should only be done with care, and with checks to identify any potential problems that may be caused by the change.

## 13.8 Interference Deliverables

The location and severity of interference can be expressed in much the same manner as for coverage predictions. Therefore, the deliverables for interference analysis can include:

- Report on interference methodologies used for each interference analysis.
- Point-to-point interference tables for fixed base stations.
- Plots of interference for mobile units.
- Statistics of interference for mobile units.

## References and Further Reading

[1] Land-Mobile Radio System Engineering, Garry C. Hess, Artech House Publishers
[2] ITU-R P.372
[3] Man Made Noise Measurement Programme (AY4119), Final Report Issue 2, Sep 2003, Produced by Mass Consultants on behalf of Ofcom (can be found on the Ofcom website)
[4] Handbook of Terrestrial Land Mobile Radiowave Propagation in the VHF/UHF Bands, ITU Study Group 3 (can be found on the ITU website)
[5] CCIR Report 945 – Method for the Assessment of Multiple Interferers

# 14

# Frequency Assignment

## 14.1 Introduction to Frequency Assignment

For any radio technology that uses frequency to distinguish between concurrent calls, it is usually necessary to assign different frequencies to base stations and mobiles that are likely to interfere with one another if they are not separated in spectrum terms. This is illustrated in Figure 14.1, which shows the interference that would be present in a network if frequency assignment is not used. Clearly, this network would be unworkable.

Frequency assignment is used to avoid this problem and also to provide additional traffic capacity in parts of the network to deal with traffic. Thus a single base station location may have, say, four frequency channels, one for each base station deployed there. Figure 14.2 shows the interference plot for the same network after frequency assignment has been performed. Now interference levels are largely eliminated. The remaining interference may be regarded as acceptable, or alternatively more design work may be performed to reduce it still further.

In general, there are two analyses of interest to the planning engineer and network designer:

- What is the minimum number of channels needed for the network in order to provide an acceptable level of service (the 'minimum span' problem) and
- How do we best assign a fixed number of channels to the base stations to minimise interference? Note that for most civilian networks working in duplex mode, there is normally a fixed frequency offset between the downlink and uplink directions, so that by assigning the downlink, we are also de facto assigning the uplink. This may not be the case in certain military networks, and in this case, the channels have to be assigned separately (but with reference to each other).

Frequency assignment methods have received considerable attention from academics proposing new and improved methods. Because these tend to be intensely mathematical in nature, we do not propose to cover them in this book, but we do provide recommended

Mobile Radio Network Design in the VHF and UHF Bands: A Practical Approach
*Adrian W. Graham, Nicholas C. Kirkman and Peter M. Paul* 

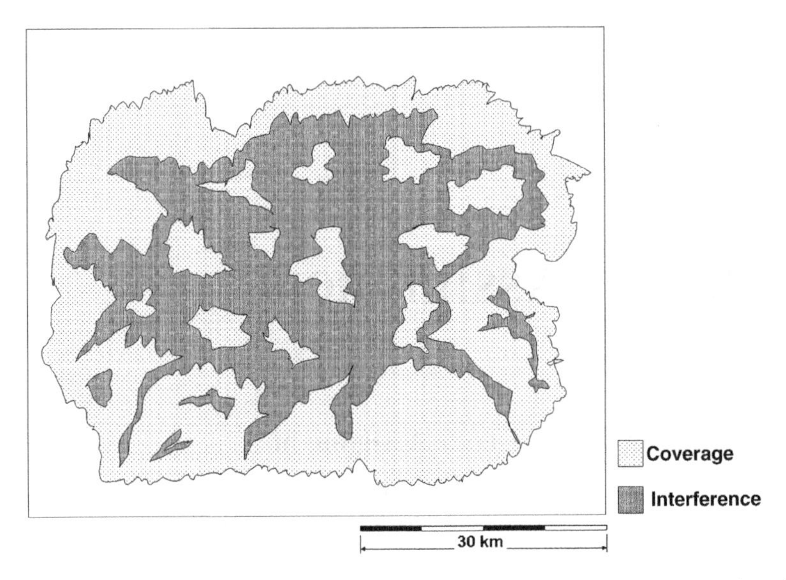


**Figure 14.1.** Illustration of Interference in a network with no frequency assignment; interference is prevalent throughout the service area, and subscribers would find it difficult to use the network.

reading at the end of the chapter for those interested. In general, each of the methods will work in a broadly similar manner as shown in below:

- Step 1: Set up initial candidate design using a set of rules or randomly.
- Step 2: test the design to see if it meets the performance criteria.

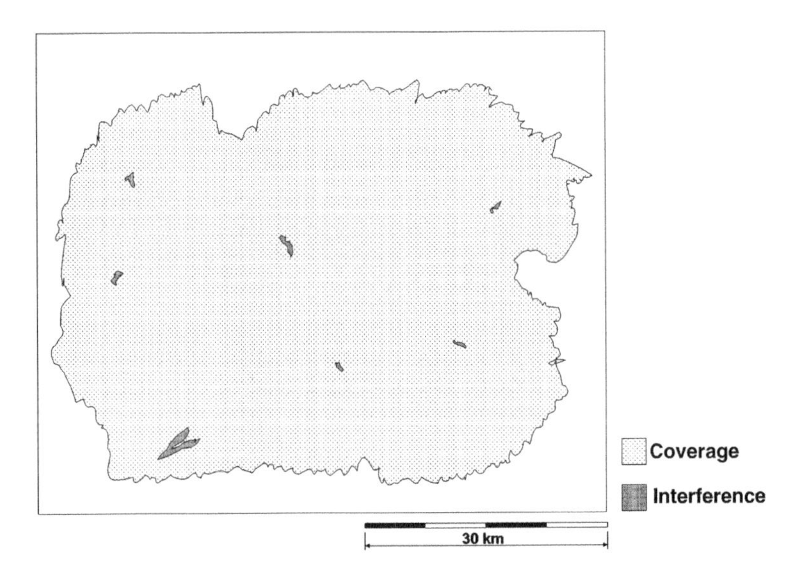


**Figure 14.2.** Illustration of Interference in a network with after frequency assignment; interference is much reduced, and ideally is in little used parts of the service area.

- Step 3: if not, revise the design by some set of rules.
- Step 4: test the revised design.
- Repeat Steps 3 and 4 until the acceptable criteria are met.

The general method is likely to be the same in each assignment type; only the rules applied in Steps 1 and 3 are likely to differ. For the initial condition, a sequential or 'greedy' algorithm is often used. This will provide a solution that is likely to be sub-optimal but acts as a starting point for the rest of the process. A number of algorithm types for Step 3 have been proposed, such as evolutionary, simulated annealing and taboo search. These apply iterative steps to refine the design based on the starting conditions and rules as to how the design can change in each step. In general, the aim of assignment algorithm designers is to provide a system that always leads to an acceptable solution whenever one is possible, in the shortest number of steps. For the engineer, it is sometimes acceptable to allow mathematically sub-optimal approaches so long as the design needs are met. This means that many of the approaches adopted in various planning tools are based on pragmatic design that sacrifice solution efficiency and processing time in order to get a result that works. In this section, we will illustrate the frequency assignment process using methods that are sub-optimal but are sophisticated and can handle the complex process of modelling real world scenarios in a timely manner and produce results that are usable.

The testing process, identified as Steps 2 and 4, will normally consist of the same type of analyses described in Chapter 13 for interference analysis, such as percentage of interference area or a quality of service metric. In some cases, the approach will be a simplified version of the methods shown in order to speed up the assignment process; a full interference analysis is then performed at the end to validate the results. For each test, the number of violations of the interference rules are added together to provide an overall 'cost' for the candidate. The aim is for the cost to be zero, thus there are no interference violations. Some algorithms apply different scores for different types of violation, thereby allowing designs that breach the test constraints in the least worst way to be identified in the case that the zero cost condition cannot be met.

For the fixed frequency case, the a graph of the process outlined above might look similar to Figure 14.3, which shows two different processes; one is based on random assignment, and the other uses sophisticated methods to 'seek' the best solution based on iterative rules. The best solution has a cost score of zero. The aim is that the second method will achieve an acceptable result before the purely random case. However, as stated before, as long as the random method finds an acceptable solution within an acceptable length of time, then the engineer can live with the less complex method. In addition, some of the more complex methods can come unstuck in certain circumstances and not provide an answer at all. In this case, the system needs to be operated by a highly trained user who can recognise the problem and perform tweaks to overcome it, or the system has to feature automatic methods of doing so.

For the discussions in most of this chapter, we will be focussing on the steps that a radio engineer might take to complete an assignment activity. We will look at some necessary theory and practical considerations in addition to this. As special cases, we will look at two complex assignment problems. The first of these is assignment for co-sited elements. This can be particularly complex to solve. The second is the situation where elements of a network are located in an area but their actual location is unknown. This is typical for military assignment. We will be looking at both these issues later in the chapter, but we start with a look at where frequency assignment fits into the spectrum management domain.

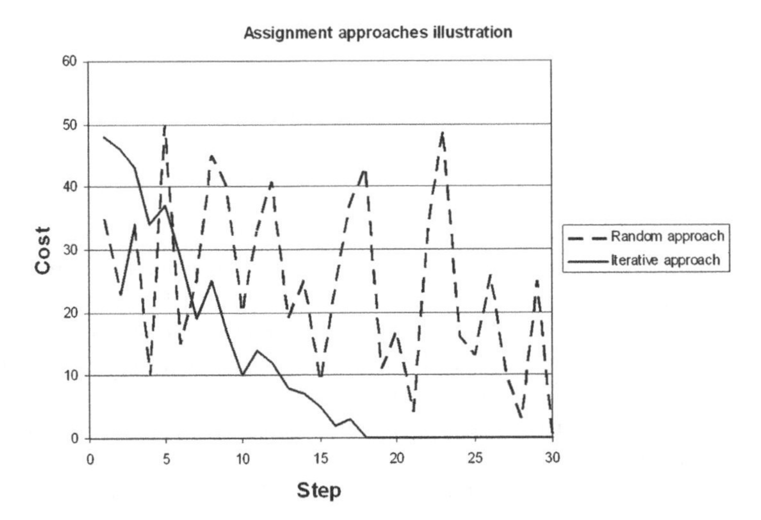


**Figure 14.3.** Illustration of random and iterative approaches to frequency assignment. A cost of zero is ideal. The iterative approach shows continual improvement, the random approach does not.

## 14.2 Frequency Assignment in Context

The frequency assignment process fits into a global structure ultimately set by the ITU table of allocations. This is a list of frequency bands together with acceptable uses for each band. These are defined either for worldwide use or for an ITU 'Region', of which there are three. 'Region 1' covers Europe including the former Soviet states, the Middle East and Africa; 'Region 2' covers North and South America and 'Region 3' covers Southeast Asia and Australasia. Individual countries can define their own variations from the general plan, and these are also recorded in the table of allocations. A view on part of the table of allocations is shown in Figure 14.4. It can be seen that individual parts of the spectrum can be used for fixed, mobile and other services depending on the desires of each nation within a region.

The figure also shows a footnote associated with a particular service. Such footnotes are used to provide additional information about frequency band usage for specific services. Each nation has its own table of allocations to be used for the allotment of spectrum to individual licensees. Although this does not have to follow the international table of allocations, it does need to ensure that it does not support services that will cause interference to services protected by the table. In practice, this means that there will be a great deal of commonality between national and regional allocations.

Domestically, the allotment of spectrum to specific users will typically be decided on the following types of criteria:

- Financial: In most countries, the national regulator will be responsible for maximising the economic benefits of spectrum, and thus the financial gain to the economy arising from a particular organisation being issued a licence will be an important issue.
- Technical: The proposed systems use that an organisation wishes to make of parts of the spectrum must be technically viable, otherwise offering a licence would be a waste.

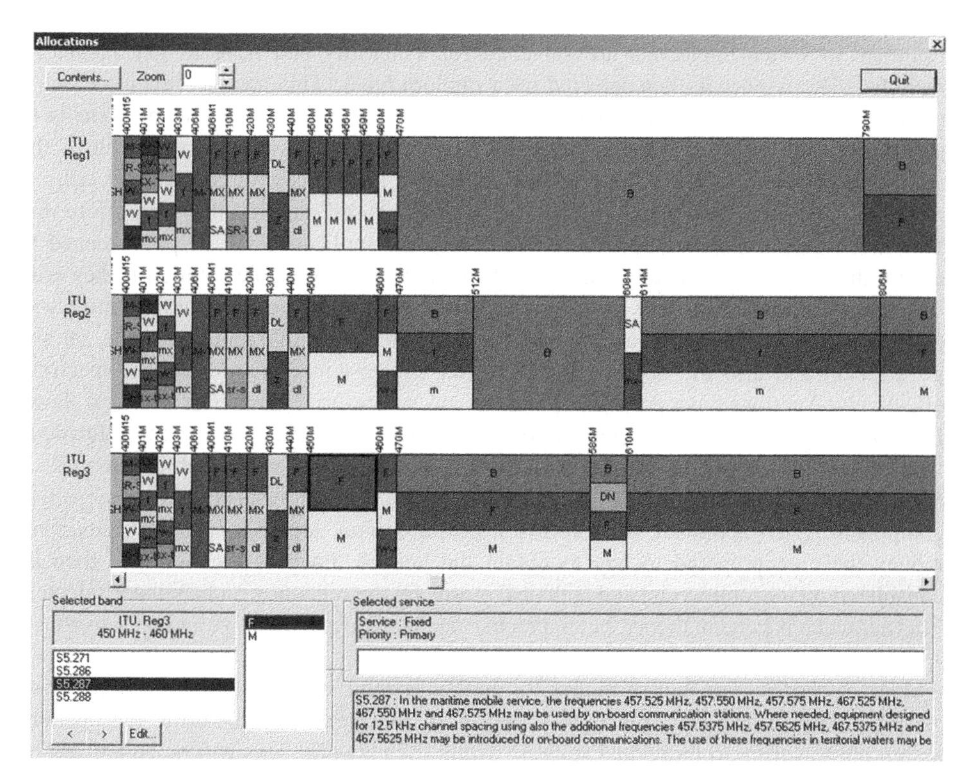


**Figure 14.4.** ITU Allocations; an example footnote is shown at the bottom.

- Efficiency: Since spectrum is a finite resource, it must be used efficiently. This must, therefore, also be used as a metric for the decision process.
- Public interest: In many case, spectrum needs to be assigned in the public interest rather than for maximum financial benefit. This will include spectrum for emergency services, military, aeronautical and other official applications.
- Minimisation of interference to other users to an acceptable level. Proposed systems must be capable of operating without adversely affecting other users.

Depending on the intended use of portions of the spectrum, different methods of deciding which operator should be given a licence when there are competing applicants can be used. Some methods that have been used include:

- Comparative hearings in which an arbiter listens to the arguments of each applicant and then makes a decision based on the facts. This is applicable to official uses of the spectrum and issues in the public interest, such as broadcast. There is the risk of decisions being based on a nonobjective basis with this approach.
- Negotiated solutions in which applicants are asked to negotiate between themselves for equitable use of the available spectrum. Again, this is typically applicable for official use of the spectrum. Ideally, this can be achieved without acrimony, but there is the risk that the parties will not be able to achieve a consensus.

- Lotteries in which applicants are screened for suitability and have to pay fee to be included. The winner is then selected on a random basis. This has the benefit of being random, but it can lead to commercial abuse, where companies are set up as shells to participate and then sold on if a licence is obtained. Also, it may not achieve the maximum revenue possible. Lotteries are applicable to licences for commercial use.
- Auctions, in which applicants will bid for spectrum based on what they can afford and how much they think the spectrum is worth. Again, applicants will be pre-screened to ensure that they are financially viable and have a technically viable solution if they win. The intention behind auctions is that the operator that wants the spectrum the most will pay the highest. This will benefit the public by providing maximum revenue for the government and also the operator will have the greatest incentive to use the spectrum effectively. Auctions have proved popular but are subject to a number of practical flaws. In some cases, spectrum has been auctioned for such high values that the entire financial viability of the winning organisation is threatened. This is not in the public interest.
- So called 'beauty contests', in which there is a financial aspect and a subjective component. Each applicant will prepare a business plan, together with the amount of money they are prepared to pay to obtain the licence. Each application will then be considered on its commercial and technical merits by the regulator, and a winner selected. This approach is good in theory, but in practice can be seen as subjective and open to abuse.

The block of spectrum or pair of blocks for duplex systems will be allotted to the winning applicant, whatever decision criteria has been used, and then it will be up to the operator to assign the spectrum to individual stations within the network. This will be completed by an appropriate frequency assignment process. If the operator wants their stations to be protected against international interference, then they will send details of the completed network design, including frequencies assigned, to the regulator so that they can be submitted to the ITU via the BR IFIC process.

For military assignments, the same process is generally followed, but allotment will typically be based on the comparative hearing and negotiated settlement approaches. This can be complex, particularly, if the force includes allies from different nations and if it is necessary to resolve issues quickly to meet operational needs.

## 14.3 Network Frequency Plans

For frequency assignment, it may be appropriate to use frequencies covering the entire allotment or it may be that only a subset is considered either for an assignment or for

**Table 14.1.** Example of equipment constraints on frequency plan for assignment.

| Radio | Minimum frequency (MHz) | Maximum frequency (MHz) | Tuning step (kHz) | Forbidden frequencies (MHz) |
|---|---|---|---|---|
| Type 1 | 30 | 80 | 25 | |
| Type 2 | 30 | 50 | 50 | |
| Type 3 | 30 | 68 | 50 | 40–41 |
| Type 4 | 35 | 60 | 25 | 30.5, 42.1 |

**Table 14.2.** Channels available after constraints applied. The allowable channels and number of channels available are shown as each additional constraint is applied.

| Constraint | Allowable channels | Channels available |
|---|---|---|
| Allotment | 30–80 MHz | n/a |
| + Type 1 radio constraints | 30–80 MHz, 25 kHz spacing | 2000 |
| + Type 2 radio constraints | 30–50 MHz, 50 kHz spacing | 400 |
| + Type 3 radio constraints | 30–40, 41–50 MHz, 50 kHz spacing | 380 |
| + Type 4 radio constraints | 30–30.450, 30.550–40, 41–42.05, 42.15–50 MHz, 50 kHz spacing | 378 |

individual sites within the assignment process. It may, for example, be necessary to reserve spectrum for future assignments to be performed separately, such as ground-to-air assignments for networks that have both terrestrial and airborne mobiles. It may be that there are limits imposed by the equipment to be used in an individual network. This is illustrated in Table 14.1, which shows four hypothetical radio types that will be used for a network assigned in a single frequency.

The available frequencies to be used as an input into the assignment process are constrained to the set of frequencies that can be achieved by each radio. This is illustrated in Table 14.2.

The list of channels that can be used after the constraints have been applied can then be used as the input to the assignment process.

There may instead be constraints linked to individual sites in the network or it even might be necessary to define a frequency plan for every station in the network. This can be used as a mechanism to refine the automatic assignment process or to handle issues that cannot be automatically accounted for.

Once complete the frequency plan or plans are then used in the assignment process.

## 14.4 Overall Assignment Process

Figure 14.5 shows a simplified frequency assignment process capable of performing both far-site and co-site assignments for networks based on individual sites or for networks ('nets') in the case of a multiple net assignment. The process starts with an allotment to be used in the assignment, a candidate network design configured to provide the wanted coverage and traffic capacity, pre-assigned parts of the network that have already been processed and third-party assignments that will need to be taken into consideration. This model is appropriate to both civil and military frequency assignments.

The raw allotment is processed if required, to filter for channels to be included in the current assignment. If necessary, individual frequency plans are associated with radio sites or for particular networks, for a multiple network assignment. The elements to be assigned are combined in a model with the fixed elements that are not to be assigned but must be considered in the design. The operator may have to set assignment rules for co-site and far-site assignment algorithms and once entered these are passed through the system into the actual assignment engines. For some methods, it will be necessary to generate an initial seed design for the frequency plan to be refined later. In others, this step will be omitted.

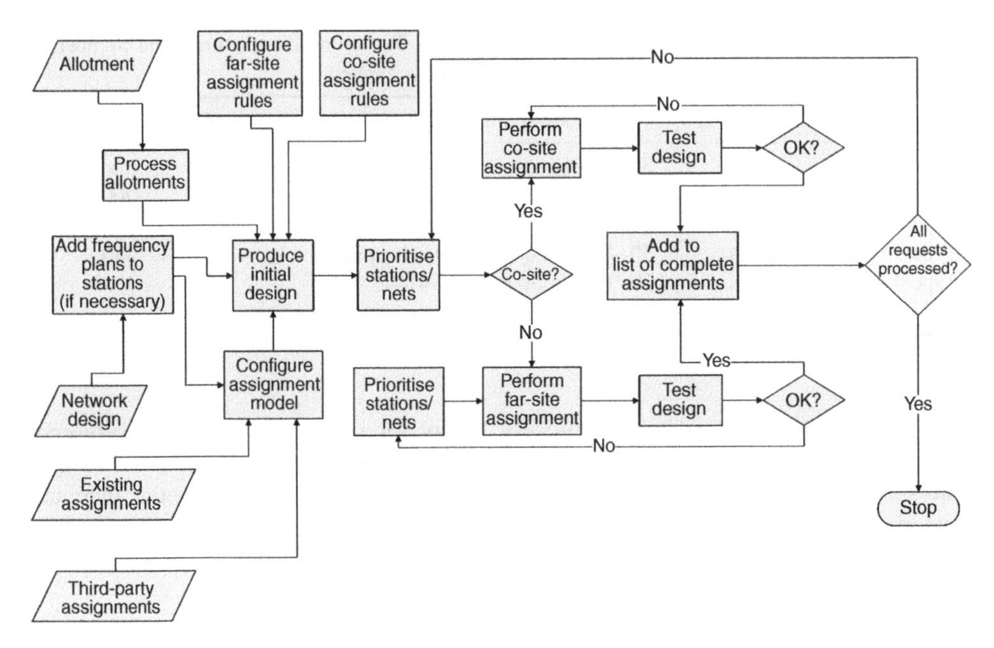


**Figure 14.5.** Outline assignment process, including both far-site and co-site assignment.

Some methods will prioritise requests for processing and the assignments related to co-site locations will be combined for assignment by the specific co-site assignment process. The far-site requests will also be prioritised before being input to the far-site assignment process. In both cases, the resulting design will be checked against the acceptance criteria. If the candidate design does not meet the acceptance criteria, the process is repeated. In general, there will be a maximum number of times that the assignment process will be completed to avoid the system locking up; in this case, it will report failed assignments and present as an output the best scheme produced. In other cases, the assignment will succeed and once all requests have been processed, the system will provide a report of the assignments made.

We will now go on to look at some of the components of this overall model.

## 14.5 Far-Site Assignment Methodologies

### *14.5.1 Prioritisation of Requests*

In many assignment methodologies, individual assignment requests will be prioritised for processing. This may be based on a most constrained first assigned (MCFA) methodology. An example of this is based on an analysis of the available channels for each frequency request. This is illustrated in Table 14.3, which shows a number of military nets to be assigned. For each net, the number of channels available to the assignment is calculated, based on the frequency range supported, tuning interval and so on. Also calculated for each area is the number of channels within that range that are already assigned to other applications. This is dynamically calculated as each new assignment is made, because the available spectrum will change as assignments are made.

**Table 14.3.** Network assignment requests and the channels available to them.

| ID | Net name | Available channels | Occupied channels | Remaining |
|---|---|---|---|---|
| 1 | Divisional Command Net | 300 | 125 | 175 |
| 2 | Logistics Net | 253 | 57 | 195 |
| 3 | Artillery Command Net | 87 | 37 | 50 |
| 4 | Air Defence Command Net | 2000 | 1550 | 450 |
| 5 | EW Command Net | 2000 | 1875 | 125 |
| 6 | Airfield Defence Net | 500 | 199 | 301 |
| 7 | Port Control Net | 500 | 250 | 250 |
| 8 | Main Supply Route Net | 1000 | 925 | 75 |

Once the number of actual channels available have been calculated, the list can be re-ordered to put the most constrained networks at the top of the assignment list, sequentially numbering them in order, as illustrated in Table 14.4.

This is the order in which the requests are processed for a sequential system. For a nonsequential process, it is still important to know which channels are available for assignment, to ensure the system assigns only valid channels.

The equivalent process is followed in mobile networks with fixed base stations, where the base station ID replaces the Net name entry shown in the tables.

### *14.5.2 Manually Seeding Assignments*

In some cases the automatic assignment processes can be assisted by deliberately setting up fixed assignments before starting the automatic assignment process. This can be achieved manually, in which case the operator manually selects some sites and manually assigns them. An interference prediction can be performed to ensure that the manual assignments do not interfere with each other. In other cases, a subset of the assignment process can be performed before the main assignment to seed the start process. As long as the assignment process itself does not change these initial assignments, the seeding will influence the frequency plan produced. This mixture of manual and automatic assignments can be used

**Table 14.4.** Assignment requests order in MCFA sequence.

| Order | ID | Net name | Available channels | Occupied channels | Remaining |
|---|---|---|---|---|---|
| 1 | 3 | Artillery Command Net | 87 | 37 | 50 |
| 2 | 8 | Main Supply Route Net | 1000 | 925 | 75 |
| 3 | 5 | EW Command Net | 2000 | 1875 | 125 |
| 4 | 1 | Divisional Command Net | 300 | 125 | 175 |
| 5 | 2 | Logistics Net | 253 | 57 | 195 |
| 6 | 7 | Port Control Net | 500 | 250 | 250 |
| 7 | 6 | Airfield Defence Net | 500 | 199 | 301 |
| 8 | 4 | Air Defence Command Net | 2000 | 1550 | 450 |

to speed up the overall assignment process and also allows the operator to tune the automatic assignment process itself. The benefits for this will depend on the circumstances of the project.

### *14.5.3 Illustration of an Assignment Process*

In this section, we will illustrate an assignment process by following through a hypothetical example based on a radio-planning tool. We will assume that the network has been designed for traffic and capacity. The resulting network is shown in Figure 14.6. Each site has three sectors to allow for traffic demand.

We will assume that a number of channels in the frequency range 380–385 MHz are available, but not as a contiguous block. Individual channels have been selected from an available allotment and these are entered into the tool as shown in Figure 14.7. This is one of the options for expressing channels to be used in this particular tool; available spectrum can also be described by bands and mixtures of bands and channels. The channels selected have been chosen at random and are not meant to be indicative of a good channel plan.

Having selected the channels available, the task is now to configure the planning tool assignment engine. The tool under consideration has a complex assignment system capable of modelling a wide variety of technologies. The configuration box is shown in Figure 14.8. There are options to select from bands of frequencies, lists of frequencies and frequency plans; in this case, the list assignment is selected. There are a variety of assignment rules available to determine how the system behaves, and there are four available assignment methods. In this case, an iterative method has been selected.

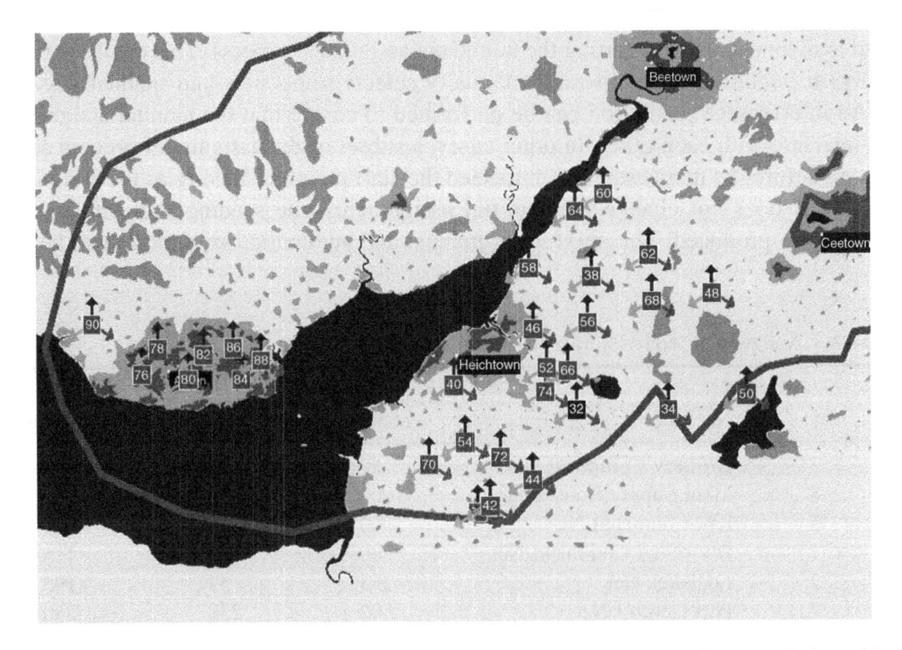


**Figure 14.6.** Network design for frequency assignment activity. Reproduced by permission of P & D Missud.

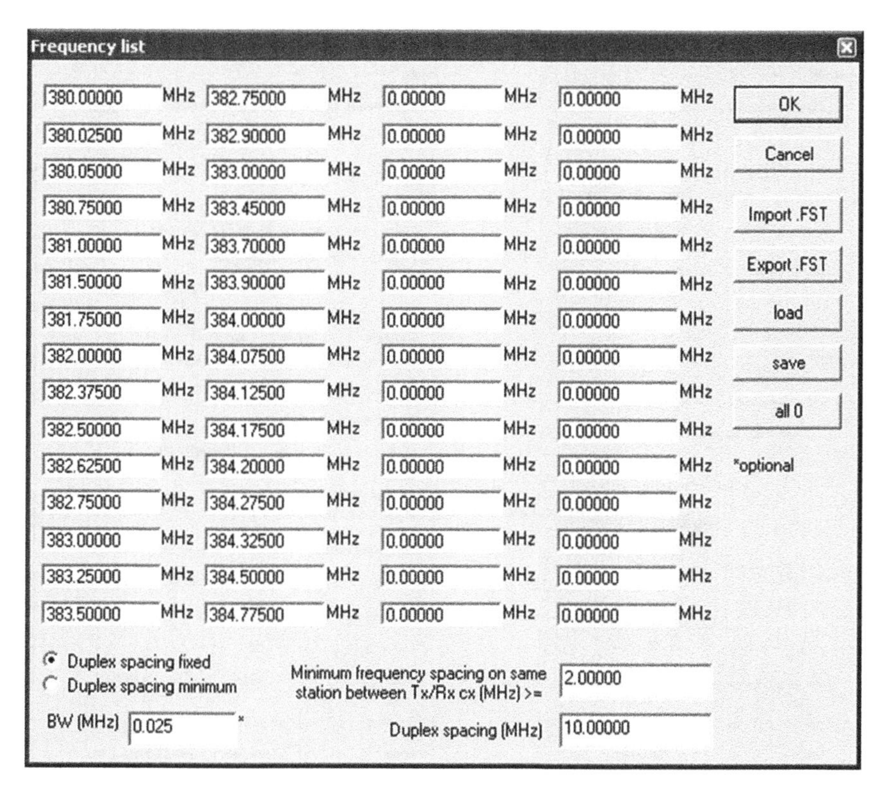


**Figure 14.7.** Channels available for frequency assignment process. Reproduced by permission of P & D Missud.

Once the system has been configured, the operator selects 'Start' and the assignment process begins. This will take some time. At the end of the process, a table of the results can be obtained, as illustrated in Table 14.5.

To ensure that the results meet the performance criteria, an interference calculation can be run on the new network design. If this is successful, then the round of frequency assignment is complete.

In many networks design activities, it may be necessary to perform many frequency assignment processes, and this may continue through the network life as new sites are added to fill in coverage gaps or to add extra capacity. The same process can be used each time, however.

## 14.6 Co-Site Assignment Methodologies

### *14.6.1 Introduction*

When transmitters and receivers are positioned in close proximity to one another, effects that can be ignored at longer ranges need to be taken into account. These are caused by unavoidable, unintentional radio emissions along with the wanted transmissions, and the vulnerability of receivers to interference on particular frequencies. In both cases the unwanted frequency characteristics often vary with the frequency of use.

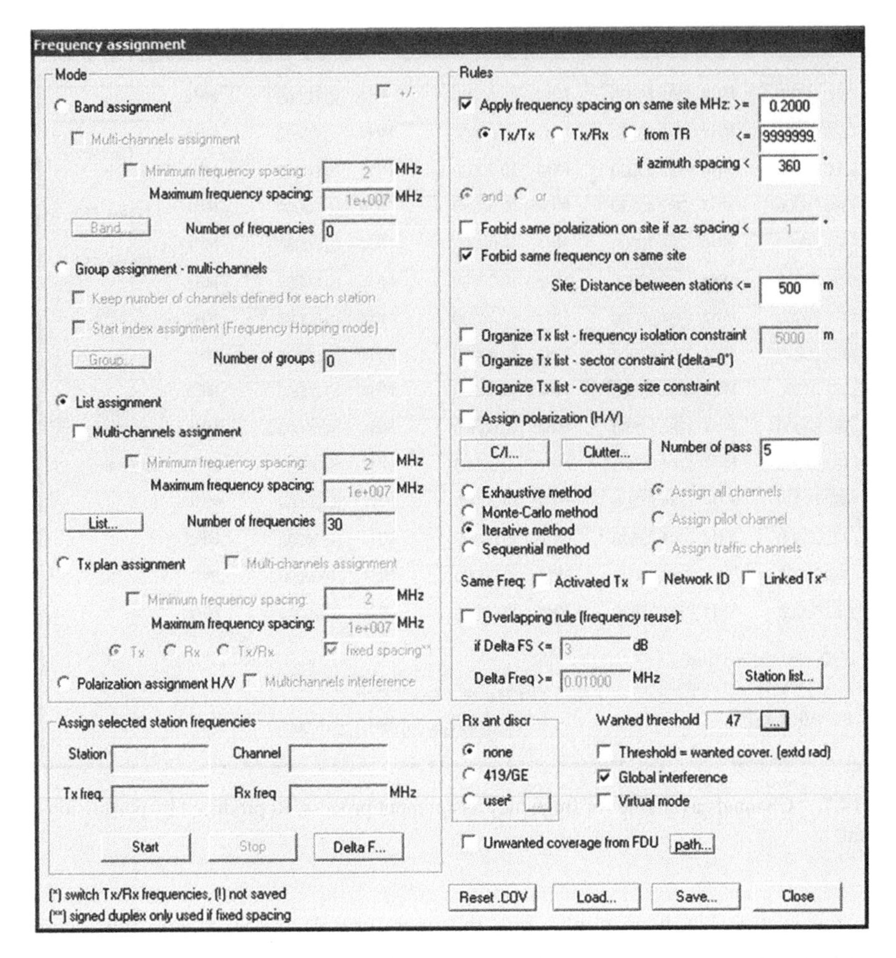


**Figure 14.8.** Frequency assignment dialog box. Reproduced by permission of P & D Missud.

In this regard we can consider the following types of mechanism that need to be accounted for:

- Harmonics of the fundamental transmission frequency.
- Intermediate frequencies used within radios.
- Image frequency.
- Minimum frequency separation.
- Inter-modulation products.

#### 14.6.1.1 Harmonics

Unwanted emissions from a single transmitter can be caused by harmonics of the wanted frequency, as illustrated in Figure 14.9. The relative strength of each harmonic is illustrative only, since it will vary due to the filtering in the radio, but it does show one

**Table 14.5.** Assignment results.

| Callsign | Sector | Longitude | Latitude | Frequency (MHz) |
|---|---|---|---|---|
| Site 1 | 0 | −2.31489 | 51.23144 | 380.0000 |
| | 120 | −2.31489 | 51.23144 | 380.7500 |
| | 240 | −2.31489 | 51.23144 | 384.3250 |
| Site 2 | 0 | −2.20104 | 51.23105 | 381.5000 |
| | 120 | −2.20104 | 51.23105 | 382.0000 |
| | 240 | −2.20104 | 51.23105 | 383.2500 |
| Site 3 | 0 | −2.44165 | 51.15145 | 380.7500 |
| | 120 | −2.44165 | 51.15145 | 380.0000 |
| | 240 | −2.44165 | 51.15145 | 384.5000 |
| Site 4 | 0 | −2.29566 | 51.33332 | 382.0000 |
| | 120 | −2.44165 | 51.15145 | 380.0000 |
| | 240 | −2.44165 | 51.15145 | 384.5000 |
| Site 5 | 0 | −2.4722 | 51.25057 | 382.3750 |
| | 120 | −2.29566 | 51.33332 | 383.5000 |
| | 240 | −2.29566 | 51.33332 | 384.1250 |
| Site 6 | 0 | −2.42441 | 51.15442 | 383.0000 |
| | 120 | −2.4722 | 51.25057 | 380.7500 |
| | 240 | −2.4722 | 51.25057 | 380.0000 |
| Site 7 | 0 | −2.37259 | 51.1749 | 383.7000 |
| | 120 | −2.42441 | 51.15442 | 382.5000 |
| | 240 | −2.42441 | 51.15442 | 383.7000 |

important aspect, which is that odd harmonics are usually far stronger than even harmonics.

Harmonics can usually be reduced in strength by good filtering, but often the design process is simply to avoid placing receivers on channels likely to suffer from the effects of

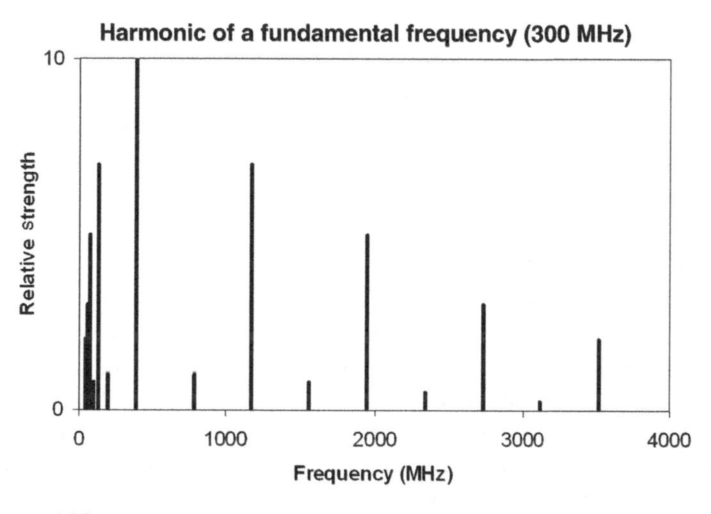


**Figure 14.9.** Harmonics of a fundamental frequency (amplitude not to scale).

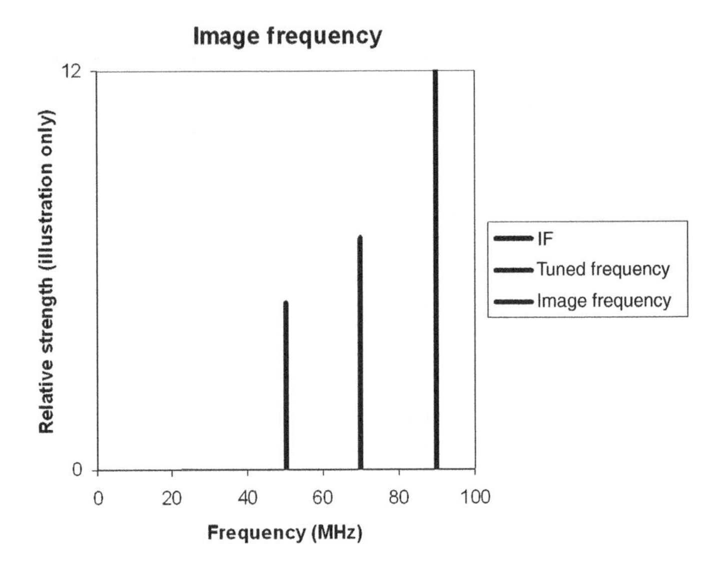


**Figure 14.10.** Example of an image frequency (amplitude not to scale).

harmonics. Note that as the number of the harmonic increases, so does the band over which it is present increase, so a harmonic of fifth order may be present over several channels (five times the original occupied band).

### *14.6.2 Image Frequencies*

Transmitters can also emit radio energy at other frequencies including intermediate frequencies used within the set and also image frequencies, which are generated offset to the tuned frequency as illustrated in Figure 14.10.

Image frequencies and intermediate frequencies can be reduced in strength by providing shielding in the radio box and transmission cables to reduce emissions into the rest of the system. Again, another approach is simply to avoid assigning receivers to channels where potential images may be present, and to avoid placing transmitters on bands where they may interfere with intermediate frequencies of other radios.

### *14.6.3 Frequency Separation*

We have already seen in Chapter 13 that co-channel interference is not the only interference mechanism between radios; there will also be energy caused by adjacent channels and well beyond. This is particularly true when the radios are sited close together since even if there is significant attenuation caused by filtering of off-channel energy, the initial levels will still be high and there will be relatively little propagation loss. This means that often it will be necessary to ensure that radios are assigned frequencies well separated from one another, either in terms of numbers of channels separation, a value in MHz or in percentage terms of the radio transmission frequency. This means, for example that if five radios are to be sited within co-site constraints, and each one needs at least 2 MHz of separation, then the minimum span of the frequencies selected will be a little over 10 MHz.

### *14.6.4 Inter-Modulation Products*

Inter-modulation products (IMPs) are caused by nonlinear interactions between elements at a site. The three possible causes are:

- Coupling between transmit antennas, radio equipment or feeders.
- Overloading of receiver input stages, causing nonlinearity.
- The 'rusty bolt' effect, where bad metallic contacts produce a rectifying action over the join, setting up spurious signals in the surrounding metalwork.

Inter-modulation effects are difficult to eliminate in practice because some generating mechanisms may be likewise hard to avoid. It can also be difficult to predict the strength of the inter-modulation products (although there are mathematical models available to estimate likely values), and so often the design approach taken is to avoid those channels that might be subject to it.

Inter-modulation products are classed in order with the higher order products above 5th often considered to be negligible. For many applications, even products can be ignored because the frequencies of the generated products are far from the band of initial interest, but this is not always the case, particularly when an individual site is used for radios covering a very large range of frequencies.

To examine inter-modulation products, we can initially consider two transmissions, operating on frequencies A and B, as shown in Figure 14.11.

We have chosen A = 380 MHz and B = 384 MHz.

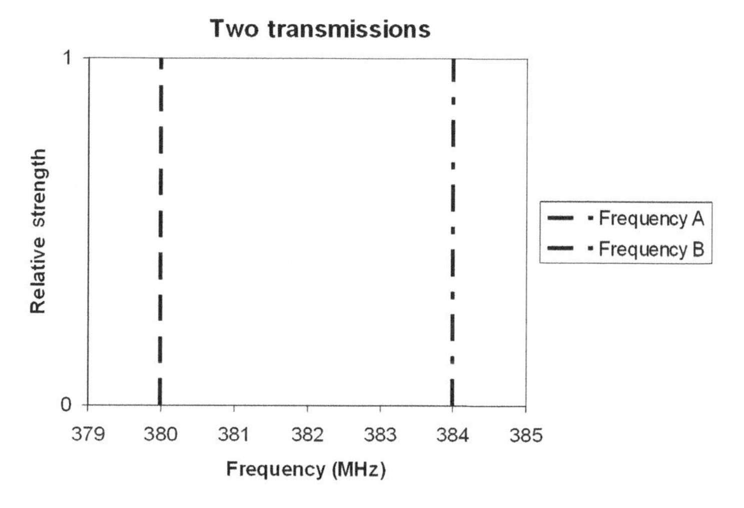


**Figure 14.11.** Two transmissions, at frequency A and Frequency B.

The two order products are:

$$A + B = B + A = 764\,\text{MHz}$$
$$A - B = A - B = 4\,\text{MHz}$$
(the fact that $A - B = -4\,\text{MHz}$ is not relevant; it still works out at 4 MHz)

These frequencies are well out of the initial band of interest and may well not cause a problem. The same process can be used to generate even products of higher order, but these are seldom a problem in practice.

The third order products include all combinations of the two frequencies involving a total of three elements:

$$2A - B = 380 \times 2 - 384 = 376$$
$$A - 2B = 380 - 2 \times 384 = 388 \quad \text{(again, negative sign can be ignored)}$$
$$2A + B = 380 \times 2 + 384 = 1144$$
$$A + 2B = 380 + 2 \times 384 = 1148$$

In general, the third-order elements are likely to be the strongest and most difficult to eliminate.

Likewise, fifth order elements include all terms with five elements in them.

$$3A - 2B = 3 \times 380 - 2 \times 384 = 372$$
$$3A + 2B = 3 \times 380 + 2 \times 384 = 1908$$
$$2A - 3B = 2 \times 380 - 3 \times 384 = 392$$
$$2A + 3B = 2 \times 380 + 3 \times 384 = 1912$$

And so on for seventh and ninth order, although often these are considered to be insignificant.

The products can be manually calculated, and there are simple computation methods of determining inter-modulation products for two transmitters (as we have just covered) and for more than two transmitters (which we will look at next), but there are many computer-based tools for helping the engineer discover the products computationally. An example of one is shown in Figure 14.12.

Computing products for two frequencies is not very difficult, and there are few terms involved. However, IMPs can be caused by any combinations of two or more frequencies, not just two. Thus, if there are three frequencies (with the third one being termed 'C'), then the third order terms include, in addition to those already identified:

$$A + B - C$$
$$A - B + C$$
$$A - B - C$$
$$A + B + C$$
$$B - A - C$$
$$C - A - B$$

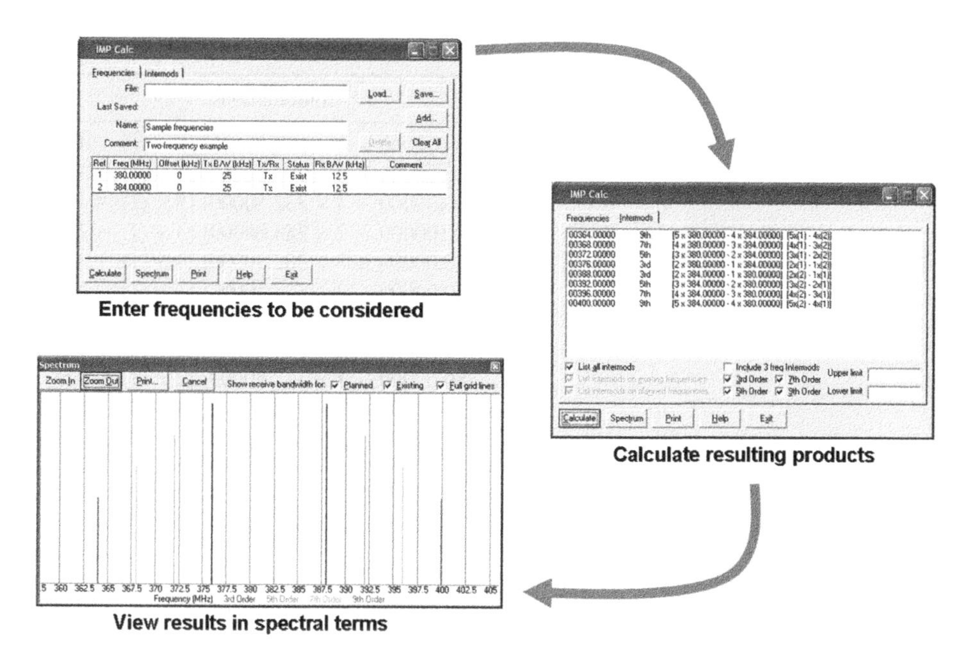


**Figure 14.12.** Example of an inter-modulation calculation tool. Reproduced by permission of ATDI Ltd.

Fifth order for three frequencies includes:

$$
\begin{aligned}
&A + B - 3C \\
&A + 2B - 2C \\
&A + 3B - C \\
&A - B - 3C
\end{aligned}
$$

and so on.

Both the number of terms and the number of products increase greatly as more channels are included in the analysis because each combination of each two or three frequency combinations must be considered. In general, combinations above three frequencies are not normally considered. An example of an analysis of the products of three individual frequencies of 380.0, 382.5 and 384 MHz is shown in Table 14.6, restricted to those products that fall close to the initial band of interest.

The problem gets rapidly worse with more frequencies under consideration, as illustrated by the spectrum analysis of five frequencies shown in Figure 14.13. In this case, the spectrum occupancy of each channel has been exaggerated to make the diagram clearer, but it should be obvious that IMPs cause a major source of spectrum pollution when they are present.

Calculation tools such as the one shown in Figures 14.12 and 14.13 are useful to identify free channels when there are existing assignments at a co-site location, but they do not help the assignments for a new co-site configuration. A more computationally complex method

**Table 14.6.** Extract of IMP calculation for three frequencies.

| Frequency (MHz) | Order | Combinations |
|---|---|---|
| 364 | 9th | [5 × 380.00000 – 4 × 384.00000] [5 × (1) – 4 × (3)] |
| 368 | 7th | [4 × 380.00000 – 3 × 384.00000] [4 × (1) – 3 × (3)] |
| 370 | 9th | [5 × 380.00000 – 4 × 382.50000] [5 × (1) – 4 × (2)] |
| 372 | 5th | [3 × 380.00000 – 2 × 384.00000] [3 × (1) – 2 × (3)] |
| 372.5 | 7th | [4 × 380.00000 – 3 × 382.50000] [4 × (1) – 3 × (2)] |
| 375 | 5th | [3 × 380.00000 – 2 × 382.50000] [3 × (1) – 2 × (2)] |
| 376 | 3rd | [2 × 380.00000 – 1 × 384.00000] [2 × (1) – 1 × (3)] |
| 376.5 | 9th | [5 × 382.50000 – 4 × 384.00000] [5 × (2) – 4 × (3)] |
| 377.5 | 3rd | [2 × 380.00000 – 1 × 382.50000] [2 × (1) – 1 × (2)] |
| 378 | 7th | [4 × 382.50000 – 3 × 384.00000] [4 × (2) – 3 × (3)] |
| 379.5 | 5th | [3 × 382.50000 – 2 × 384.00000] [3 × (2) – 2 × (3)] |
| 381 | 3rd | [2 × 382.50000 – 1 × 384.00000] [2 × (2) – 1 × (3)] |
| 385 | 3rd | [2 × 382.50000 – 1 × 380.00000] [2 × (2) – 1 × (1)] |
| 385.5 | 3rd | [2 × 384.00000 – 1 × 382.50000] [2 × (3) – 1 × (2)] |
| 387 | 5th | [3 × 384.00000 – 2 × 382.50000] [3 × (3) – 2 × (2)] |
| 387.5 | 5th | [3 × 382.50000 – 2 × 380.00000] [3 × (2) – 2 × (1)] |
| 388 | 3rd | [2 × 384.00000 – 1 × 380.00000] [2 × (3) – 1 × (1)] |
| 388.5 | 7th | [4 × 384.00000 – 3 × 382.50000] [4 × (3) – 3 × (2)] |
| 390 | 7th | [4 × 382.50000 – 3 × 380.00000] [4 × (2) – 3 × (1)] |
| 390 | 9th | [5 × 384.00000 – 4 × 382.50000] [5 × (3) – 4 × (2)] |
| 392 | 5th | [3 × 384.00000 – 2 × 380.00000] [3 × (3) – 2 × (1)] |
| 392.5 | 9th | [5 × 382.50000 – 4 × 380.00000] [5 × (2) – 4 × (1)] |
| 396 | 7th | [4 × 384.00000 – 3 × 380.00000] [4 × (3) – 3 × (1)] |
| 400 | 9th | [5 × 384.00000 – 4 × 380.00000] [5 × (3) – 4 × (1)] |

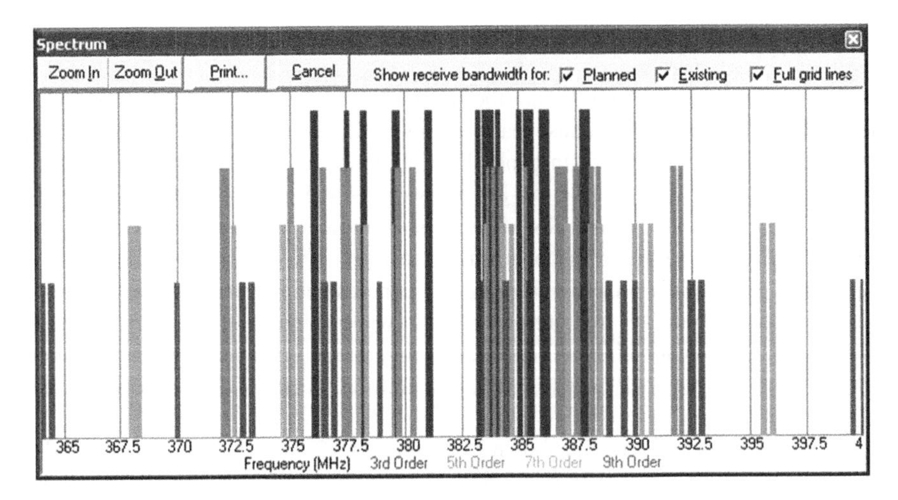


**Figure 14.13.** Inter-modulation products of five frequencies, restricted to band of interest.

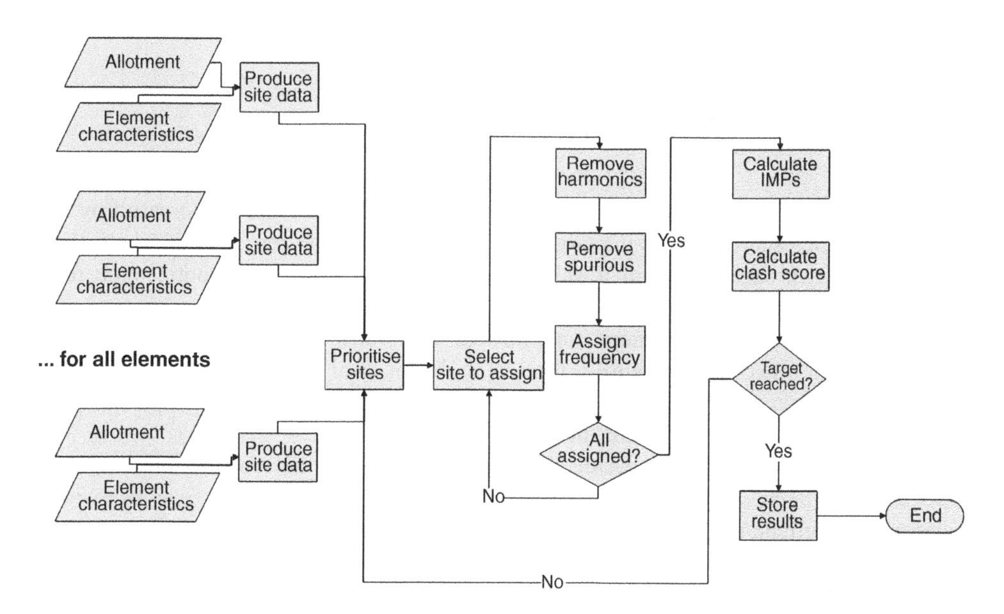


**Figure 14.14.** Simplified automatic co-site assignment process.

has to be used in order to allow automatic assignment in this case, as illustrated in Figure 14.14 which shows a simplified process diagram.

Initially, allotments will have to be generated on a per-assignment basis, to include the channels available to each assignment after a far-site analysis has been carried out. The characteristics of each element are also included, to identify frequencies that cannot be used with that type of radio. This may further constrain the channels in the individual allotment, and it also adds rules for the later calculation of the clash score. This is performed for every element in the assignment. For elements present at the co-site location but not to be re-assigned, the allotment is essentially the channel or channels occupied but marked as not to be changed. Again, their equipment characteristics also have to be considered in the assignment process.

Sites are prioritised for assignment if needed by the assignment process, and then each station is assigned. Where frequencies have already been assigned to elements, the harmonics and resulting spurii can be removed from the list of available channels. The process will also prevent elements being assigned frequencies already given to another element. It is also possible to consider the inter-modulation products of elements already assigned at this point and remove them from the available list, but this will add significantly to the overall computation time of the algorithm. An assignment is made from the list of available channels in the allotment for that site, and this whole process is repeated until each element on the site is assigned. After this, the IMPs are calculated and then the clash score is calculated for the candidate solution.

The clash scoring system will depend on the method adopted, but may be along the following lines:

- Odd harmonics clash up to 3rd order =10.
- Even harmonics clash to up 4th order =1.

- Third-order IMP clash =5.
- Fifth order IMP clash =3.
- Even IMP clash up to 4th order =1.

and so forth.

Each time an assignment violates one of the scores, the total score is increased by that amount. If the total score exceeds the target score, which would be zero in an ideal situation, then the process is repeated until a valid configuration is obtained or until some other termination condition arises.

A practical process may be more complicated, with rules for frequency selection and candidate scheme selection to optimise the search for an ideal solution, and rules covering different levels of co-site based on system separation, but the overall form of the solution is likely to be similar to that shown in the diagram.

Co-site assignment and engineering is a complex process that can tax the engineer, particularly for large systems. Useful, but somewhat dated, information can be seen in reference [1], particularly with regard to inter-modulation products.

## 14.7 Assignment Methods for Mobile Units

For the assignment of networks with fixed infrastructure, the base station locations and technical parameters are always known explicitly. This is not the case for mobile networks with no fixed infrastructure, or where elements will move during the life of the assignment. This is often the case for military networks and also for systems that are temporarily established for a large event or incident. If it is not possible to know where elements will be, then traditional coverage plots, interference analysis and assignment cannot be done, and an alternative approach must be taken.

Some examples of where this situation may arise include the following examples:

- Troops are to be used to protect civilian convoys operating through a range of routes. Should bandits attack the convoys, then the troops must be able to communicate to one another to call in reinforcements. There is no way of knowing where the bandits might attack along the route, so available spectrum must be protected for the whole length of each route.
- Special Forces troops are to be deployed in a particular operational area, and the frequencies assigned to them must be protected from other users. Operational security will preclude other forces knowing where the Special Forces are operating, thus their frequencies must be protected over their entire operational area.
- Forces are deployed to protect a large airfield and associated fuel dumps, hangers and the surrounding hills. The troops must be able to investigate activity over any of the area, including the nearby hills, and their communications links must support this.
- Tank battalions are in a known location and have target locations to achieve within given timescales. Their actual route may vary according to contact with the enemy and other circumstances, but they have a system that reports their locations constantly. Thus their positions are always known, but the assignment has to be valid for the entire mission.

- An unmanned airborne vehicle (UAV) is to provide video imagery over an area devastated by a natural event. The position of the UAV is known at each instant, but the assignment for the control channels must be protected over the entire operation.
- Kidnappers have taken several people hostage and are demanding a ransom to be paid. Police are desperately trying to find the location of the kidnappers. When they do, one option will be to storm the location to free the hostages, given that the group that has taken them hostage is known to be very violent. If so, this will be a major operation that cannot be supported by the normal police radio network. Temporary infrastructure will be necessary for all of the people to be involved, possibly including police officers from many forces, specialist police officers, special forces, ambulance and fire services. Planning for this event has to be carried out, and spectrum reserved for the operation. Since the location will not be known until the kidnappers are located, the spectrum must be reserved over the entire possible range of their location.

Many other examples can be imagined. In each case, it is clear that an area rather than a specific location must be protected. First, a method must be used to define network element behaviour in some constrained way. This can be achieved by linking force location to a polygon rather than an individual location. An example of this is shown in Figure 14.15. Although units are identified by markers in specific locations, they only indicate that they are somewhere within the associated polygon. Three polygons are shown; a large one surrounding the whole operational area, one delimited by a dashed line for the marines and another delimited by a dashed line for the recce (reconnaissance) platoon.

Any of these groups can be anywhere within their operational areas, and their use of assigned frequencies must be protected across the whole area. This means that interfering signals must be lower than the interfering level anywhere within the polygon, so it is not just important that competing networks on the same frequency are not in the same polygon, but

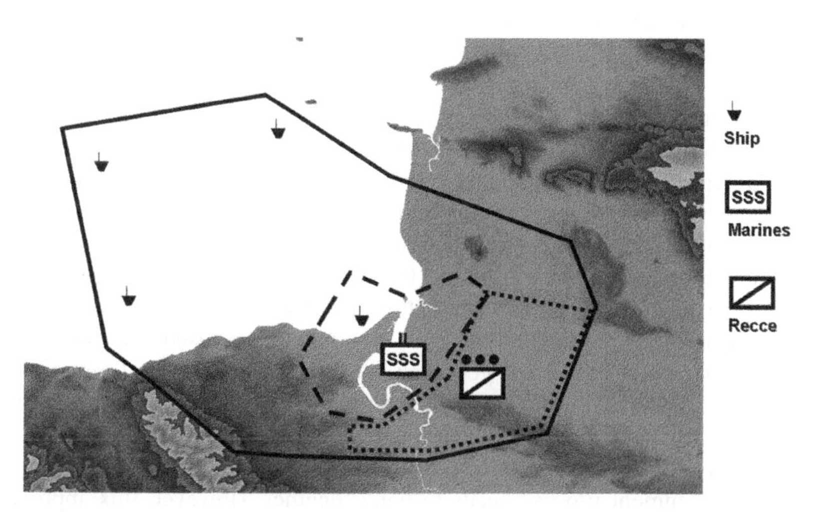


**Figure 14.15.** Mobile subscribers who may move anywhere within a polygon during an assignment, linked to the polygons for the assignment process, rather than fixed locations. Reproduced by permission of P & D Missud.

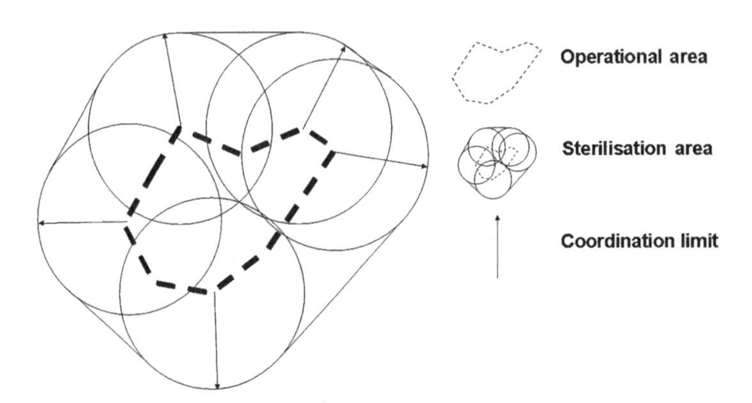


**Figure 14.16.** Sterilisation area and coordination limit; the sterilisation are is the operational area plus the area within the coordination limit of its boundaries. If the coordination limit is large enough, so long as interferers on the same channel are kept out of the sterilization area, then the wanted system should be protected from interference.

they are also far enough away that their transmissions will not interfere even on the edges of the polygon. This leads to the concepts of 'sterilisation area' and 'coordination limit'. This is illustrated in Figure 14.16. This shows the initial operational area for the Marines unit shown in the previous figure.

Around it a 'coordination limit' has been established. This is calculated from the network parameters based on a link budget between the interferer and the maximum tolerable interference limit at the edge of the operational area polygon. It is effectively the link budget solved for distance, to give the minimum distance separation necessary between an interfering transmitter and the edge of the wanted operational area. The coordination limit is used to derive a larger polygon than the operational area, and it is the region enclosed by the original polygon plus all points within the coordination limit distance. This is then the area that must be protected, or 'sterilised' from the wanted assignments being re-used.

Using this approach, the assignment problem is reduced to being a comparison of overlapping polygons; if a new assignment request overlaps the sterilisation area of an existing assignment, that frequency cannot be re-used, otherwise it can.

The comparison of polygons approach is very simple in theory, but very complex in application. For example defining an entire battlefield in this way involves thousands of polygons for the operational areas of all the units within it, plus thousands of fixed locations. Some nets may require that the same assignment is re-used in different polygons in different locations and not linked geographically. The coordination limit is not static, but has to be re-calculated for every interaction between network types, where the parameters change and thus the coordination limit likewise changes. Co-sited elements also need to be assigned according to the co-site methods discussed in the previous section, and the whole assignment process needs to hang together. However, this approach does offer the ability to solve large assignment problems in a timely manner while ensuring that interference is managed appropriately, and it has been implemented in some assignment systems.

## 14.8 Frequency Assignment Deliverables

Frequency assignment deliverables are relatively straightforward, once the work has been completed. The list will normally include:

- A frequency plan, listing all frequencies used, possibly along with a figure for the number of times the frequency is re-used in the assignment.
- List of sites, together with their assigned frequencies.
- Interference statistics, as a metric to identify how good the assignment is in terms of minimising interference or quality of service.

## References and Further Reading

[1] Frequency Engineering in Mobile Radio Bands, William M Pannell C. Eng, MIERE, Granta Technical Editions in association with Pye Telecommunications Ltd

[2] Methods and Algorithms for Radio Channel Assignment, Ed. Robert Leese and Stephen Hurley, Oxford University Press, ISBN-0-19-850314-8.

# 15

# Verification

## 15.1 Validation and Verification

Radio network design and implementation must include a range of checks to ensure that the activity is being performed correctly and that the system delivered at the end of the project will satisfy the original need. There are two main aspects to this activity:

- Validation (are we building the right network?). This is effectively a theoretical exercise to ensure that the design properly reflects the user's expectations (as captured in the URS and downstream of this (see Figure 15.1).
- Verification (are we building the network right?). This is a practical activity to go out and measure actual network performance. Ideally, this will be measured against metrics that best reflect actual user expectation (a formalised form of the following type of questions; can I make successful call? Can I hear the person at the other end?).

Both of these are essential in the network design process. A typical example of how validation and verification fits into the design and implementation process is shown in Figure 15.1.

As discussed in Chapter 6, customer aspirations must be processed into a formal user requirements specification, against which the rest of the design documentation is built. The URS must be fully validated against user aspirations, because it forms the only formal description of what the network is required to provide; customer aspirations not captured in the URS may well be ignored in the rest of the design process. The functional specification (FS) produced must be thoroughly validated against the URS, and the detailed design document and test specification must be thoroughly validated against the FS. It is only if all these steps are completed correctly that the network is likely to meet the customer's eventual satisfaction.

Design metrics and the results of physical measurement surveys are used to verify that the FS is met. Meeting the FS and proof of completion are the only metrics used to judge whether the network meets its requirements in a formal sense. Of course, if the customer then uses the network and find it fails to deliver as expected, even though it has passed the

Mobile Radio Network Design in the VHF and UHF Bands: A Practical Approach
*Adrian W. Graham, Nicholas C. Kirkman and Peter M. Paul* 

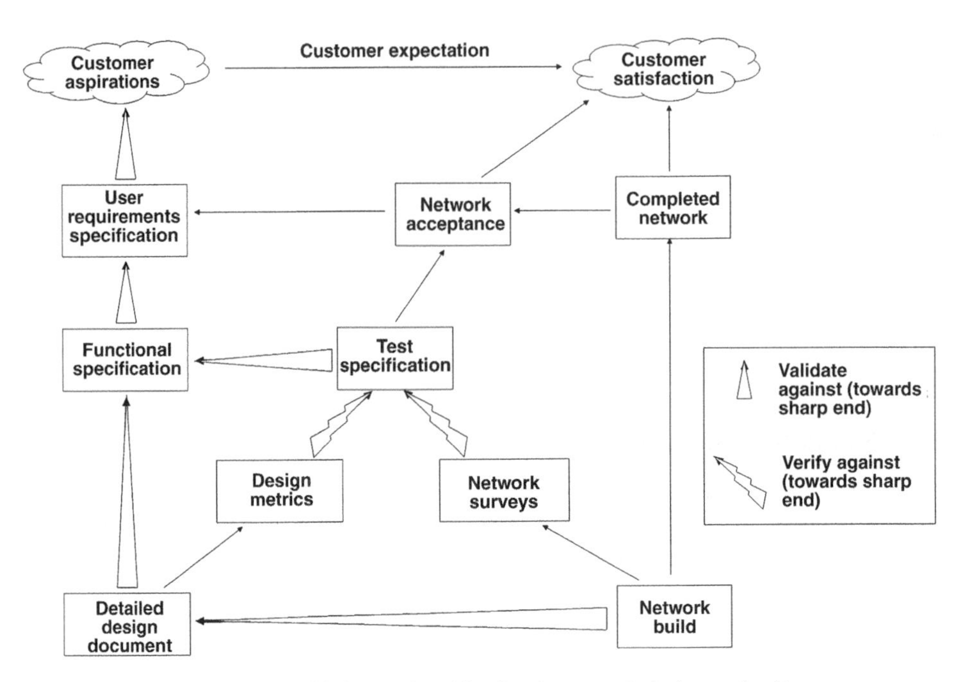


**Figure 15.1.** Validation and verification in network design and rollout.

test specification, then this points to a failure in the process to capture the requirements, convert them into an effective FS, test specification or design. In most cases, it is the first step of capturing all the requirements that proves the sticking point – but this is discovered all too late. This shows the importance of validation to ensure that each step is carried out correctly. Verification should ideally be a simple case of ensuring that the network has indeed been performed correctly.

The methods of validation have been discussed throughout the book, and so has the verification of the design of the network. We will now go on to look at the methods of physically verifying network performance, which normally involves producing physical surveys. We start by introducing the different types of survey that can be performed.

## 15.2 Introduction to Surveys

A survey is a physical measurement campaign used to verify one or more performance metrics of an active network by measuring it within the service area. Typically, a survey system consists of the elements shown in Figure 15.2. For many applications, a vehicle offers the most cost-effective method of surveying large areas of a network. For aeronautical or maritime networks, an aircraft or ship may be necessary to host the survey equipment.

The network need not be in service at the time of measurement for a survey to be carried out, but it must have been built – otherwise there will be nothing to survey. Because of this, surveying typically happens late in a project, unless the project has been structure in such a way that part of the network is available for test before the full system. The point at which a system will be ready for a survey will depend on the project strategy, and some approaches

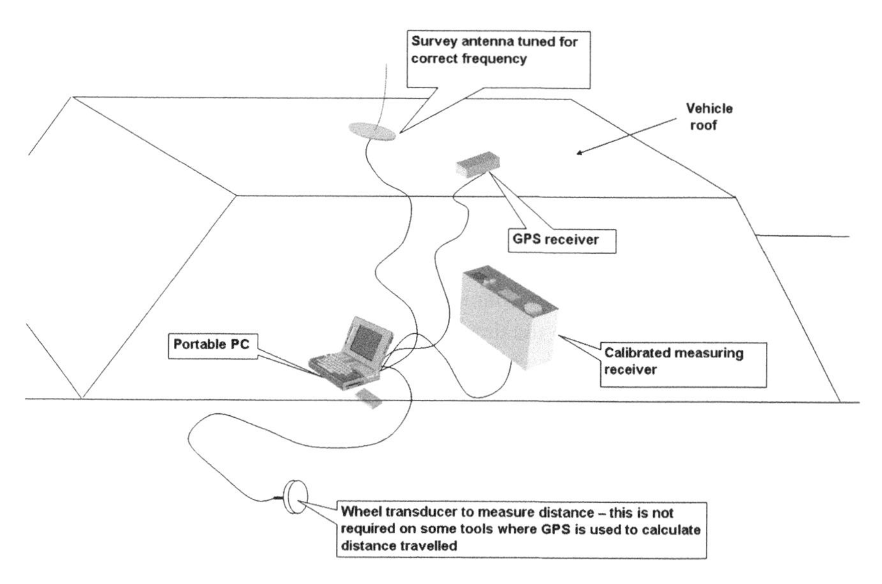


**Figure 15.2.** Typical survey set up, shown against a stylised survey vehicle. The radio and GPS antennas must be on the exterior of the vehicle, the other elements will normally be inside.

will provide measurable results before others. For example the following project strategies are possible:

- Test system; in which a test system is built very early in the project, and the results of surveys are fed into the rest of the design process. This can be used to calibrate propagation models, evaluate some aspects of the actual performance of the proposed technology and identify some project risks early in the project. It is likely that the test system will be limited in functionality (perhaps to engineering functions only) and thus not all risks will be identified. This is a reasonable strategy for fairly high-risk designs, since it can help to define the risks involved so that mitigating actions can be taken. However, since the test system will normally be abandoned, it is typically higher cost and thus it is not so desirable approach for less-risky network designs.
- Pilot system; in which a fully functioning subset of the eventual network is built, with the intention either of abandoning the pilot once it has served its purpose or, ideally, building the rest of the network around it. This is more expensive than producing a test system, but it will typically allow more of the system performance to be measured. This will usually not be limited to engineering data but will ideally include real-life users and their traffic. The pilot system approach can allow the measurement of the system on an end-to-end basis, which is highly useful. Pilot systems are expensive and if they take time to build and evaluate, they can delay overall network rollout. They are appropriate for particularly high-risk networks but in all other cases, the justification for their use is limited.
- Traditional design; in which the network rollout commences after design has completed, and the network is rolled out as late as possible with respect to its in service date (or, put conversely, the in service date is as soon as possible after the network has been rolled out). In this approach, a survey is not possible until late in the design.

The type of survey possible will be dependent on the system available to be surveyed. Surveys fall into three general categories:

- Calibration survey.
- Engineering survey.
- Network performance surveys.

A calibration survey is carried out as early as possible, using a test or pilot system or, if available, another system with similar characteristics. A calibration survey is often used to tune propagation models and inform the design process, and they are best performed as soon as possible in the project. Often, a calibration survey will measure the received signal strength indication (RSSI) value rather than more sophisticated metrics; this is far from ideal but has to be tolerated on occasion. Figure 15.3 illustrates a typical scenario. The entire service area is shown using the outline dashed area. The start and end points of the survey are shown by a cross, and the survey route taken is shown as the solid line. For this particular survey, only a part of the network has been tested. It is likely that this survey would be part of a survey campaign to cover the whole service area.

Also shown in the figure is an example of the typical type of results likely to be obtained from measurement. This shows the processed measured data (see Section 15.3 below) rather than raw figures; hence the variations are not as variable as instantaneous measurements would be. The processed data can be compared to predictions carried out in a planning tool, and the overall mean error and standard deviation can be calculated between the measured and predicted values. This can be used to tune the propagation model directly, but if

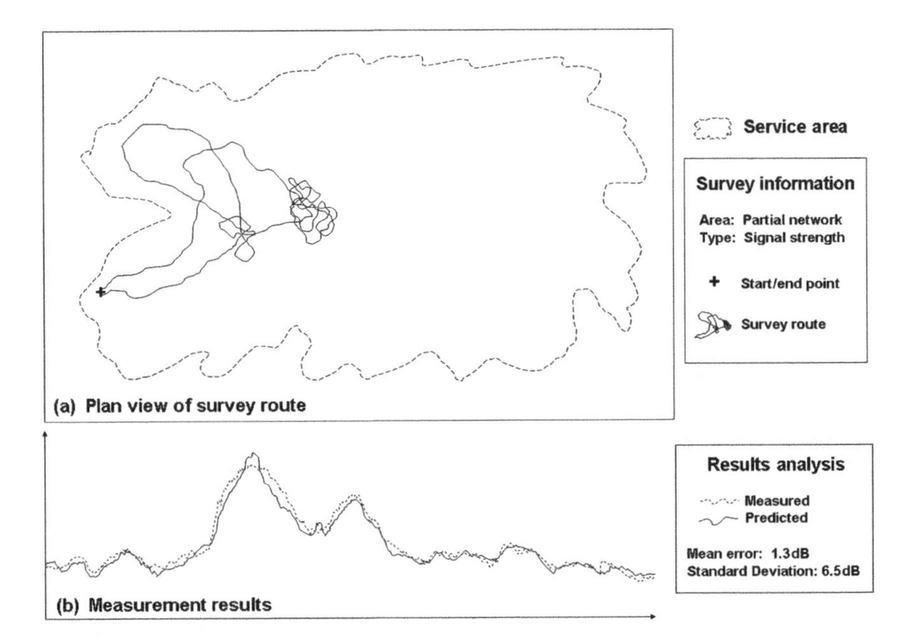


**Figure 15.3.** Survey information and results. Reproduced by permission of Radio Systems Information Ltd.

**Table 15.1.** Illustration of effects of tuning a propagation model using data from a calibration survey.

| | Before tuning | | After tuning | | | |
|---|---|---|---|---|---|---|
| Category | Mean error (dB) | Standard deviation (dB) | Mean error (dB) | Mean error improvement (dB) | Standard deviation (dB) | Standard deviation improvement (dB) |
| Rural–open | 1.5 | 5.6 | 1.5 | 0 | 5.0 | 0.6 |
| Suburban | −2.5 | 7.3 | −0.4 | 2.1 | 6.9 | 0.4 |
| Urban–light | −3.1 | 6.6 | −2.2 | 1.1 | 6.1 | 0.5 |
| Urban–mid | −4.1 | 5.5 | 1.1 | 3.0 | 5.7 | 0.2 |
| Urban–heavy | −1.0 | 7.0 | 0.9 | 0.1 | 6.2 | 0.8 |
| Forest | 3.7 | 4.3 | 1.2 | −0.6 | 4.5 | −0.2 |
| **All** | **1.3** | **6.5** | **0.5** | **0.8** | **4.9** | **1.6** |

information about the different mobile environments can also be identified, then it is possible to tune further options, such as the attenuation applied to each clutter category. This can be modified to improve the mean error and standard deviation in each category as well as for the overall accuracy.

Engineering surveys are used during the build process to ensure that the network is functioning as expected when it is being rolled out, and also when the network is in service to probe problems and prove modifications, enhancements and extensions to the original network. Engineering surveys will often measure metrics such as received signal strength and handovers. Figure 15.4 shows an example where a survey has been performed to

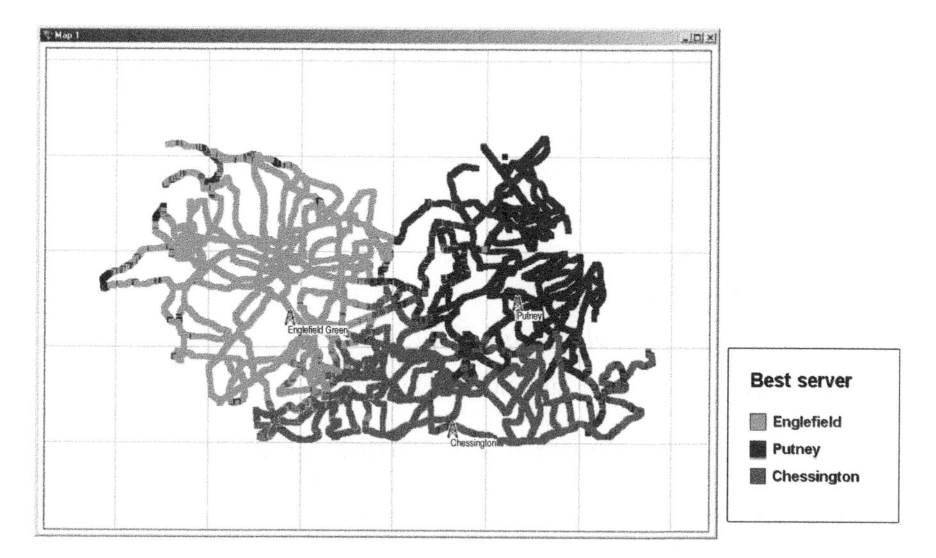


**Figure 15.4.** A typical engineering output – a best server plot based on measurement. Reproduced by permission of Radio Systems Information Ltd.

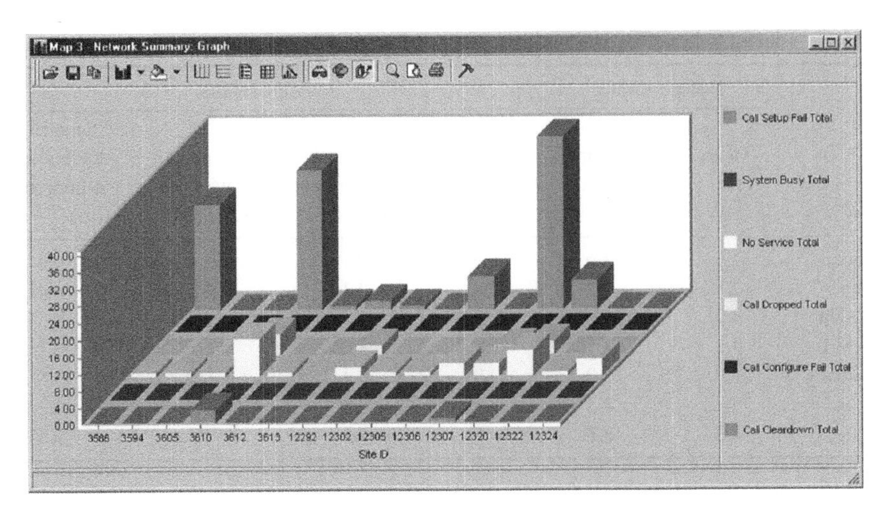


**Figure 15.5.** Results of a network performance survey, in this case call metrics. Reproduced by permission of Radio Systems Information Ltd.

determine which base station provides the strongest signal in each area surveyed. The survey route shown has been produced by driving along many roads in the survey area.

Network performance surveys are used to prove the network performance for contractual purposes and can also be routinely carried out as a health check during the life of the network. These types of surveys will typically be aimed at high-level parameters such as speech quality, bit or packet error rates (or an even higher level data metrics) and call success rates. An example is shown in Figure 15.5, which shows the results of a survey to determine call statistics such as call set-up failure, dropped calls and so forth. This type of survey can be meaningful only when the network is in operation because it relies on the presence of other users, whose use of the network will affect the survey results.

Whatever type of survey is being performed, it is crucial that the survey is statistically valid and that the results are repeatable. Unfortunately, many surveys are carried out without an understanding of the methods that must be adopted to ensure that the survey is valid, and we look at some of the key areas next.

## 15.3 Survey Fundamentals

### *15.3.1 Capturing Statistically Valid Data*

There are three aspects to capturing statistically valid data:

- Capturing enough data to represent the median field strength in the short sector effectively and not just the instantaneous value during a single sample.
- Surveying a sufficient amount of the network to be able to extrapolate the survey results to the performance of the entire network.
- Ensuring that the assessment of the network represents it on all (or a stated number of) days in the year.

We will look at capturing valid data and the processes of setting up an effective survey route shortly. Time variations will occur for radio propagation, particularly for networks with long links, such as maritime and aeronautical systems, and a single survey will not capture this variation. This means that it may well be necessary to perform many surveys before an understanding of network performance is achieved. Traffic will also be time variable, and this also has to be taken into account in network performance survey strategies.

We now go on to look at the mobile signal and how it can be measured accurately.

### *15.3.2 The Mobile Signal*

As we saw in Chapter 5, radio signals very rarely travel from a base station to a mobile by a single direct route. In all practical situations signals bounce off terrain and clutter and arrive at the receiver from different directions after taking many different routes. This situation causes fast fading and produces deep fades at half-wavelength intervals so that by moving by only 1 m might cause a receiver to experience a 20 dB change in signal strength. To make matters worse, the pattern of fading is not constant and varies with local conditions such as vehicles moving nearby. If we measure instantaneous received signal levels, then the level will change according to these factors even without moving the receiver. Even if the measuring equipment is sufficiently fast to detect every fade, the conditions that created the fade will have changed. This means that it will be impossible to repeat the measurement, and unrepeatable measurements are not statistically useful. All of this means that instantaneous spot measurements are useless on their own.

The situation is illustrated in Figure 15.6. This shows the signal level for a short sector. The percentage of time for which a particular signal level is exceeded over the entire sector is shown on the right hand side. The purpose of the survey is *not to measure the level of the signal at each instant* but rather to identify the level at which the signal is exceeded for 50 %

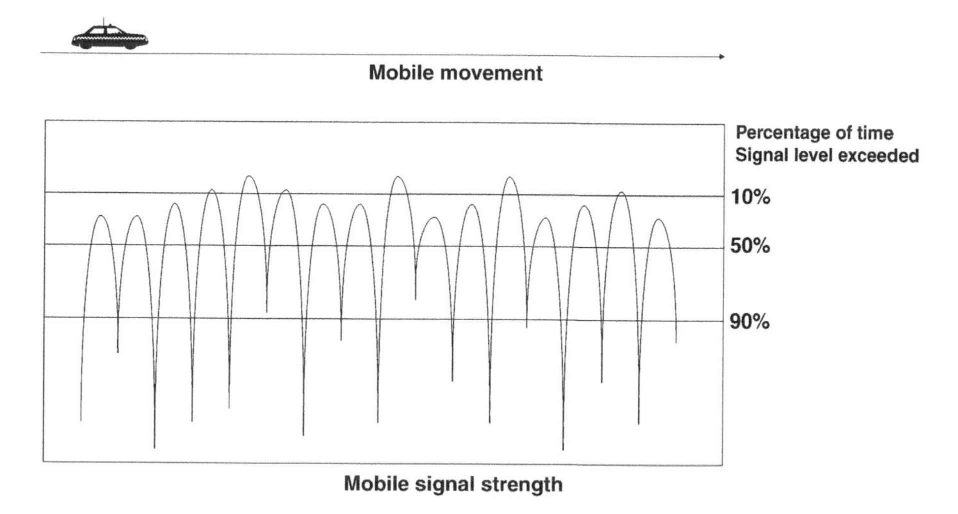


**Figure 15.6.** Mobile signal as the receiver moves, showing typical fades. The percentage of time a particular signal level is exceeded is shown on the right hand side.

of the time over the length of this path. This is a far more stable metric and, since as we saw in Chapter 5, if we know the 50 % value, we can determine the availability level for different percentage of times or locations for the mobile since the signal variation will typically behave according to a known statistical distribution. For mobile networks, Rayleigh statistics are typical.

The measurement metric wanted will depend on the activity being performed. For example for direct comparison to median field strength predicted by radio propagation predictions, a value of 50 % (the median value) would be appropriate. For a usable signal level for subscribers, the value may be 90 or 95 %, and for interference, the value may be 10 % or less. For whatever the wanted case, the level can be expressed in terms of a field strength value wanted for a given percentage of locations and times. For example it might be '20 dBμV/m at a 95 % locations/50 % time confidence level'. This means that there would be a 95 % probability of measuring a signal level of 20 dBμV/m or higher at any location fulfilling this requirement for 50 % of the time. This can be further extended to areas by defining a percentage of area that must meet this requirement, such as '95 % of the service area should receive a signal level of 20 dBμV/m at a 95 % locations/50 % time confidence level'. This is then the type of metric that can be generated during a survey campaign and compared against the FS.

## *15.3.3 Sampling Rate*

Consider the situation where only one instantaneous measurement is taken in an attempt to characterise the median field strength in a given short sector. A single sample value will be obtained. If we repeat this process, other sample values will be obtained. Figure 15.7 shows four separate instantaneous measurements being taken in the same short sector. Each measurement is different, but they all share one common characteristic; they do not in

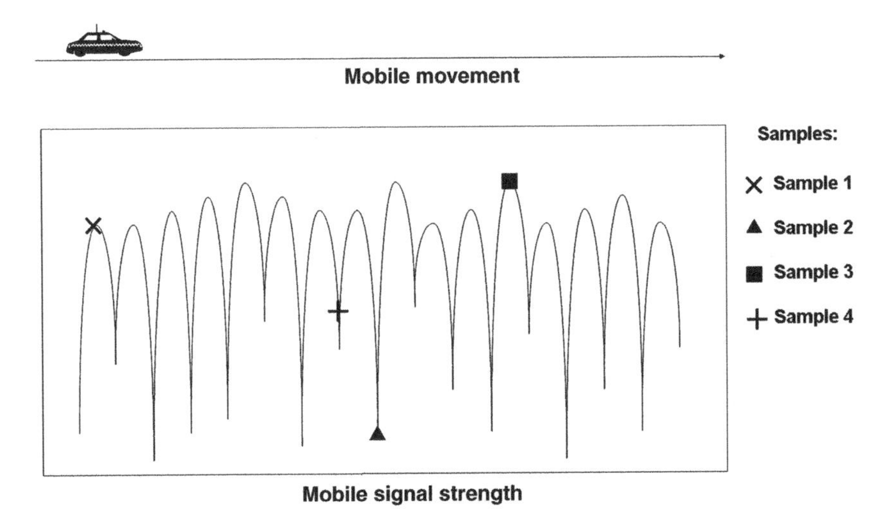


**Figure 15.7.** Inadequate sampling or signal; the measurements taken to not capture the characteristics of the signal.

any meaningful way represent the median value. Measurements done in this way are misleading and worse, flat wrong. The real value can only be derived by averaging a number of random samples within the short sector.

The question is how many spot samples must be taken over the short sector in order to characterise it to the level of accuracy? If too few samples are taken then the calculation will be inaccurate for the reasons given, while the receiver scan speed sets an upper limit to how many samples can be taken at a given vehicle speed and sampling distance. Also, since the number of samples that can be taken per second may be a metric that affects the cost of the receiver used, we need to determine what values are reasonable to avoid having to buy or build an over specified receiver system.

W.C.Y. Lee performed some very useful work that demonstrated that, as expected, the more samples that are taken, the more accurate the results are. His work included the following theoretical figures for 90 % confidence:

This graph shows that, for example, taking 50 samples over the sample gap will give a 90 % statistical confidence of the measurement being within 1.0 dB of the true median value. This is probably good enough for most network design activities, and it is difficult to improve upon.

As long as a reasonable number of samples are taken over an appropriate sampling distance (by not driving the survey vehicle too fast) then the results obtained will be valid and, most importantly, will be repeatable as the effect of individual fades are not affecting the results. Note also that taking more than about 50 samples over a sample gap distance does not significantly improve the accuracy, although it does no harm. Note also that taking as few as 12 samples over a sample gap distance is not wrong – it merely reduces the accuracy of the result. As long as that is taken account of in the analysis, the results can still be meaningful.

An important point to note is that the samples used to calculate the signal level over the sample gap distance must be instantaneous signal levels and must not have been averaged by the measuring receiver, as this will remove any information on the depth of the fading.

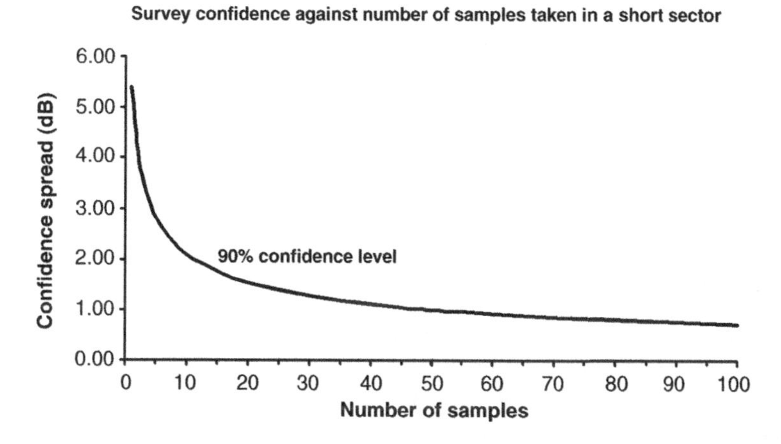


**Figure 15.8.** Measurement accuracy against number of samples taken. Reproduced by permission of Radio Systems Information Ltd.

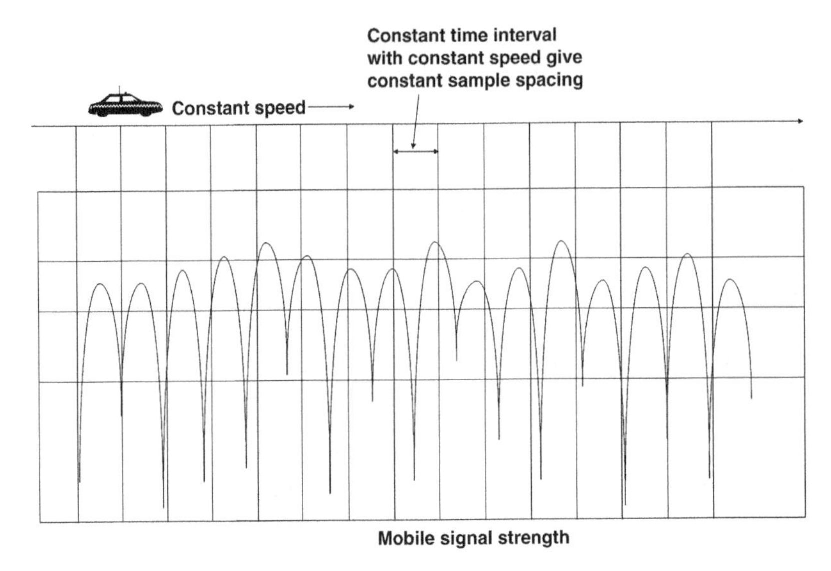


**Figure 15.9.** If a survey systems samples at a constant rate and the vehicle is travelling at a constant speed, then the sampling interval along the short sector is constant, as shown.

Additionally, the number of samples taken is not sufficient to guarantee the performance shown; the samples must be randomly selected and not correlate with one another.

Suppose, for example that we are using a survey system that measures one sample in a defined time slot. In this case, vehicle speed is not considered. If the vehicle is travelling at a constant speed, as shown in Figure 15.9, then the samples are taken over an equally spaced distance. This is good because it means that the individual measurements are de-correlated and thus will allow the median value to be derived through averaging.

Imagine, however, that the survey vehicle comes to a halt at a traffic signal, as illustrated in Figure 15.10. In this case the survey system keeps measuring even though vehicle is stationary; it effectively measures the same point (or at least points too close together to be de-correlated) over and over again. This will affect the average value obtained and thus the median value will be wrong.

This kind of problem can be overcome only in one of two ways; either the survey vehicle must move at a constant speed or the survey equipment must be capable to determining the movement of the survey vehicle and only taking a sample once the vehicle has moved far enough, however, long this takes.

### *15.3.4 Sample Gap Distance*

In Figure 15.6 we showed an example of signal variation over a short sector. If we extended the measurement beyond short sector distance, we can expect the median field strength to change because the propagation path will change as will the effects of terrain and clutter. The question is, therefore, how long should that distance be? The survey distance must be a large enough area to ensure that individual fades are not affecting the result, but small enough to ensure that local conditions do not change. W.C.Y. Lee has shown that

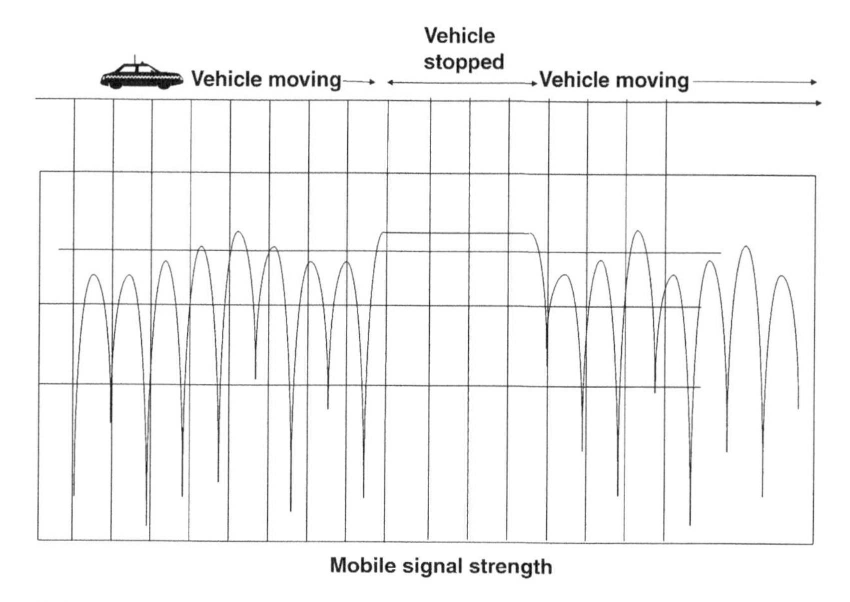


**Figure 15.10.** If a survey systems samples at a constant rate and the vehicle slows down or stops, Then the sampling will become skewed.

this distance (also known as the sample gap distance) affects the accuracy as shown in Figure 15.11.

This graph shows that measuring the median signal strength with a sample gap distance of, say, 40 wavelengths will give a 95 % probability (2 sigma) of a measurement being within 2.1 dB of the true median and a 68 % probability (1 sigma) of it being within 1.0 dB.

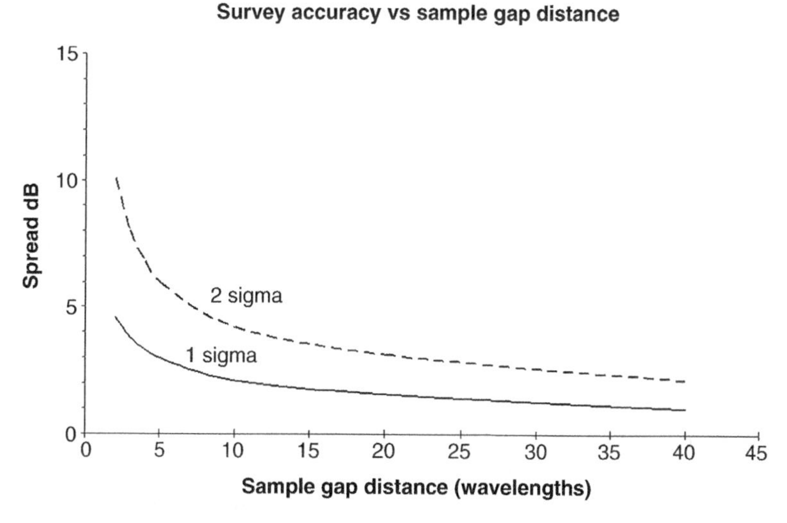


**Figure 15.11.** Sample gap distance.

This is the inherent inaccuracy caused by an unpredictable fading signal superimposed on the background median signal strength.

There is little improvement in accuracy by choosing sample gap distances greater than about 40 wavelengths, while below 20 wavelengths the uncertainty spread starts to increase rapidly. Therefore a measurement interval of 40 wavelengths is a good compromise between the result being affected by individual fades and smoothing out the long-term fading by using too large distance. A sample gap distance of 40 wavelengths corresponds to the following distances for the frequencies shown:

150 MHz - 80 m

450 MHz - 27 m

900 MHz - 13 m

In general the sample gap distance should be kept below about 100 m to avoid changes in terrain affecting the median signal level for VHF and above.

### *15.3.5 Selecting a Measurement Receiver*

It should be clear that surveying must be performed correctly in order to provide useful data. This not only includes consideration of the sampling gap distance, number of samples to take within that distance and distribution of those samples, but also ensuring that the measurement receiver is capable of performing the required measurements. Unfortunately, not all receivers at the market are ideal for survey work. It is worth considering a few key points about measurement receivers, so that statistically valid measurements can be taken. These include the following:

- In all cases, the receiver should be fully calibrated. This includes periodic assessment to ensure that calibration has not been lost over time.
- The receiver should take instantaneous samples as fast as possible, but note that more than 50 samples per sample gap distance results in little improvement in accuracy. A good speed is 1 sample per frequency every 5 ms; faster receivers allow more frequencies to be surveyed simultaneously while still surveying each frequency with sufficient sample speed.
- The trigger point for taking a sample may be may be time based or distance based. Distance triggering is better as it avoids biasing due to vehicle speed; however the effects are minimal when time triggered samples are used (e.g. every 5 ms) with typical distance sample gaps (e.g. every 50 m).
- Some receiver systems use wheel transducers to measure distance. These can be used to trigger instantaneous samples (e.g. every 5 cm) or to measure the sample gap distance (e.g. every 50 m). Wheel transducers have the advantage that they still work even if GPS is not available (e.g. for indoor surveys or in tunnels). However, if a time triggered receiver is being used with a GPS receiver then the wheel transducer is not required as the sample gap distance can be derived from the GPS.
- The limiting factor in survey tool design is often the communications link between the receiver and PC. Values for every raw sample are passed to the PC for calculation of

the percentage confidence levels – leading to data transfer rates of megabytes per second in some situations. Therefore some receivers will calculate the percentage confidence levels internally and only pass the final value for each sample gap. This reduces the amount of data needing to be analysed at the PC but limits the potential for further post-survey analysis.

If the receiver passes the raw data to the PC for analysis then it should *never* carry out any averaging or smoothing of the samples. This is because averaging will remove information on the depth of fading, making it impossible to derive high percentage confidence levels from the raw data.

### *15.3.6 Surveying Fundamentals Summary*

Surveying is not a trivial task and it requires knowledge of planning and performing surveys that are statistically valid. Unfortunately, there is still a significant lack of understanding of these principles in the radio-planning domain, and this does lead to inaccurate and misleading surveys being performed on a fairly regular basis. This is not only bad in itself, but it often undermines genuine information obtained from good network design principles; some people prefer bad measurements to good theory, with disastrous results for the project.

In this section we have highlighted some of the key aspects of surveying. We close this section by examining some actual survey data in the UHF band, and looking at different sampling methods. This is illustrated in Figures 15.12 and 15.13, both of which refer to the same short period within a particular survey.

Figure 15.12a shows the survey vehicle speed during this part of the survey. The survey was taken in an urban environment with traffic. There are two points in this section where the survey vehicle is stopped. Figure 15.12b shows the raw samples taken during this period. Notice the number of deep fades, and the general signal variability due to fast fading. The receiver used for the survey was capable of taking a large number of samples in the sample gap distance and therefore the data is very useful for performance analysis of the network.

Figure 15.13a shows the effect of trying to take a single sample per second. If the value taken during the sample is compared to the raw data, it is easy to see that the samples do not track the raw data; in fact, they are random snapshots with little real meaning. It is impossible to reconstruct the important characteristics of the raw data from these samples.

Figure 15.13b shows a comparison between the raw data and calculated median values based on averaging over 1 s and over 50 m. Both versions track the raw signal characteristics far better than the example shown in 15.12b. Of the two, the distance-based method is preferable because the processed data can be normalised in distance – in other words, the speed of the vehicle is irrelevant and it is possible to store one median measurement per 50-m distance. This is easier to use when comparing to predicted values and it is also more useful for characterising performance over the entire survey route.

To illustrate this important concept, think of a slightly ridiculous example; suppose a distance of 500 m is to be measured. The survey system features the ability to trigger samples both by time and by distance (50 m). If the vehicle starts, gets half way, stops for several minutes and then starts again to complete the survey. The distance-triggered system is fine; it has 10 average values, equally spaced along the survey distance and each measurement is equally important. The time-based-triggered measurements feature similar

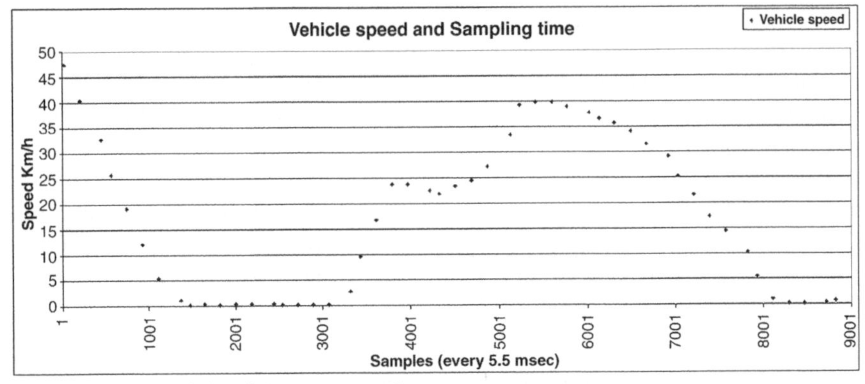


(a) Vehicle speed during data measurement

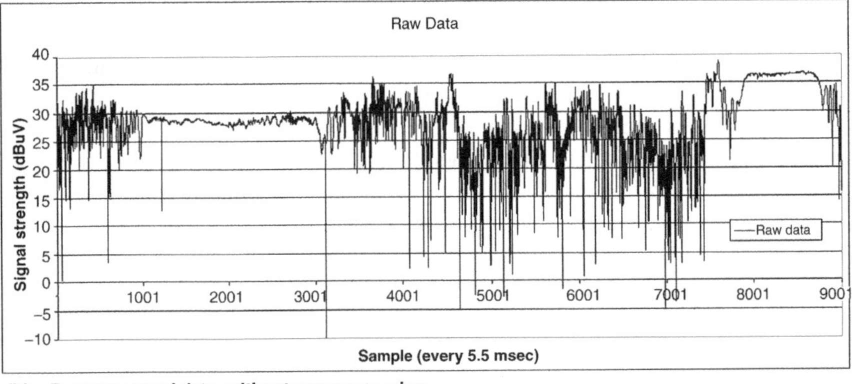


(b) Raw measured data, without any processing

**Figure 15.12.** Measurement example. Reproduced by permission of Radio Systems Information Ltd.

measurements for the time when the vehicle is moving, and also lots of measurements taken when the vehicle is stopped; these are not only superfluous, but must also be removed to avoid skewing the survey results. This is clearly not as desirable as the results of the distance-normalised measurements taken by the distance-triggered system.

The important point is surveys should not be undertaken lightly and great thought must be put into making sure they will provide useful, accurate data.

## 15.4 Digital Surveying

Digital networks need a different approach to surveying from analogue networks because the basic signal strength is not always a good indicator of network performance. Secondary effects such as multi-path propagation or badly configured handover algorithms can often become the main factor affecting network performance. Even bit error rate (BER), if it can be measured, is not always a good indicator of network performance. The general solution is to measure the actual parameter that the end-user perceives as quality, e.g. speech on a

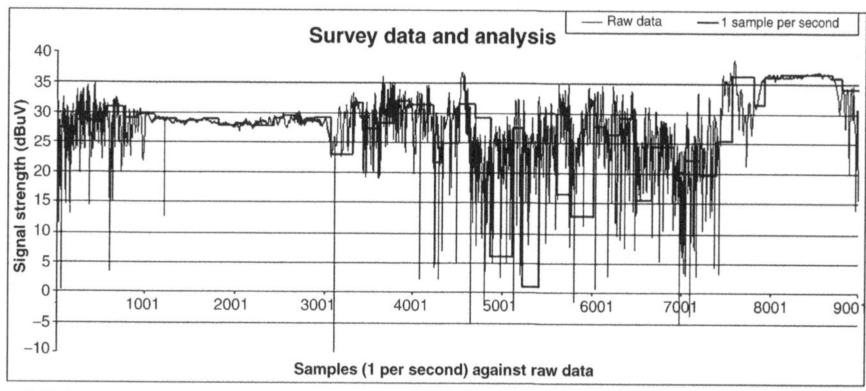


**(a) Data sampled at 1 sample per second against raw data. Note that the samples do not properly represent the raw data, due to insufficient sampling rate.**

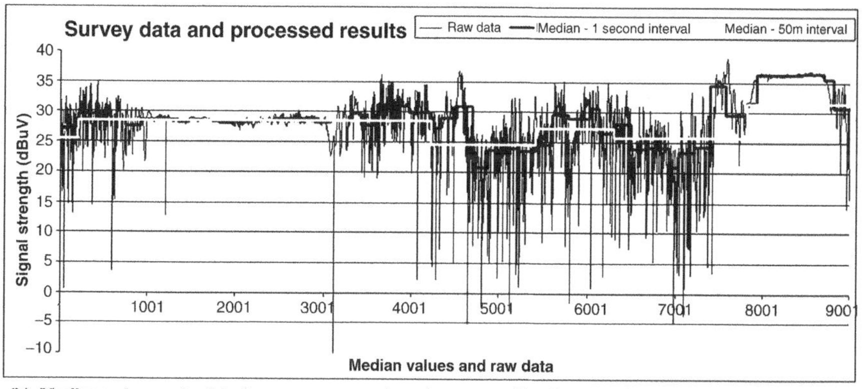


**(b) Median values calculated once per second, and once per 50m interval. Both are better than the data shown in (c), but the 50m interval version is better because it accounts for changes in vehicle speed and more easily allows results to be further processed.**

**Figure 15.13.** Sampling comparisons. Reproduced by permission of Radio Systems Information Ltd.

speech network. This then accounts for the digital coding flywheeling over short duration fades.

The key parameters that affect the user's perception of a network are the following:

- Speech quality.
- Reliable call set-up without call dropping.

Speech quality can now be measured objectively using algorithms that compare a received speech sample with a master copy of the sample and report the result as a mean opinion score (MOS) value on a scale of 0–5. The best known of these algorithms is PESQ (perceptual evaluation of speech quality) which is now an ITU standard (recommendations ITU-R P.862 and ITU-R P.862.1). This approach involves transmitting short sections of natural speech over the network repeatedly while performing the survey, and then using

**Table 15.2.** Typical PESQ scores.

| System | Typical maximum PESQ MOS |
|---|---|
| PSTN | 4.0 |
| GSM | 3.5 |
| TETRA | 3.1 |
| Typical limit of acceptability | 2.1 |

automatic methods to determine speech quality. The quality is represented by a score, and the scores shown in Table 15.2 are typical.

The MOS threshold of acceptability will vary from user to user and will also depend to some extent on the language and specific speech sample used for testing. Therefore a calibration exercise should always be carried out to determine the acceptable threshold for speech quality for each network. Perhaps surprisingly, speech quality is not always directly related to RSSI or BER. This is illustrated in Figures 15.14 and 15.15. These shows the relationship between RSSI and PESQ MOS as measured on a real TETRA network using 4-s speech samples in an area of bad multi-path propagation.

Figure 15.15 shows the relationship between BER and MOS. The data is based on short 4-s speech samples that have been passed through a TETRA channel simulator set for TU50 and TU5 (i.e. simulating a typically urban environment at 50 km/h and 5 km/h). Both male and female voice samples have been used at a limited number of percentage BER levels (0, 1, 2, 3, 4, 6 and 8 %).

These graphs show that whilst there is the loosest of relationships between RSSI and BER against the PESQ MOS – for any given BER or RSSI there is a wide range of MOS values. The most that can be said is that the lower the RSSI or BER, the higher the likelihood of a poor MOS value. The most likely explanation for this is that the equaliser and speech vocoder on a digital network are not linear systems with a predictable response – random errors and burst errors produce different responses, with some errors being corrected and

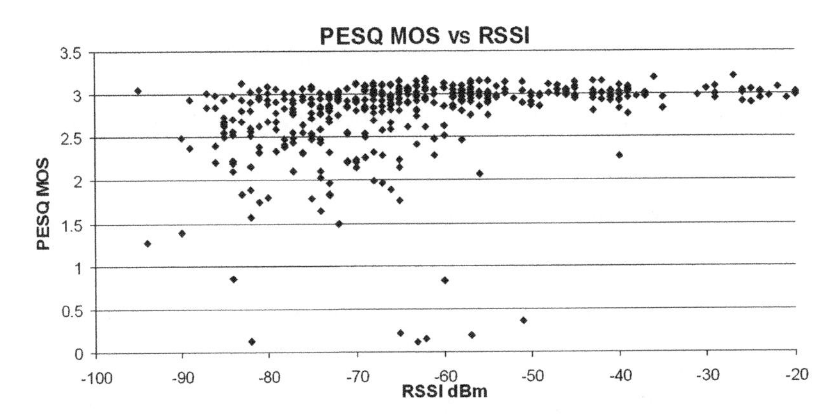


**Figure 15.14.** Comparison of speech quality against RSSI – the relationship is not clear. Reproduced by permission of Radio Systems Information Ltd.

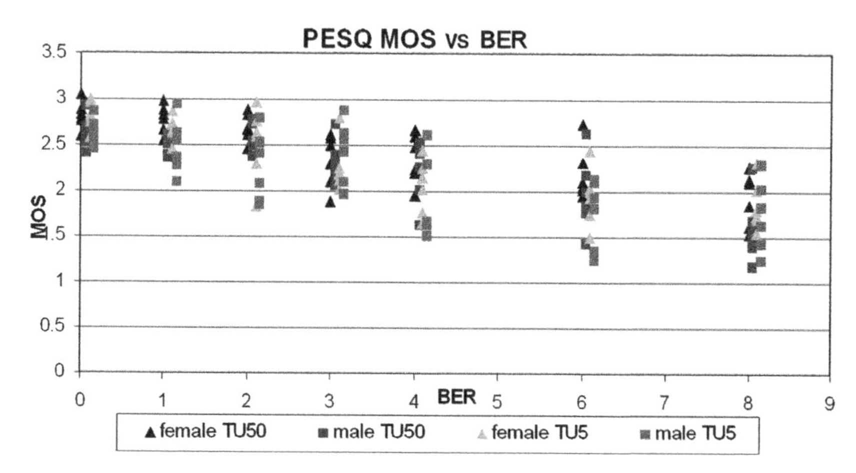


**Figure 15.15.** Comparison of speech quality and BER – only a loose relationship. Reproduced by permission of Radio Systems Information Ltd.

other errors being passed onto the vocoder with different responses depending on the importance of the errored bit. Some errors will produce virtually no audible distortion while others will produce a complete dropout. This implies that judging overall network performance as perceived by the user on either RSSI or BER is not a good idea although it is still important to measure these parameters for an engineering survey. They can also be used to probe network performance in which poor performance has been reported. Engineering surveys are needed when first setting up and configuring the network. The following parameters are usually logged:

- RSSI or signal strength.
- Site ID.
- Handovers.
- Message error rate (MER).

In general, BER is difficult to measure in practice on digital speech networks, as it needs to be measured at the same rate as the raw speech data. This is difficult to do in practice as; for example on TETRA networks it is not possible to gain access to the data transport layer to pass a known bit sequence for testing. Some radio terminals give access to the 'training sequence' set of bits (for TETRA this is 21 bits repeated every frame) but this is a coarse measure and only a broad indicator of channel BER.

In conclusion, measuring speech quality for digital networks is as follows:

- Simple.
- It produces repeatable results.
- It is more statistically valid than using the radio terminal's internal BER measure.
- It requires no special radio terminals.
- It directly relates to user-perceived network quality.
- It produces a metric that is easily understood by nontechnical end-users.

## 15.5 Digital Network Performance Surveys

The previous section explained why we use speech quality as prime metric for network quality measurements. When performing such surveys, the speech sample should be:

- Sufficiently long to provide enough speech sounds for accurate quality assessment.
- Sufficiently short so that the network performance can be assumed to be constant over the distance covered by the survey vehicle.

In practice, a speech sample length of 4 s is about as short as can be used with PESQ while still producing reliable and repeatable results. A typical test call manages to send eight speech samples per 55 s giving an average time of 6.9 s per sample. For statistically valid and repeatable results, a speech sample should be taken every 100-m distance or less – over this distance the network coverage and performance can be assumed to be reasonably constant. This implies that the survey vehicle will need to be limited to a maximum speed of 52.2 km/h (32.4 mph). This is quite slow, particularly when surveying major roads. One simple technique to increase the number of samples taken is to use two or more radio terminals each operating independently (but in this case it is necessary to be careful of mutual interference between the radio terminals). The test call can be configured to test uplink and downlink alternately every other speech sample, to more closely simulate the way a real user makes a call, or can just test in one direction.

## 15.6 Planning a Survey Campaign

Any survey must be constructed against a suitable test plan, otherwise the results will not be as useful as they could be. For formal surveys, the characteristics of the survey will effectively have been established by the contractual requirements or if not, the main characteristics should have been set by the conditions in the TS.

For any other survey, a survey plan to gain an understanding of the full network performance should address the following aspects:

- The survey route should be representative of the whole service area. This will mean the survey should include some measurements from all areas.
- The survey should include measurements from all types of environment that subscribers are likely to encounter, including urban environments and representative rural environments.
- The percentage of the total service area will depend on the confidence required in the results. For high degrees of confidence, and for contractual surveys, it may well be necessary to survey a high percentage of the roads in the service area; perhaps all major ones.

It is also necessary to determine the fundamental aspects of the survey, such as:

- The number of frequencies to be measured and the number of radio terminals to be used in the survey.
- The sample gap distance to be used for analysis and for configuration of the receiver.

- The number of samples to be taken per sample gap distance.
- The speech sample length for digital voice networks.
- The distance to be covered per speech sample.

This can be used to calculate the maximum vehicle speed, as illustrated in the previous section. Other practical aspects will also have to be considered, and some of these are described next.

### *15.6.1 Vehicle Surveying*

When fitting the survey system to a vehicle, there are some important points to consider:

- Normally, a centre-mounted (typically the middle of the vehicle roof) 1/4-wave whip will provide the most even radiation pattern. The performance of this system against the actual antennas to be used by network subscribers should be well understood.
- All other antennas on the car should be removed.
- The test radio terminals must be correctly isolated (see below).

All these will improve the performance of the survey and the quality of the results.

### *15.6.2 Pedestrian/Indoor Surveying*

When considering the requirement for either pedestrian or indoor surveys, the first question should always be, 'can this be emulated by a vehicle survey instead?' Pedestrian and indoor surveys are time consuming and require significant effort, so if it is possible to make corrections to the vehicle configuration that will provide similar results, this is generally desirable. Methods of doing this are described in the next section.

It should also be borne in mind that if portable survey equipment is used, then even if a trolley-mounted antenna is used there will probably still be a correction factor to normalise data to a body-worn antenna.

For indoor surveys, tracking the receiver location within the building is always difficult, but some potential solutions are:

- The survey technician can mark locations on the map in real time as he or she moves during the survey – the software can then interpolate positions between these known locations.
- An approach is to treat individual rooms as having more-or-less constant coverage in the room, and aggregate all readings taken in the room to produce a single result per room.

### *15.6.3 Emulating Hand-Portable Performance from a Vehicle Radio*

If there is no requirement to survey the coverage in specific buildings, just to obtain a general figure for hand-portable coverage in an area, then it is often possible to make corrections to the results of vehicle surveys to represent indoor coverage. This typically involves applying an attenuator between the antenna and the receiver. This reduces the signal level by a set number of dB, and it provides a correction to allow for:

- The difference in antenna efficiency for body-worn hand-portable to vehicle-mounted 1/4-wave whip.
- The difference in transmit power from the mobile (when measuring speech performance in the uplink direction).

As an illustration of this, for TETRA systems, the antenna difference factor typically used is 11 dB, with the power difference factor being 5 dB (power difference between Class 4 1 W hand-portable to Class 3 3 W vehicle radio). Strictly speaking the power difference attenuation should only be applied in the transmit path. The 11-dB attenuation can also be applied to a conventional calibrated signal strength survey if using a vehicle antenna to measure hand-portable performance. This factor is applicable for TETRA at 400 MHz – other factors will apply for other frequencies and other technologies.

### *15.6.4 Operating Multiple Radio Terminals in a Survey Vehicle*

For a network performance survey, there is often the need to operate more than one radio terminal at a time during a survey. In this situation, all of the radios will be both transmitting and receiving and therefore there is the potential for interference between them.

It is important not to use separate antennas for each radio terminal because:

- If this is done, none of the antennas will have a good radiation pattern; each will affect the other and thus skew the radiation patterns.
- If separate antennas are used, then the physical antenna separation requirements will prevent more than one being fitted to a typical survey vehicle.

Typically, about 60 dB of isolation between antennas is needed. This can be achieved with about 3 or 4 m of separation between the antennas; this is not usually possible on a vehicle roof. It is necessary to add a second antenna; one potential solution is to use a trailer for the 2nd antenna. In this case, the antenna must be mounted on the trailer at the same height as the antenna on the vehicle roof to ensure the measurements are consistent.

The alternative approach is to use an antenna combiner unit, which allows multiple radios to be used with a single antenna. Typical antenna combiner units will allow four radios to be used with a single antenna.

## 15.7 Network Analysis and Network Acceptance

A survey is normally required to determine if a network has met the contractual performance.

As previously described, an acceptance statement might be along the following lines:

*95 % of the entire region should receive at least 20 dBμV/m at a 98 % locations for 50 % time availability level, using a 50 m sample gap*

For a digital network this might be re-phrased as:

*95 % of the entire region should have speech quality above a MOS threshold of 2.1, taking 1 sample every 100 m*

However, a further qualification is needed to make sure the 5 % that does not meet this requirements is evenly spread over the entire region and not just in one area. Therefore an additional statement might be:

*Each 1 km square shall have a maximum of 5 % failed samples*

Collecting one sample every 100 m means that each 1 km$^2$ will have 20 samples and that a maximum of one sample may fail and still meet the acceptance criteria. One option is to allow the possibility of further surveys in a square that fails initially to see if more good samples can be collected to achieve the pass. The pass or fail criterion means that it is only the individual square that is important, not the individual measurements inside it. This is illustrated in Figure 15.16, which shows a number of such squares in a network together with the pass or fail criteria.

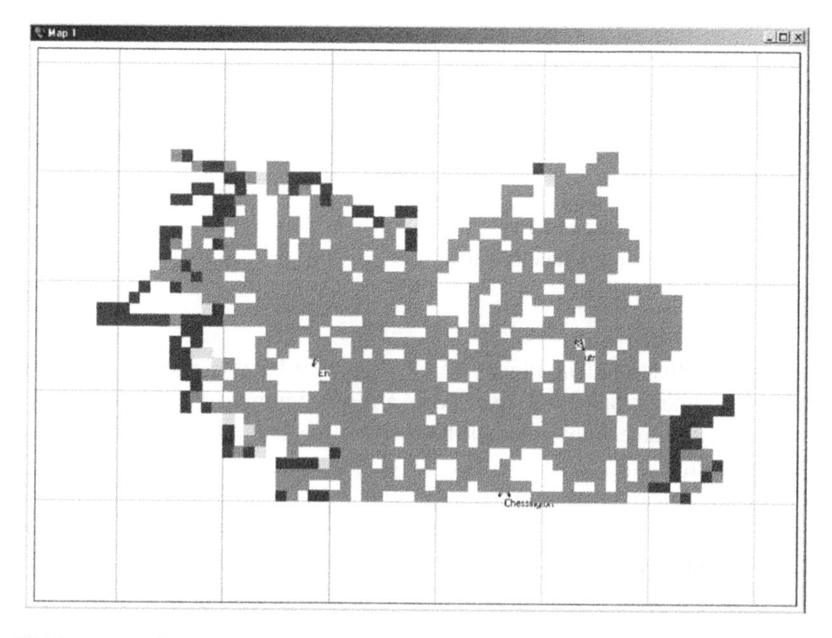


**Figure 15.16.** Example of pass/fail criteria expressed in square kilometre terms. The light shaded Squares are passes, and the darkly shaded ones have failed. Reproduced by permission of Radio Systems Information Ltd.

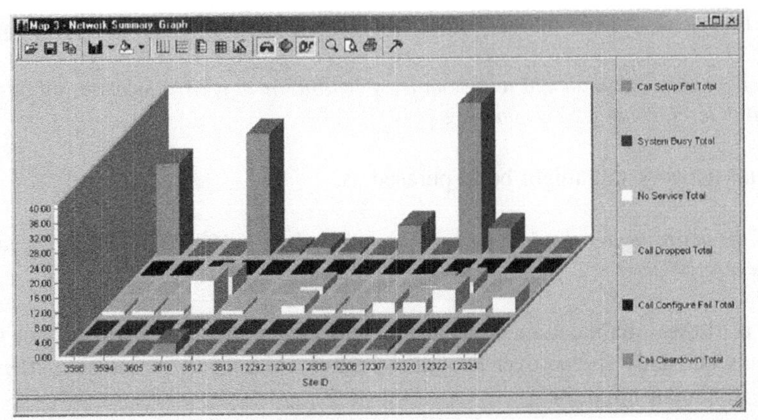


(a) Example of call statistic analysis, looking at the reasons for failed calls

Map 3 - Network Summary: Summary Report

Management Summary | Level 2 | LEC Summary | LEC Squares Failed | LEC Squares Passed | LEC Squares Insufficent Samples | LEC Squares Not Visited

| Coverage Class | RSSI Threshold | RSSI | RSSI Pass/Fail | VQ Threshold | UL Actual | DL Actual | UL Pass/Fail | DL Pass/Fail |
|---|---|---|---|---|---|---|---|---|
| Vehicle Major | 99.0% >= -103dBm | 99.8% | Pass | 96.0% >= 2.2 | 97.3% | 96.6% | Pass | Pass |
| Vehicle Minor | 96.0% >= -103dBm | 100.0% | Pass | 96.0% >= 2.2 | 96.8% | 98.3% | Pass | Pass |
| Hand Portable | 96.0% >= -87dBm | 98.9% | Pass | 96.0% >= 2.2 | 96.4% | 95.4% | Pass | Con Pass |
| E1 Brentford | 96.0% >= -82dBm | 100.0% | Pass | 96.0% >= 2.2 | 100.0% | 100.0% | Pass | Pass |
| HandPortable R... | 87.0% >= -87dBm | 100.0% | Pass | 87.0% >= 2.2 | 96.7% | 98.3% | Pass | Pass |

(b) Network summary of a number of characteristics, with pass or fail against network requirements shown

**Figure 15.17.** Typical additional network performance criteria and survey results. Reproduced by permission of Radio Systems Information Ltd.

Further network level pass criteria include:

- Percentage calls set-up successfully.
- Percentage calls clear down successfully.
- Percentage handovers failed.

Some of these are illustrated in Figure 15.17. Importantly, this survey system not only records the measurements taken but also shows a direct comparison with the requirements. This makes it easier to determine whether the network meets the requirements or not, which is always a good thing.

## 15.8 A Case Study

To conclude this chapter, we can examine a case study showing how analysis of different aspects of survey data can point to problems in network design or implementation. This is shown in Figure 15.18.

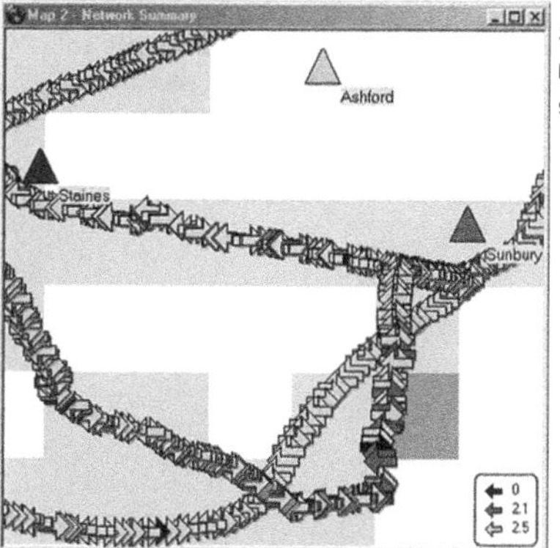


**Step 1: The dark square indicates a failed 1km square near to the Sunbury base site, where the speech quality seems to be a problem. This needs further investigation.**

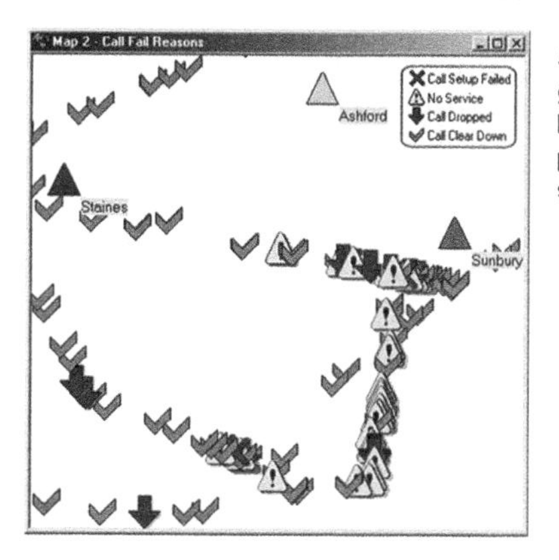


**Step 2: Looking at the call failure reasons for this area gives much more information. In particular, close to the base site there are a number of 'Dropped Calls' and plenty of 'No Service' events. These are probably significant.**

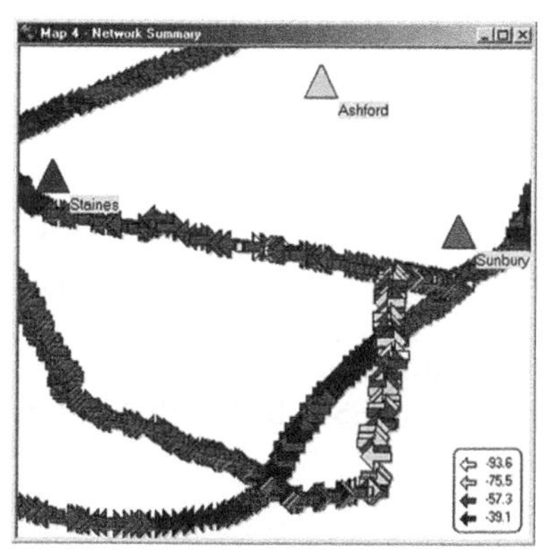


**Step 3: In this map, both call events and audio events are shaded according to the signal strength. The problem road does show poorer signal level (lighter shade), but certainly not low enough to cause a problem. Further analysis is required.**

**Figure 15.18 (a).** A case study. Reproduced by permission of Radio Systems Information Ltd.

## 15.9 Verification Deliverables

The deliverables required will depend on the contractual requirements of each individual project, and thus the points below are illustrative of the type of deliverables that are typical.

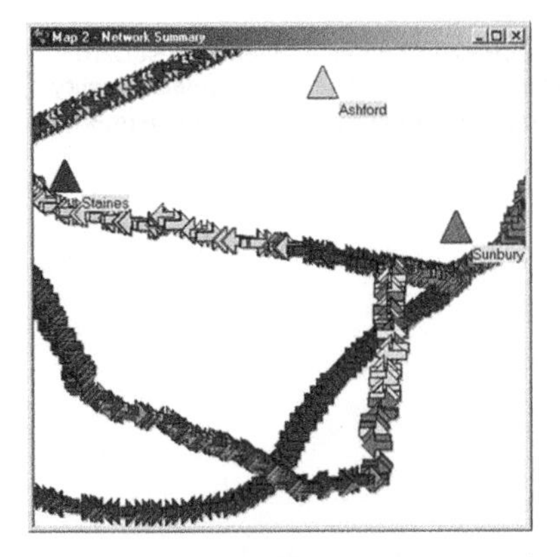


**Step 4: A further view of the data allows events to be shaded according to the base station site that was used – and this is where the cause of the problem starts to become clear; the radio terminal is simply not handing over correctly from one base site to another.**

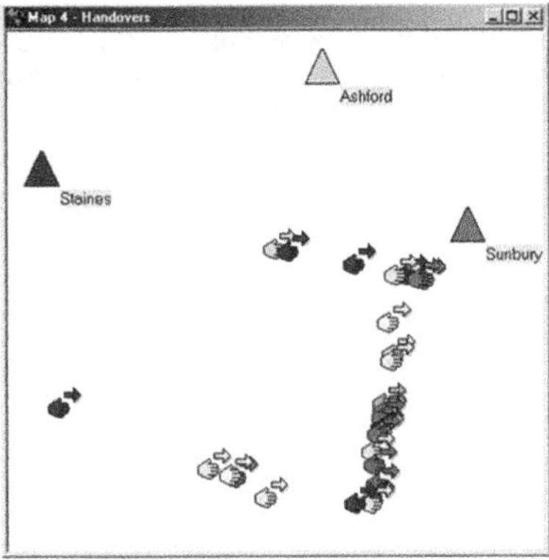


**Step 5: Handover analysis provides the final part of the story. The handover events are shaded according to the site that the radio attempts to register with. The problem area has an excessive number of handovers, with up to 5 base sites being involved.**
**The cause is that mobile radio terminals are using the wrong base site. This means that cell boundaries are being stretched to the point where the radio terminal cannot find the preferred neighbour cell and then resorts to multiple handovers in an attempt to find a usable site.**

**Figure 15.18 (b).** A case study. Reproduced by permission of Radio Systems Information Ltd.

- Report on how survey was carried out and how the survey system was configured.
- Survey data in raw form, to allow later analysis if needed.
- Survey route and condition information, recorded by survey technician during the survey.
- RSSI statistics for each area and pass/fail results according to network performance criteria.
- Additional agreed statistics such as call success rate and associated metrics.
- For digital networks, statistics for the agreed metrics such as PESQ, BER, MER or other.
- Plots, graphs and tables to support survey analysis.

# 16

# Mobile Network Development Cookbook

## 16.1 Introduction

This final chapter summarises some of the material covered in the rest of the book and can also act as a checklist for setting up and running projects and associated activities. It is in the form of bulleted lists and diagrams, with references to the parts of the book where more information can be found.

## 16.2 Pre-Project Activities

### *16.2.1 Stakeholder Analysis (Chapters 6 and 8)*

It is vital to understand the stakeholders involved in any project; they are the people who will determine whether the project succeeds or not; stakeholder analysis is all about defining *risks* (and sometimes opportunities) arising from the involvement of various groups. The purpose of spending effort to understand the stakeholders is to:

- Help understand the background to the project.
- Aid in understanding the motives behind actions that stakeholders take during the project.
- Identify any risks that specific stakeholders pose.
- Prepare mitigation methods to counter those risks where possible.

### *16.2.2 Technology Analysis*

It will not be possible to design a network without understanding the design aspects of the technology. Remember to focus on those elements that can be modified in the design. Other aspects that cannot be changed but may influence the design (such as timing limits) should also be taken into account.

Mobile Radio Network Design in the VHF and UHF Bands: A Practical Approach
*Adrian W. Graham, Nicholas C. Kirkman and Peter M. Paul* 

Specifications for open standards such as TETRA, GSM and UMTS UTRA can normally be found on the Internet at the website of the sponsoring organisation (ETSI in these cases). Specifications for other technologies are normally available from the manufacturers. They may also supply planning guidelines and typical link budget figures.

Remember that it will be necessary to add extra time to research planning methods for technologies you are unfamiliar with, and that if necessary, it may be cost-effective to bring in external consultants with specialist knowledge. This is often the quickest way to learn the techniques for designing new networks.

### *16.2.3 Pricing the Design and Preparing a Project Plan*

Pricing for consultancy tasks is often a matter of experience and will depend on the requirements of the task. It is important to remember to include effort and time for the following activities as well as the (obvious) core design activities:

- Project management (creating and maintaining the project plan; managing the project; maintaining the risk register).
- Quality management (creating and maintaining the quality plan; peer reviews etc).
- Interacting with the customer (meetings; reporting on progress; ad hoc tasks).
- Importing customer data.
- Consider whether some re-working of the design may be necessary, in response to customer requests.
- Do not forget to include the price of any data that may need to be bought for the project.
- The project plan must include all of the normal design activities required, plus the time required for other activities such as:
  - Delivery timescales of data supplied by environmental data suppliers;
  - Delivery timescale of data supplied by the customer/network operator;
  - Potential delays due to querying of information or re-supply following detection of errors in the original data sent (happens frequently);
  - Time for the customer to review documents provided as milestone deliverables;
  - Printing numerous plots at large scale can take a long time.

It might also be possible to consider the following questions when putting the price together:

- If the project is particularly risky, will the customer accept some of the risk by accepting a nonfixed price (a daily rate for example, with milestones to constrain the activity)?
- If project progress depends on obtaining data from the customer, what will happen if the data is delayed or delivered in a form that cannot be used? This can delay the project so the customer must be made aware of this and must commit to performing what is required of them.
- Other data suppliers must be capable of performing in the required timescale; cheapest is not always necessarily best. What fallback options are there if the supplier fails to perform?

## 16.3 Project Phases and Documentation (Chapter 6)

Once the project starts, the first activity is to create the MS and get this agreed with the customer. A typical project structure, showing where documentation arises and is used, is shown in Figure 16.1. This will vary from project to project, but creating a diagram on the lines of that shown is a useful reminder on when documents should be produced and where they should be used.

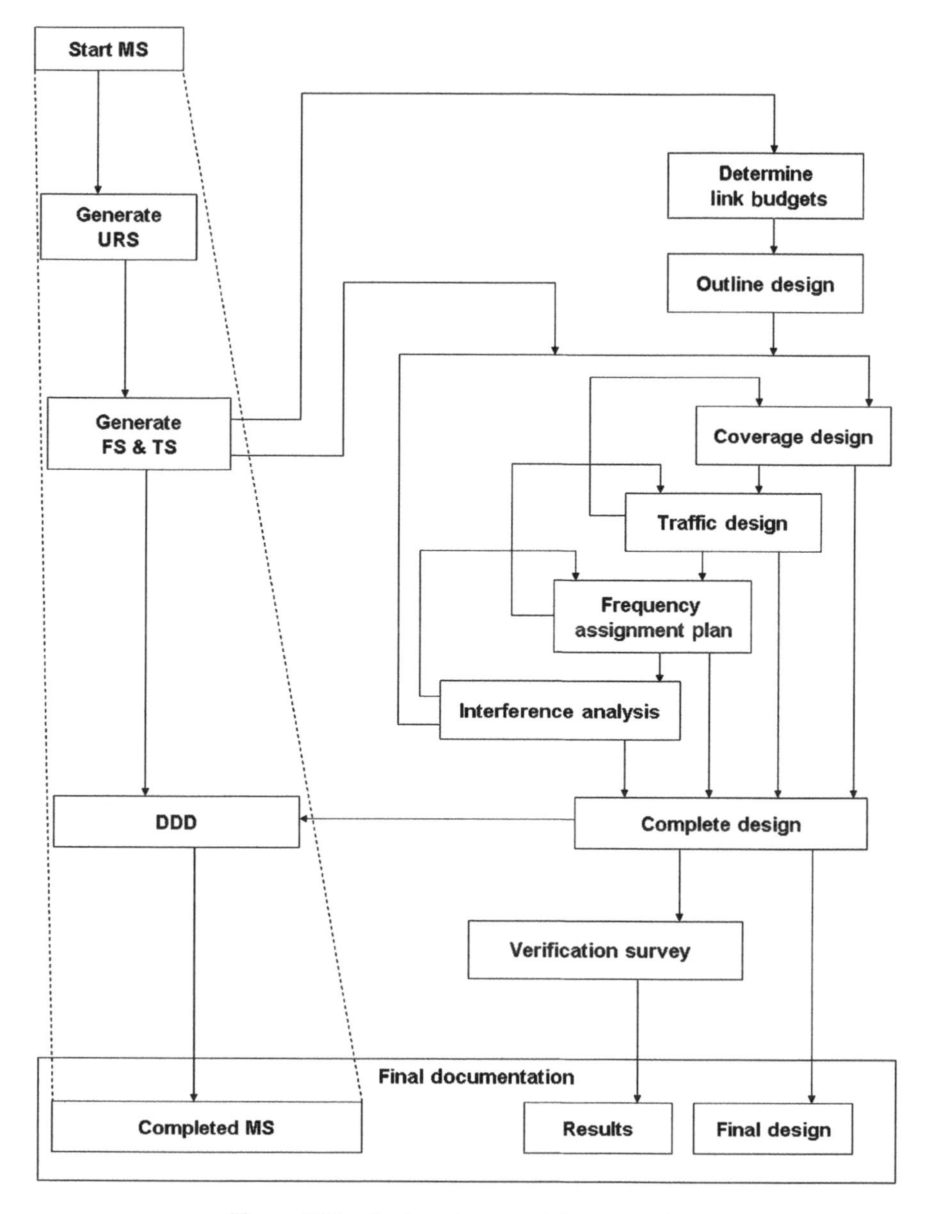


**Figure 16.1.** Project phases and documentation.

## 16.4 Setting Up the Project (Chapter 8)

A number of issues will need to be addressed as soon as the project starts. Remember that the timescale to get each component in place will affect the ability to deliver the project design, and so it is vital to consider when the following stages need to start:

- Assignment of Project Manager.
- Acquisition of project staff.

- Activity to define and refine planning methods to be used (may need a consultant or senior designer).
- Acquiring a suitable project area to host staff and equipment.
- Acquisition of, or arranging access to, a suitable planning tool; if acquiring, it may take time to determine the right system and to get it delivered and installed.
- Acquisition of, or arranging access to, suitable hardware required; e.g. computers and ancillaries, networking equipment, file sharer, printers (may need to acquire printers capable of printing at A3 or larger in colour).
- Staff training on planning tool and/or planning methods to be used.
- Acquisition of, or access to, suitable environmental data; it may take time for data to be delivered and processed for use, so ensure this activity is launched as soon as possible.
- Initial meetings with customers to launch project and refine communication interfaces.
- Creation of MS.
- Initiate design of other documents (see previous section).

It is important that issues that will take time to resolve, such as the acquisition and processing of data are addressed as soon as possible to prevent possible delays later. The project manager must keep on top of all the items initiated in this section, and those additional elements required for a specific project.

## 16.5 Data Requirements (Chapter 7)

### *16.5.1 Environmental Data*

The environmental data required for a project may include some or all of the following:

- Digital terrain map of project area.
- Digital map extending beyond the project area for interference analysis.
- Digital terrain map for coordination purposes, covering an area far larger than the project area.
- Detailed clutter map for project area.
- Outline clutter map for interference and coordination.
- Map images for analysis of whole project area.
- Map images for detailed analysis of small parts of the project area.
- Map images for interference analysis and coordination.
- Building outline maps for project area (or rooftop maps).
- Road maps.
- Rail maps.
- Maps of features of special interest.
- Population maps or statistics relating to postcodes (or zip codes).
- Postal or zip code area maps.
- Other subscriber modelling maps (possibly generated specifically for the project).
- Rainfall maps (for backhaul planning).
- Atmospheric refractivity maps for long range planning and interference analysis.

A detailed analysis of all the tasks to be performed should be conducted to identify all data requirements at the beginning of the project. Remember that different data may be required at different phases of the project (e.g. low-cost/low-resolution for outline planning, detailed, high-resolution data for main planning activity – see Section 7.6). A cost-benefit analysis should be conducted to identify the characteristics of data required for each task.

### *16.5.2 Radio Equipment Technical Parameters*

Radio equipment technical characteristics for planning purposes can typically be obtained from manufacturer data sheets or from specifications for radio technologies (e.g. TETRA, GSM, and UMTS). It can be difficult to obtain sufficiently detailed and accurate data, especially for older radios and occasionally it will be necessary to estimate some parameters based on the properties of similar radios. The characteristics required will depend on the tasks to be carried out during the design (see Section 7.7).

### *16.5.3 Antenna Characteristics*

Antenna characteristics will include antenna gain and polar response in the horizontal and vertical plane. The resolution and accuracy should be appropriate to define the response adequately. As a typical starting point, resolution of about 5 degrees horizontally and 1 degree vertically, to a resolution of 0.1dB is typically more than adequate. The antenna manufacturer normally provides antenna characteristics to a sufficient degree in the data sheets provided. Some manufacturers produce electronic catalogues that can greatly assist in the selection of antenna types (to the benefit both of the planner and the manufacturer!).

## 16.6 The URS (Chapters 6 and 8)

The User Requirement Specification document is the definitive description of the required performance from the customers' point of view. It should be created with the following in mind:

- It should capture all of the requirements of the customer (user).
- It should identify each specific requirement in an unambiguous manner that can be tested in a binary fashion (criterion met: yes/no).
- It must capture the spirit of what the customer intends as well as the detail; high level requirements must also be included – but again in a manner that allows an unambiguous test.
- It must be consistent; individual requirements must not contradict one another.
- Ideally, each requirement should be given a priority to allow trading where necessary.
- It must be *thoroughly* validated against the actual intentions of the customer.

The production of the URS will normally involve at least some aspect of iteration, so that the customer's requirements and the stated requirements are brought fully into line. The customer must signify that the URS is a valid representation of the requirements.

## 16.7 The FS and TS (Chapters 6 and 8)

The FS and TS should ideally be created at the same time. The FS is created against the requirements identified in the URS and nothing else. The TS should have a one-to-one correspondence with the FS and should also consider the following:

- Ideally, the FS should have a one-to-one or one-to-many correspondence with the requirements identified in the URS. It may be necessary for a number of features and tests to be carried out to meet one requirement, and if so, this should be reflected in both documents.
- Each functional point must be testable by a process that is physically realisable within the context of the project and which produces an unambiguous result.
- Successful completion of an identified test is the sole arbiter of whether a requirement has been met or not. This is why both the FS and TS must be very detailed.
- Successful completion of all elements (or an agreed percentage) is the sole arbiter of whether the network meets the design criteria or not.

It should be clear that the role of the URS, FS and TS are absolutely fundamental to the project and must be designed to properly reflect the intentions of the customer, but in a measurable manner that can prove (or disprove) compliance without ambiguity.

## 16.8 Coverage Design (Chapter 9)

A typical coverage design approach is illustrated in Figure 16.2. This particular example is based on the principle that an initial design is produced (which may only be a single site or may be a partial design), and then there are iterative steps to improve a number of base station sites in order to improve the performance of the design. In this case, the aim is to first engineer each site using the equipment already deployed by modifying power, antenna height, azimuth and tilt. If this fails, then different antennas are tried (again modifying the site parameters). Only if all the options fail, then a different site is considered. Once a site has been designed, the coverage achieved by the whole network is tested against the requirements. The process continues until the requirements are met (in practice, they it may stop because it becomes apparent there is no solution. This type of diagram is useful to inform the work instructions for coverage design, albeit that it must be tailored for the requirements of individual projects.

## 16.9 Traffic Design (Chapter 10)

A typical traffic design process is shown in Figure 16.3. This is essentially the process of determining where traffic demand will exist within the network and then matching each base station to the coverage within its service area, by adjusting the service area. This is an iterative process that involves re-designing coverage using the coverage design process.

Traffic has to be considered for the different distribution of calls typical over the day, week and longer cycles, and thus the traffic design must be tested against as many potential scenarios as possible to ensure that it meets all expected demand.

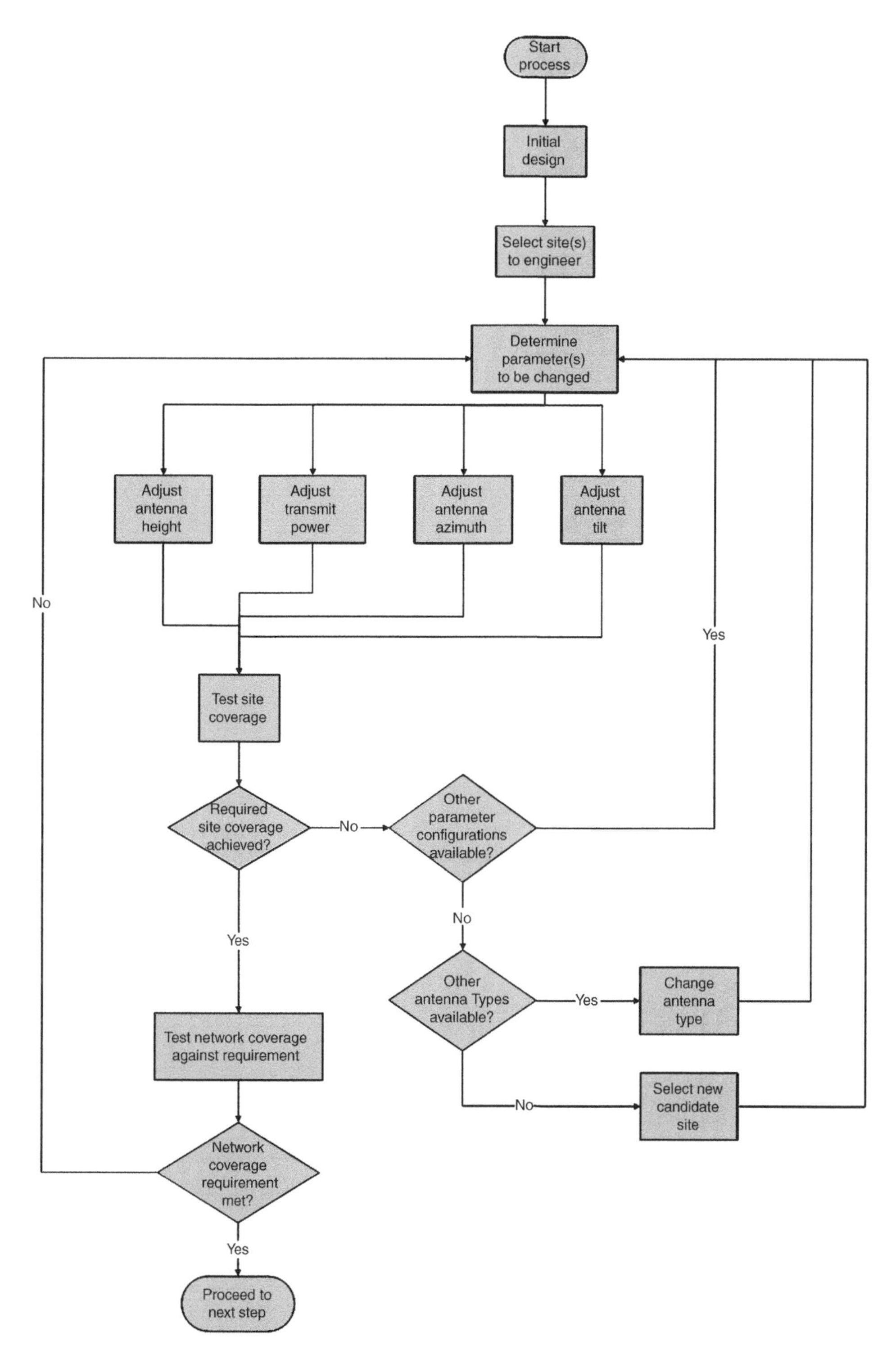


**Figure 16.2.** Coverage design process example.

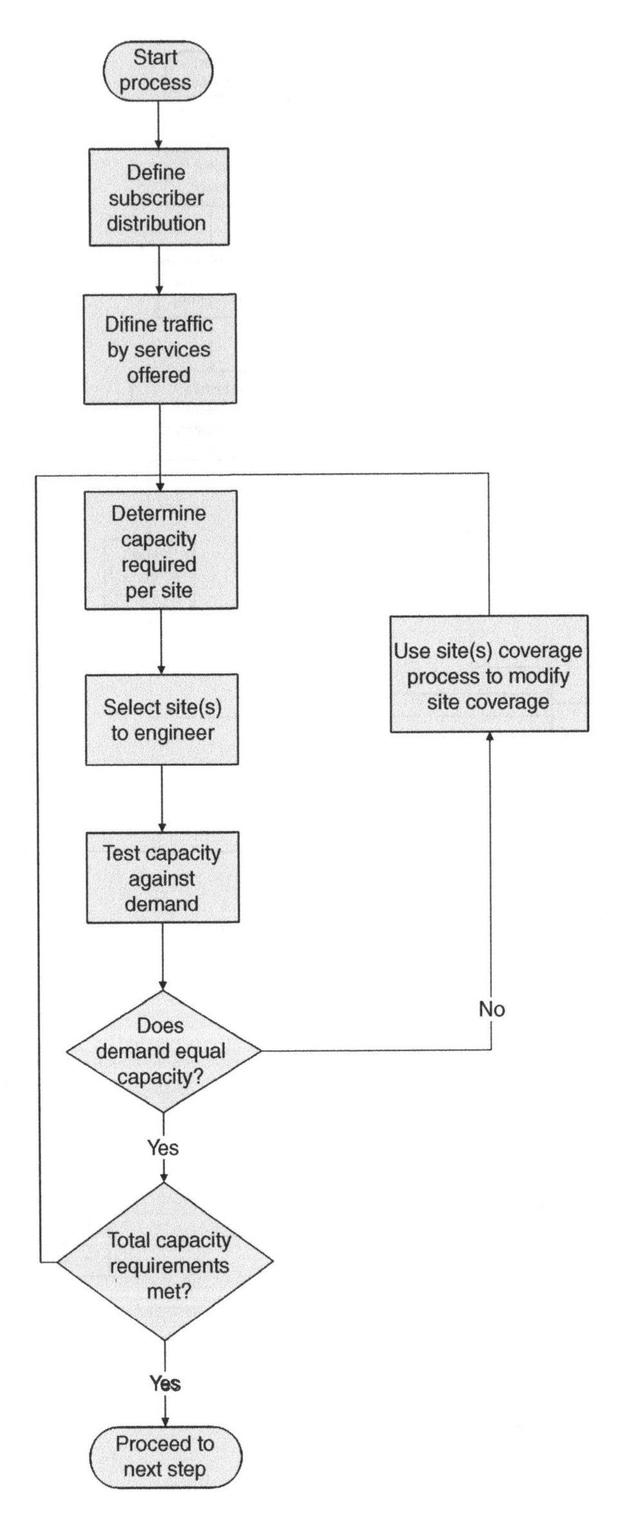


**Figure 16.3.** Traffic design process example.

## 16.10 Interference Mitigation (Chapter 13)

The methods to use to mitigate interference will depend on the circumstances. In general, it will be necessary either to reduce the effect of the interfering transmitter or increase the effectiveness of the wanted transmitter. For interference between unwanted base station transmissions and mobiles, the coverage can be adjusted using the methods shown in Section 16.8.

Practical measures include:

- Reduce interferer effect.
  - Reduce interferer antenna height
  - Reduce interferer power
  - Increase interferer antenna tilt
  - Use directional antenna to reduce interferer effect in a given direction
  - Change interferer frequency
  - Move interferer (try to use terrain or clutter shielding)
  - Improve network design by changing several sites
- Reduce effect of interferer on victim.
  - Increase victim antenna height (to increase received signal level at mobile)
  - Increase victim transmit power
  - Change victim antenna to reduce gain in direction of interferer
  - Change victim frequency
  - Move victim (try to use terrain or clutter shielding)
  - Improve network design by changing several sites

Ideally, the interference mitigation effects that have the least impact on the network design should be used in preference.

## 16.11 Frequency Planning (Chapter 14)

The frequency assignment process will depend on the complexity of the task and the tools used to perform it. Typical approaches for full assignment and co-site assignment are shown in Figures 16.4 and 16.5, respectively.

## 16.12 Verification (Chapter 15)

Verification and validation are both vital parts of the project. Their involvement in the project can be summarised as shown in Figure 16.6. Validation is covered by the previous sections, and is performed against the FS.

The key points to be borne in mind for verification are that:

- Verification for design and implementation are carried out according to the requirements of the TS.

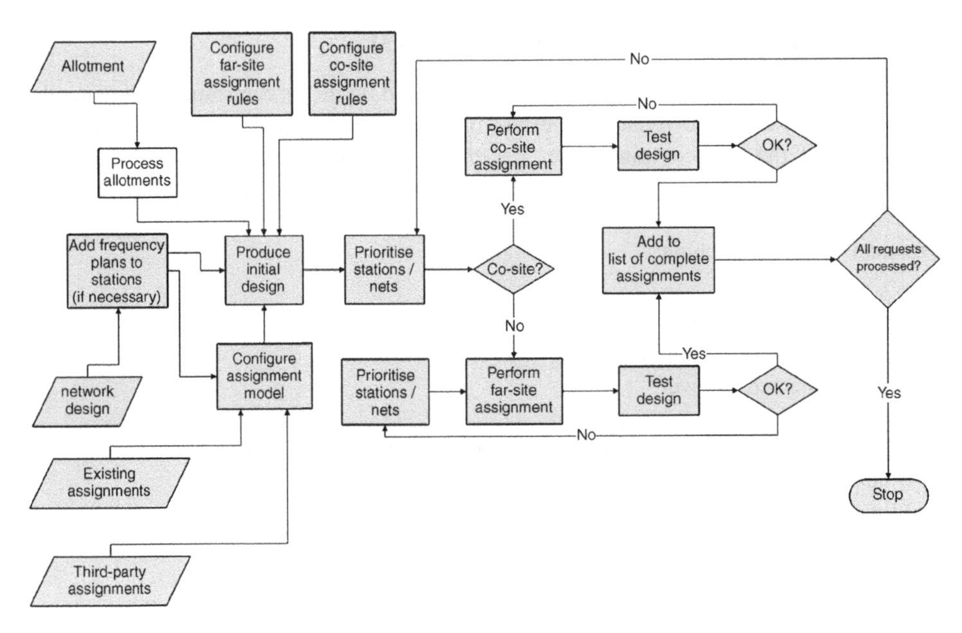


**Figure 16.4.** Typical full assignment process.

- For surveys, the following are crucial:
  - The survey must be performed against a test plan, not just on an ad hoc basis. For formal survey, this will be against the TS.
  - The survey must produce *Statistically Valid Results* otherwise it is worse than useless.
  - Read the above bullet point again; it is not only vital, it is often not adhered to.

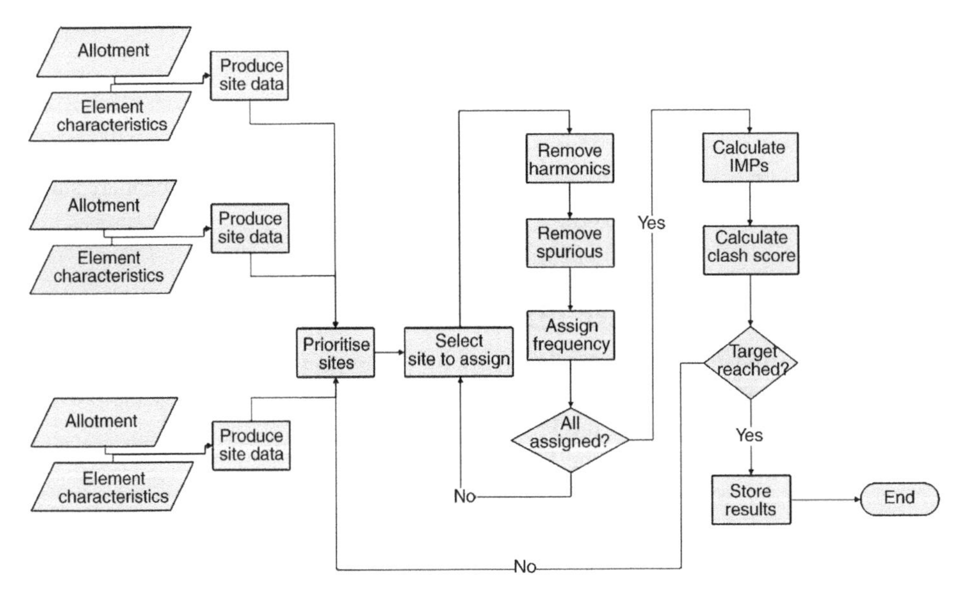


**Figure 16.5.** Typical co-site assignment process.

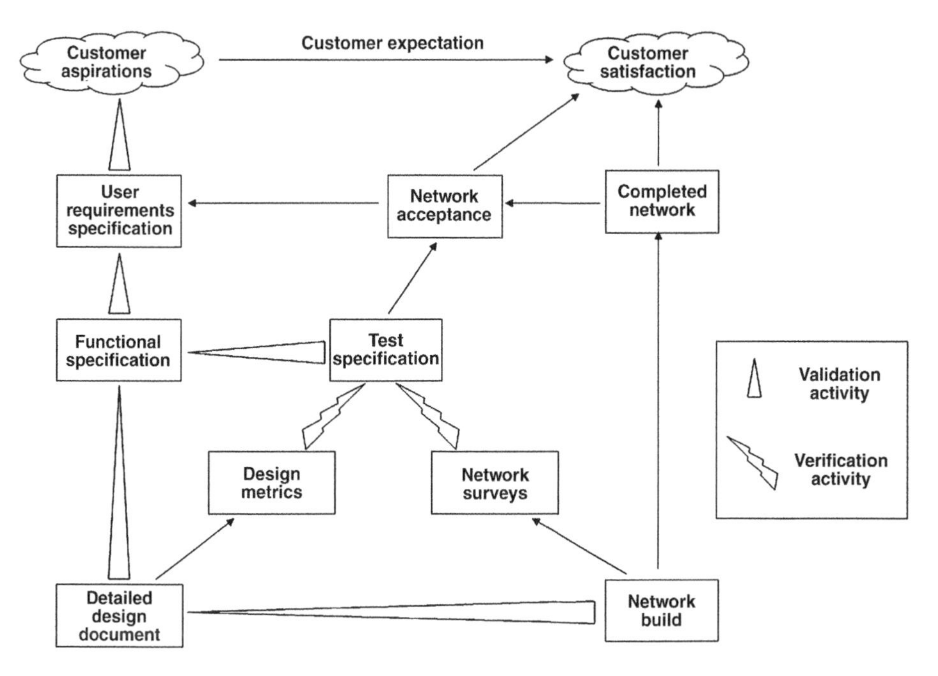


**Figure 16.6.** Validation and verification.

## 16.13 Typical Project Deliverables

### *16.13.1 Project Documents*

- Method Statement (MS).
- User Requirements Specification (URS).
- Functional Specification (FS).
- Detailed Design Document (DDD).
- Test Specification (TS).

### *16.13.2 Design Deliverables*

#### 16.13.2.1 Coverage Design Deliverables

- A list of the candidate sites, including their locations.
- Antenna height, orientation and tilt for each antenna to be used at the site.
- Power to be used for each channel.
- Coverage plots per site, for each different category of service in the requirements.
- Coverage plots for the whole network, for each different category of service in the requirements.
- Best server plots for the whole network, and optionally for sub-areas of it.
- Statistics for each requirement, provided in tabular format.
- Statistics for the contribution of each site to overall coverage, based on best server coverage.
- Other information as agreed by the customer. This may include comments on difficult areas to cover; potential options for network re-design and so on.

#### 16.13.2.2 Traffic Design Deliverables

- A list of the candidate sites, including their locations.
- The number of sectors to be used.
- Antenna height, orientation and tilt for each antenna to be used at the site.
- The grade of service estimated for each sector.
- The number of frequencies required for each site (to be used in the frequency assignment process).
- Power to be used for each channel.
- Other, technology-specific information as required.

#### 16.13.2.3 Interference Deliverables

- Report on interference methodologies used for each interference analysis.
- Point-to-point interference tables for fixed base stations.
- Plots of interference for mobile units.
- Statistics of interference for mobile units.

#### 16.13.2.4 Verification Deliverables

Apart from design verification deliverables (described in the previous sections), the following survey deliverables are typical:

- Report on how survey was carried out and how the survey system was configured.
- Survey data in raw form, to allow later analysis if needed.
- Survey route and condition information, recorded by survey technician during the survey.
- RSSI statistics for each are and according to network performance criteria.
- Additional agreed statistics such as call success rate and associated metrics.
- For digital networks, statistics for the agreed metrics such as PESQ, BER, MER or other.
- Plots, graphs and tables to support survey analysis.

#### 16.13.2.5 Frequency Assignment Deliverables

- A Frequency plan, listing all frequencies used, possible along with a figure for the number of times the frequency is re-used in the assignment.
- List of sites, together with their assigned frequencies.
- Interference statistics, as a metric to identify how good the assignment is in terms of minimising interference or quality of service.

## 16.14 Final Thoughts

These are our final thoughts that we often keep in mind when approaching projects.

Unless you have an understanding of the involvement and intention of all-important stakeholders, you do not have a full understanding of the project risks.

If you profess to be a radio network-planning engineer, you must understand propagation mechanisms, propagation models and fading characteristics.

Projects should be 'business-centric' not engineering-centric; engineering should always be performed on behalf of the customer, the dependence should not be the other way round.

Information flow between departments in a business is key; no Chinese walls, no 'us and them'.

Good engineering is critical engineering; neither above nor below the requirements.

Documentation; MS, URS, FS, TS and results.

Validation; URS against the user expectation; FS against URS, everything else against FS.

Verification; against TS.

Surveys; against TS or other test plan – STATISTICALLY VALID, repeatable.

When writing reports, remember to write down why, what, where and when; not just how. Too many engineers simply write down what has been done, not why they have done it nor when it should be applied. This does not allow others to follow the reasoning behind this, nor gain a full appreciation of the *process*.

*Radio network design: If it is not enjoyable, you are not doing it right!*

# Index

Printed and bound by CPI Group (UK) Ltd, Croydon, CR0 4YY

Printed and bound by CPI Group (UK) Ltd, Croydon, CR0 4YY
16/04/2025
14658474-0002